Art, Technology and Nature

Since 1900, the connections between art and technology with nature have become increasingly inextricable. Through a selection of innovative readings by international scholars, this book presents the first investigation of the intersections between art, technology and nature in post-medieval times.

Transdisciplinary in approach, this volume's 14 essays explore art, technology and nature's shifting constellations that are discernible at the micro level and as part of a larger chronological pattern. Included are subjects ranging from Renaissance wooden dolls, science in the Italian art academies, and artisanal epistemologies in the followers of Leonardo, to Surrealism and its precursors in Mannerist grotesques and the Wunderkammer, eighteenth-century plant printing, the climate and its artistic presentations from Constable to Olafur Eliasson, and the hermeneutics of bioart.

In their comprehensive introduction, editors Camilla Skovbjerg Paldam and Jacob Wamberg trace the Kantian heritage of radically separating art and technology, and inserting both at a distance to nature, suggesting this was a transient chapter in history. Thus, they argue, the present renegotiation between art, technology and nature is reminiscent of the ancient and medieval periods, in which art and technology were categorized as aspects of a common area of cultivated products and their methods (the Latin *ars*, the Greek *techné*), an area moreover supposed to imitate the creative forces of nature.

Camilla Skovbjerg Paldam is Associate Professor of Art History, Department of Aesthetics and Communication, Aarhus University, Denmark.

Jacob Wamberg is Professor of Art History, Department of Aesthetics and Communication, Aarhus University, Denmark.

Science and the Arts since 1750

Series Editor:

Barbara Larson is Associate Professor of Art History at the University of West Florida, USA.

This series of monographs and edited volumes explores the arts - painting and sculpture, drama, dance, architecture, design, photography, popular culture materials - as they intersect with emergent scientific theories, agendas, and technologies, from any geographical area from 1750 to now. It welcomes studies on the aesthetic conditioning of scientists as well as those that explore the influence of technologies, medicine, and science on visual culture either in a specific cultural or social context or through webs of influence that cross national, political, or imperial boundaries. Projects additionally might address philosophies of mind, brain, and body that changed the way visuality and aesthetic theory were understood or how new theories can be used to reinterpret the past.

Art, Technology and Nature

Renaissance to Postmodernity

Edited by
Camilla Skovbjerg Paldam
and Jacob Wamberg

Routledge
Taylor & Francis Group
LONDON AND NEW YORK

First published 2015 by Ashgate Publishing

Published 2016 by Routledge
2 Park Square, Milton Park, Abingdon, Oxon OX14 4RN
52 Vanderbilt Avenue, New York, NY 10017, USA

First issued in paperback 2017

Routledge is an imprint of the Taylor & Francis Group, an informa business

British Library Cataloguing in Publication Data
A catalogue record for this book is available from the British Library.

Library of Congress Cataloging-in-Publication Data
Art, technology and nature: Renaissance to postmodernity / Edited by Camilla Skovbjerg Paldam and Jacob Wamberg.
pages cm
Includes bibliographical references and index.
ISBN 978-1-4724-1172-3 (hardcover: alk. paper)
1. Art and technology. 2. Nature (Aesthetics) 3. Technology and civilization. I. Paldam, Camilla Skovbjerg, editor. II. Wamberg, Jacob, editor.

N72.T4A775 2015
701'.05—dc23

2015007094

ISBN 13: 978-1-138-31032-2 (pbk)
ISBN 13: 978-1-4724-1172-3 (hbk)

Contents

List of Illustrations

Black and White Figures

14.1 Olafur Eliasson, *The New York City Waterfalls* (2008), installation. East River, New York. Julienne Schaer/Courtesy Public Art Fund. © 2008 Olafur Eliasson/Photo credit: Olafur Eliasson

Note: Every effort has been made to secure all reprint permissions. However, in the case of Figures 10.1, 12.1, 12.2 and Colour plate 4 this has not been possible. The editors apologize for any inadvertent errors or omissions and, if notified, will endeavour to correct these at the earliest opportunity.

Notes on Contributors

Christian Ulrik Andersen is Associate Professor at Aarhus University, Chair of the Digital Aesthetics Research Centre and part of the Participatory Information Technologies Centre. He researches digital art and culture, generally in relation to the practice of interface critique, and is a frequent speaker at media art festivals and conferences. He is co-editor of *Interface Criticism: Aesthetics Beyond the Buttons* (2011, with Søren Bro Pold) and the journal *aprja.net* (with Geoff Cox).

Hans Dickel, Ph.D., is Professor of Art History at the University of Erlangen-Nuremberg. His research interests include modern and contemporary art, drawing, photography and art theory. He has published widely on modern art with a particular focus on its relation to nature and technology: *Kunst als zweite Natur. Studien zum Naturverständnis in der modernen Kunst* (2006), *Künstlerbücher mit Photographie seit 1960* (2008).

Olafur Eliasson is a Danish-Icelandic artist whose sculptures, paintings, photography, film and installations have been exhibited widely throughout the world. Not limited to the confines of museum and gallery, his practice engages a broader sphere through architectural projects and interventions in civic space. His works include *Green river* (2000, Stockholm), *The Weather Project* (2003, London, Tate Modern) and *The New York City Waterfalls* (2008).

James Elkins is E.C. Chadbourne Chair at the Department of Art History, Theory and Criticism at School of the Art Institute of Chicago. His research interests include art theory, art history and scientific images. Recent publications include *Artists with PhDs: On the New Doctoral Degree in Studio Art* (2009) and *What Photography Is* (2011).

Marion Endt-Jones teaches at the University of Manchester. Her postdoctoral research, funded by the Henry Moore Foundation and the British Academy, has focused on the cultural history of natural objects and the cabinet of curiosities in contemporary art. She is the curator of *Coral: Something Rich and*

Strange at Manchester Museum (2013–14) and the editor of the eponymous accompanying book (2013).

Claire Farago is Professor of Renaissance Art, Theory and Criticism at the University of Colorado Boulder. Her main interests are in art theory and cultural exchange between Europe and the rest of the world. She has published widely on Leonardo da Vinci and is currently completing, with an international group of scholars including Carlo Vecce, Janis Bell, Juliana Barone and Martin Kemp, a scholarly edition of da Vinci's *Treatise on Painting* published in 1651. Her most recent book is *Art Is Not What You Think It Is* (2012, co-authored with Donald Preziosi).

Maria Fabricius Hansen, Ph.D. and Dr.Phil., is Associate Professor of Art History at the Department of Arts and Cultural Studies, University of Copenhagen. Her present research focuses on the relation between art and nature, especially the grotesques in sixteenth-century Italian art. Her publications include *The Eloquence of Appropriation: Prolegomena to an Understanding of Spolia in Early Christian Rome* (2003) and 'Maniera and the Grotesque' (in *Manier und Manierismus*, edited by W. Braungart, 2000).

Hanna Johansson is Research Fellow (Academy of Finland) at the Department of Philosophy, History, Culture and Art Studies, University of Helsinki. She focuses on modern and contemporary art and their relationship to nature, natural forces, landscape and ecosystems. Recent publications include *Homes in Transformations: Dwelling, Moving, Belonging* (editor, 2009) and *Nature in Nordic Contemporary Art: From the Environment to a Common Shared World* (2014).

Caroline A. Jones is Professor of Art History in the History, Theory, Criticism section of the Department of Architecture at MIT. She studies modern and contemporary art, with a particular focus on its technological modes of production, distribution and reception. Jones has also worked as an essayist and curator. Her publications include *Sensorium: Embodied Experience, Technology, and Contemporary Art* (editor, 2006), *Eyesight Alone: Clement Greenberg's Modernism and the Bureaucratization of the Senses* (2005) and *Machine in the Studio: Constructing the Postwar American Artist* (1996).

Pernille Leth-Espensen, Ph.D., is Postdoctoral Fellow in Art History at Aarhus University, financed by the Novo Nordic Foundation. Her doctoral thesis is titled *Art, Hermeneutics, and Technoscience* (2013). As a part of her Ph.D. project she was a resident at SymbioticA, The University of Western Australia. Her research interests include art-science collaborations and the relation between biotechnology, bioart and biopolitics.

Alisa Luxenberg is Professor of Art History at the University of Georgia. She investigates Franco-Spanish artistic and cultural confrontation and exchange as well as visual modes of realism of the eighteenth and nineteenth centuries. Her most recent book examines Baron Taylor's illustrated travelogue, *Voyage pittoresque en Espagne, en Portugal et sur la côte d'Afrique* (2013).

Kristine Nielsen is Assistant Professor of Art History at Illinois Wesleyan University. Her research interests include modern and contemporary art, the historiography of art history and visual studies, Cold War media cultures, GDR monuments and the history and theory of iconoclasm. Publications include 'What Ever Happened to Ernst Barlach? East German Political Monuments and the Art of Resistance', in *Totalitarian Art and Modernity*, edited by Mikkel Bolt Rasmussen and Jacob Wamberg, 2010.

Camilla Skovbjerg Paldam is Associate Professor of Art History, Aarhus University. Her research includes avant-garde art and theory, especially Surrealism, collage techniques and Surrealist legacies in contemporary art. Her publications include *Surrealistiske collager. Underfulde billeder i kunst og litteratur* (*Surrealist Collages: Marvellous Images in Art and Literature*, 2011).

Søren Bro Pold is Associate Professor of Digital Aesthetics at Aarhus University and founding member of the Digital Aesthetics Research Center (http://darc.imv.au.dk/). He works on digital and media aesthetics with a focus on interface criticism, including electronic literature, net art, software art, urban and mobile interfaces and controlled consumption. With Christian Ulrik Andersen he edited *Interface Criticism: Aesthetics Beyond the Buttons* (2011).

Markus Rath, Dr.Phil., is Assistant Professor of Art History at the University of Basel. His research interests are articulated sculpture, visual-haptic perception and *paragone* theory. Edited volumes include *Paragone als Mitstreit* (2014, with J. v. Gastel and Y. Hadjinicolaou) and *Das haptische Bild. Körperhafte Bilderfahrung in der Neuzeit* (2013, with J. Trempler and I. Wenderholm).

Bjørn Okholm Skaarup received his Ph.D. from the European University Institute, Florence. His edited doctoral thesis, *Anatomy and Anatomists in Early Modern Spain*, was published by Ashgate in 2015. He has conducted postdoctoral research funded by The Carlsberg Foundation at the Warburg Institute and Columbia University on teaching and training in the first Italian art academies. Skaarup is also an artist whose work has been exhibited internationally.

Lisbet Tarp is Postdoctoral Fellow in Art History at Aarhus University. Currently, she is working on the project *The Visual Laboratory: Alchemy, Instrument, and Hybridity in the Early Modern Period* supported by The Novo

Nordisk Foundation. She has published articles concerning the relation between nature and art, decorative arts of the sixteenth and seventeenth centuries, and Ole Worm.

Jacob Wamberg, Dr.Phil., is Professor of Art History at Aarhus University. He works on evolutionistic theories of the visual arts, especially in relation to nature and technology. His present focus is posthuman aspects of avant-garde art. His publications include *Landscape as World Picture: Tracing Cultural Evolution in Images* (2009 [2005]) and *The Posthuman Condition: Ethics, Aesthetics and Politics of Biotechnological Challenges* (2012, edited with Kasper Lippert-Rasmussen and Mads Rosendahl Thomsen).

Introduction:
A Short History of Art, Technology and Nature

Jacob Wamberg and Camilla Skovbjerg Paldam

Postmodern Convergence, Modern Separation

Why join art, technology and nature in a constellation and turn this into the theme of a book? The question itself suggests a certain unrelatedness among these domains, but in this volume we suggest the triad occupies some significant corners of the material phenomena among which humans act. Nature comprises all phenomena produced without human intervention; technology and art encompass two dimensions of human production. Technology furnishes artefacts that are oriented towards physical and utilitarian use (instruments, utensils, machines), whereas art yields artefacts evoking extraordinary, metaphysical forces, whether primarily mental (in modern times) or spiritual or magical (in premodern times).

More specifically, art and technology appear lately to be forming a closer relationship with one another and with nature – one in which ontological categories are becoming porous, allowing for new, even epoch-making, crossovers. Whereas, until recently, our industrial civilization regarded nature as a more or less stable, passive background to human enterprise – be it as reservoir of materials (in technology), a domain of cognitive investigation (in science) or a panorama for aesthetic contemplation (in art and relaxed states of mind) – nature is now increasingly understood as something whose destiny must be judged in close contact with human history: both closely influencing and being influenced by human agency and its technological prostheses. In both 'natural' phenomena like climate change and 'cultural' ones like biotechnology, it is practically impossible to insert boundaries dividing human from natural agencies. And in emerging academic fields, such as some neurosciences, or hybrids of natural and cultural history (such as the biocultural 'Deep History' or the all-encompassing 'Big History'), cultural artefacts and the human body are treated as an ensemble, in which interaction with objects conditions the development of the brain, yielding a framework with increasingly indeterminate borders between biological and technological evolution.[1] Are we witnessing a re-emergence of the kind of history presented in Pliny the Elder's *Naturalis historia* (c. 75–79 CE), a transdisciplinary account that does not separate natural products from cultural ones, but addresses both nature and those things that humans make from her, including what we call 'technology' (useful artefacts) and 'art' (aesthetic artefacts)?[2]

In any case, the convergence of the natural and the cultural seems to have also affected those elements of visual media that we call 'art'. Just as utilitarian technology is becoming increasingly intermingled with the business of nature, so is contemplative art becoming more closely involved with the physical processes of technology and nature. A keyword here, with its roots in 1960s and 1970s avant-garde movements like Fluxus and Situationism, is 'interactivity', which – this is the utopia, at least – converts the artist into an agent who is doing something with, and in, the world, rather than merely reflecting it. At the same time, the distinction between artist/agent and viewer is broken down, as the artwork invites the viewer to physically interact with what is otherwise the exclusive preserve of those artefacts labelled 'technology', that is, things for use or, to a lesser degree, for leisure activities (such as toys and games). Specifically, in many contexts we see art engaging more directly with nature than formerly, above all by transgressing landscape painting, which used to contemplate nature at an aesthetic distance. We have seen this development in such fields as earth art, body art, and perhaps especially the expanding field of bioart, in which the triangle of art, technology and nature becomes particularly difficult to disentangle. To take a paradigmatic example, Eduardo Kac's now-famous transgenic *GFP Bunny* (2000) is a work of art but, in an Escher-like knot, it is also an organism in its own right, behaving just like a naturally created animal, except for its property of (reportedly) emitting green light when exposed to blue light, owing to the biotechnological insertion of genetic material (Green Fluorescent Protein = GFP) from *Aequorea victoria*, a luminous Pacific jellyfish, into its albino mother's egg, prior to conception.[3] Kac's rabbit can therefore be categorized not only as, firstly, an artwork and, secondly, a natural organism but also, thirdly, a technologically shaped artefact.

This book thus regards art as situated in a narrowing gap between technology and nature, claiming that in recent years, the boundaries of these three areas – art/technology/nature – have become more negotiable, and accordingly exploring possible genealogies, or contrasting periods, for this denser constellation. Since the Enlightenment, we have become accustomed to thinking of art, technology and nature as strictly separate domains: nature being 'out there', technology and art being 'in here', but segregated by their contrasting rules. An early celebration of rising industrialization, such as Turner's painting, *Rain, Steam and Speed: The Great Western Railway* (Colour plate 1) is situated at an interesting intersection of this High-Modern distinction and emerging postmodern convergence. This painting absorbs into its very centre a sphere of culture that in many preceding centuries was tolerated only peripherally in imitative art: here, technology, in the form of a black locomotive, rushes across the River Thames, over the recently constructed Maidenhead Bridge. Although it attains a new role as the central agent in a work of art, the machine still relates to the forces of nature in a very different way than the artwork mirroring both it and the whole natural panorama. Whereas the painting reflects steam power in a free, wild state (the steam and smoke blending with the turbulent atmosphere under the open sky), the steam engine at the heart of the locomotive uses this power in a subdued and controlled manner (the atmospheric pressure turned to the service of humans inside the boiler). Furthermore, even though the oily chaos of the picture mirrors the most unbridled and sublime forces of nature – and indeed, subjectively animates them, so as to capture the experience of being in their midst – nevertheless, this recreation in brush strokes amounts to no more than an imaginary, static vision bound to a flat canvas. In contrast, as Turner himself had probably recently experienced, a locomotive and its steam engine could transport your body physically, at great speed, through the misty air.[4] Nonetheless, this speed happened at a price, for the 'animation' of the steam, so vividly foregrounded in Turner's imaginary vision, was neutralized in the machine, and replaced by industrial exploitation based on the stable, mathematical laws of a new physical science. Despite Turner's desire for an ecstatic synthesis of the forces of nature, technology and art, his work remains trapped in the High-Modern distribution of roles: art (the landscape painting) communicating a quality of aliveness in nature, but at a purely imaginary level; technology (the locomotive) physically moving nature, but in a mechanical and dead manner; and nature (both terrain and atmosphere being represented and utilized), a reservoir of inscrutable forces, which may be respectably mirrored and controlled (Figure I.1).

The German philosopher Joachim Ritter, who viewed the notion of landscape as the prime example of what he termed the 'aesthetic' attitude to nature, described the distribution of roles between art and aesthetics, and technology and science, as highly symptomatic of post-medieval, urban culture:

> In the historical epoch in which nature, its forces and materials become 'object' of the natural sciences and of the technical utilization and exploitation that are based on these sciences, it is the – no less universal – mission of poetry and visual art to make the same nature 'aesthetic' and present in its relation to the perceptive human being.[5]

Basing his view in particular on Friedrich Schiller's late eighteenth-century ideas that modernity concerns humankind's separation from nature, compensated for by a sentimentalist longing for it,[6] Ritter considers this aesthetic attitude towards nature as indeed a sentimental relationship – one in which the cynicism inherent in industry's framing of nature as dead matter lying free for use is counterbalanced by a feeling for nature as if it were alive. Whereas in industry with its technological apparatus, backed by objective mechanistic science, nature is conceived of as a huge machinery of compact,

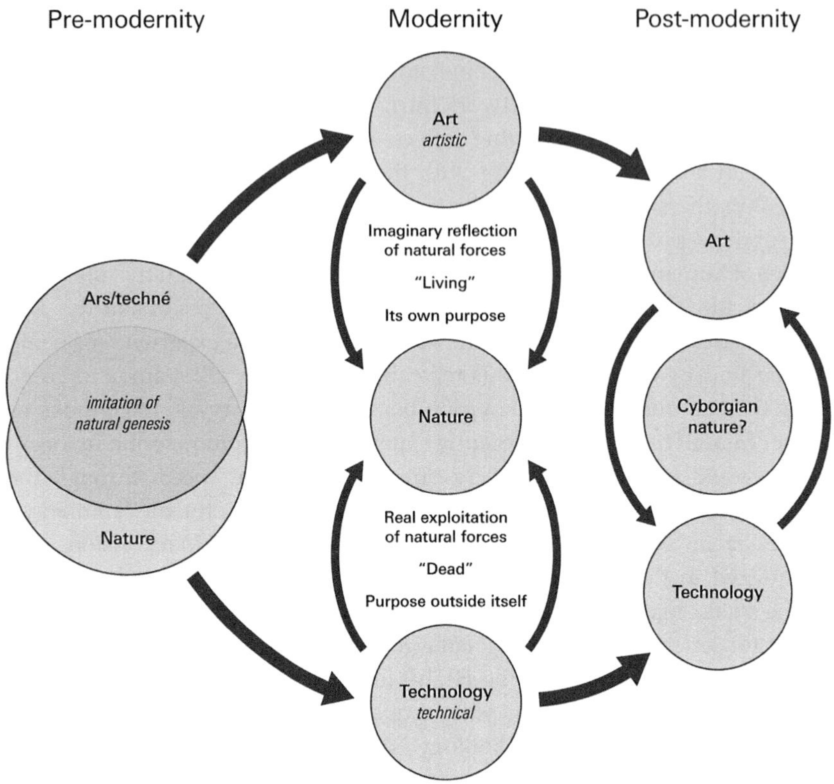

I.1 Diagram of the evolving relations between art, technology and nature. Illustration: Jacob Wamberg

indifferent objects to be manipulated; in aesthetics and its condensed form – art – subjective projection ensures that nature appears reanimated.

Both aesthetics and science, the two High-Modern observatory modes that underlie the more practice-involving regimes of art and technology, can characteristically be said to be somewhat detached from the interactive lifeworld of human interests. As Immanuel Kant shows in his first and third critique – the *Critique of Pure Reason* (1787) and the *Critique of Judgement* (1790) – though both forms of detachment are certainly dependent on sense impressions, these corporeal stimuli are nevertheless sublimated to higher, more abstracted versions in which the observer throws off any bodily or otherwise interactive engagement with the phenomena observed, whether pertaining to sex, appetites or politics. Kant's term, 'disinterestedness' – meaning detachment from physical or ethical concerns – although coined specifically to address the aesthetic experience, may be applied to both pure reason and to aesthetic judgement.[7] Science, its cognitive laws regulated by purely mental a priori assumptions, represents the most extreme form of this disinterestedness (objectivity). On the other hand, aesthetics, putting cognition into free, voluntary play with imagination (subjectivity), retains an involvement with the senses, since the imagination may be seen as a restructuring of fragmented sense impressions.

The missions of the clusters science/technology and aesthetics/art significantly part ways in the manner that each relates to the philosophical notion of purpose, or in Greek, *telos*. Thus, the detachment from corporeal engagement that identifies disinterestedness might be said to be freedom from purpose, *telos*. In the early seventeenth century, Francis Bacon, soon followed by Descartes, performed the groundbreaking, modern scientific deed of doing away with the Aristotelian heritage of explaining all nature's operations in relation to a *telos*, displacing final purpose in favor of disinterested natural laws.[8] In the modern paradigm in which these neutral laws function, nature is viewed as functioning like a huge machine, rather than – as was previously the case – an animate organism. Amazingly, when these purposeless laws are implemented in the domain of technology, with the human-produced machinery mirroring its natural counterpart, the object of the laws – the machine and its raw, natural matter – is given over to an opposing role, an absolute instrumentalization of human interests. The 'cynicism' of industrial technology is thus that it subsumes nature under a *telos* outside itself – namely, the bodily interests of one species, humankind.

In contrast to the change of function that happens when natural laws are transferred from nature's 'world machine' (*machina mundi*) to the machine constructed by humans – in the process shifting from absolute disinterestedness (the non-purpose of nature) to absolute interestedness (the purpose of human technology) – the implementation of the aesthetic experience in art retains its distance from human physical needs. For, according to Kant, the artwork establishes itself as its own *telos*.[9] This artistic autonomy mimics natural objects' lack of a pre-established *telos* outside themselves, whether divinely or naturally given. Yet, unlike technology, which fills out this lack with its humanly defined *telos* (nature as slave to technology), art construes this lack as dignity (nature as its own purpose). Thus, Kant notes that organisms are generated in a way that cannot be reduced to those general laws that govern the machinery of inorganic nature, and posits this freedom from those laws as an ideal yardstick for art's creations, stressing its freedom from an external purpose. Nevertheless, Kant emphasizes that the appreciation of art is also based on an understanding of its essential difference from nature.[10] Thus it is through a quite ambivalent sympathy that art assists nature in resisting the instrumentalization of industrial technology: on the one hand, artworks are comparable to self-contained organisms as they come into being in a freedom from generalizable rules and exterior purpose; on the other hand, artworks are essentially different from nature in that they come to 'life' only through the artist's subjective animation.

Primordial Beginnings: Art as a Magical Extension of Technology

If the strict distribution of the roles of art, technology and nature is what distinguishes the modern from the postmodern, it may also be considered a rule set that distinguishes the modern from the premodern. Although the essays in this volume focus on the post-medieval period, we hope that readers will forgive us for presenting a longer perspective, harking all the way back to the beginnings of culture in the early Stone Age; for such an evolutionary span may offer useful insights concerning the roles of art and technology, and their mutual relationship to nature, in the later periods, too – and, indeed, this is in accordance with our 'neo-Plinian' task of dissolving the absolute gap between natural and human history. Thus, one could ask, did not both art and technology emerge as parts of the same agenda: to improve upon a nature that somehow did not seem sufficient in its naked state? It seems uncontroversial to state that this was from the first the purpose of technology: by chopping trees with axes, hunting animals with spears, burning branches with torches and so on, goods and forces were wrested from nature which, but for these interventions using manufactured instruments, would have been unobtainable. Those phenomena that we today categorize as art – from ornaments, music and ritual dances to mimetic images and sculpture – were however probably at least partly directed towards the same natural goods and forces wished for by early humans. In the many cases in which technology could not independently mobilize those goods and forces, we will assume that art took over, and supplemented or completely outdistanced technology by means that we today categorize as 'magical': alleged remote manipulation, such as healing through chanting and vocal formulae, cursing through fetishes, influencing weather through rituals, increasing fertility through totem images and so on.[11] To the already large number of proposals for art's original evolutionary function, we will add the following: art as a magical extension of technology.[12]

To understand both the similarities and the differences between these oldest forms of art and technology more clearly, it is important to see that both these practices physically partake in a more abstract sort of semiosis than that found in the animal world. Whereas in the animal world, communication by means of signs always seems to refer to accessible phenomena in the immediate environment, human semiosis expands the range of referents to phenomena that are not immediately accessible or may even be inaccessible, such as past and future occurrences (e.g. birth and death), remote places, numbers and sets of instructions for practical procedures, such as the manufacture of new artefacts. As paleoanthropologist Alondra Yvette Oubré has suggested, this virtual world of not immediately present referents probably had very fluid boundaries with – and indeed may have been catalyzed by the emergence of – what we now term 'supernatural phenomena', spirits and magical agents and their collective home, a *beyond*.[13] These extraordinary realms can be seen as projections onto the environment of the surplus of images produced by developing consciousness: the matrix and imprint in the brain of the new, more abstract semiosis. As some of the more flamboyant ideas generated by this consciousness – what we moderns denigrate as just the fictitious products of 'imagination' – were homeless in relation to the immediately sensible outer world, special embeddings were produced for them that were partly material and partly retained the qualities of fleeting brain images. In short, these ideas condensed into ghosts, spirits or magical entities, including artefacts marked by 'immaterial' aspects. Together, these constituted a 'metaphysical' or religious sphere, a sphere reaching beyond purely physical qualities but nevertheless having its collective base in a remoter region of the perceptible realm.

The first materially based 'beyond', a remoteness presenting consciousness's excess of inner images as emanations from another world, was probably the subterranean domain of rocks, caves and tunnels, a hidden and massive zone, difficult to access, which, according to the Jungian theory of religion, served as the womb of the earth.[14] Such a macroscopic womb seems to be required as a source for the new, excessive semiotics, since the new, voluminous brain, spilling over with images, referential or not, emerged at a price: the human infant is born in a radically immature state, even compared to birds and mammals whose progeny already require prolonged parental care. As Lacan has suggested, the whole semiotic apparatus that the human being builds up, in order to establish an ego – from the initial, iconic self-image, to the armour of the alienating identity of developed language – may be seen as a filling out of the dehiscence accompanying birth, 'an organic insufficiency in his natural reality'.[15] We will regard precisely this 'primordial Discord' as the lack being felt upon the early expulsion from the womb, and therefore signs become substitutes for those things no longer, or not yet as present as those phenomena experienced in the womb's immediacy. Accordingly, the ultimate source of this consoling semiosis would be a compensatory, grander womb: the womb of the earth. The role of artefacts that we today call 'art' – that is, artefacts with a primarily or, from our viewpoint, exclusively semiotic identity, such as vocal formulae, ornaments and images – was probably to connect to and to mobilize, indeed to materialize, this subterranean womb, imagination's outer projection, so that it interacted with the sensible world and did marvellous, magical things with it. Thereby, it extended the immediate wish-fulfilment realized in the female womb, performing deeds that technology in its naked state, that is, un-empowered by indexical, iconic or symbolic signs, could not achieve. A notable example of this is Palaeolithic cave painting, for, according to a recent, anthropologically influenced theory, developed independently of Jungian-Lacanian theory, but clearly confirming it, Palaeolithic humans participated in shamanistic rituals in which they traced the forces of the fertile netherworld onto the cave walls, the very membrane between the here and the beyond.[16]

This initiatory function of art as a magical extension of technology does not accept the modern division that we saw unfolded in Turner's work – namely of 'doing', that is, physically moving and transforming dead matter (technology), and merely 'representing', that is, referencing to physical phenomena in a purely imaginary, detached, if vivid form (art). Rather, the initiatory function of art precedes this division through a performative category, doing through signs – which is exactly what we call magic when 'doing' allegedly concerns the physical realm, not just those socially constructed conventions such as initiation and marriage which are the basis of J.L. Austin's performative category.[17] In prehistoric cultures, in fact, the functions of 'useful' artefacts (axes, spears, clothes, torches and so on) were not completely distinguishable from those of 'aesthetic' artefacts (ornaments, songs, spectacular rituals, images, etc.); both were encompassed by rituals, decoration and their constituent magic. Here again, we see natural forces assisted by supernatural ones.

Ancient Wishfulness: Art and Technology Perfecting Nature

Such a proposed distribution of the roles of art and technology in the beginnings of human culture also allows us to better evaluate their future destiny, both in their interrelationship and in their relationship to nature. Thus, in the Western cultural sphere (the focus of attention in this volume), as late as Graeco-Roman antiquity and the Middle Ages, art and technology still constituted two parts of a common domain of artefacts and methods for creating them, extending from agriculture to medicine and from handicrafts to architecture via the 'properly artistic' domain (the mimetic arts) to the so-called liberal arts, methods of cultivating the free citizen, including grammar, dialectics, rhetoric, arithmetic, geometry, astronomy and music (the last-mentioned also being one of the mimetic arts).[18] Interestingly, this common domain – called *techné* in Greek, *ars* in Latin – not only bears witness to the originally shared purpose of all artefacts, whether functional or mimetic, but its etymology highlights the separation that was to come in the High-Modern era. Thus, *techné* and its derivatives later came to be reserved for technical and mechanical artefacts (what we call technology, today), whereas *ars* became the property of the modern artistic domain (the non-functional Beaux Arts, extracted from a broader base of aesthetically satisfying handicrafts). A similar etymological genealogy, crossing the domains of *ars/techné*, applies to the Greek *poiesis*, which originally covered the manufacture of any product pertaining to *techné* but is also the etymological root of the modern, non-utilitarian domain par excellence, artistic *poetry*.[19]

That art and technology originally coexisted under the same term – *ars/techné* – indicates that from antiquity to the Middle Ages, there existed a sustained desire to improve on nature from within, one that in many ways extended the presumed prehistoric agenda of magical art supplementing useful technology. Articulating, fortifying and partaking of the forces of nature, artefacts of all kinds – from those of carpentry to those of medicine, from those of sculpture to those of poetry – apparently sought to perfect what nature intended to, but could not, accomplish unassisted.[20] In the ancient and medieval cosmologies, this assistance rendered to nature in her striving for perfection was typically interpreted in an organic manner that extended prehistoric ideas of a living cosmos. Yet, a decisive turning point occurred in classical antiquity (or was at least urgently desired by the ancient philosophers), which complicated both the relationship between *ars/techné* and nature, and that between mimetic and functional artefacts: the displacement of nature's creative forces, from matter to spirit, from nature's bosom to an ethereal heaven lying outside nature.[21] This displacement apparently resulted from an increasing autonomization of consciousness, an intensified abstraction of semiosis, shifting the projection of the surplus brain images from the nearest perceptible remoteness, the subterranean, to the most distant perceptible remoteness, the sky. With this shift, the sky was upgraded to a superior reality, the heavens, from which not only the intellect but also all terrestrial things derived. In the most extreme version of this cosmology, we meet an ultimate inversion of brain images and material reality, in which the real is no longer comprised of things, however magical, but instead is comprised of eternal prototypes reproducing themselves as things: impressions in brittle matter of a higher, more durable, semiotic reality, whether comprised of super-icons such as Platonic ideas, or super-symbols, such as the Judaeo-Christian Logos.

In this dualistic world view, all created or existing things – humans, animals, plants and minerals and the artefacts that human hands made from these – could be conceived of as organisms, or at least their performance was to be judged by the ideal yardstick of an organism. The geocentric universe itself was the super-organism, in which the heavy, base material elements (earth and water) were animated into lower or higher degrees of organization via the influence of heavenly forces (from the lighter elements air and fire to the quintessence of the celestial spheres and on to the divine spirit) – all of which impregnated matter like semen.[22] For Aristotle and countless followers, such as Thomas Aquinas, art – comprising all cultivated products and their methods – was an activity that closely imitated and in some regards perfected this organic nature. Again, a case of art doing what nature intended but could not do herself. Echoing the celestial impregnation of nature, the carpenter, for instance, was seen as impregnating the passive material with the idea of his intellect via the work of his hands.[23] It is this close collaboration between art and nature that is reflected in Pliny's *Naturalis historia*, mentioned above. Pliny includes man-made artefacts in his vast encyclopedic description of all natural phenomena between heaven and earth, setting his pioneering history of the visual arts – architecture, sculpture, painting – in the context of the last books, which are dedicated to the materials of the earth, the stuff of which these arts are made.[24]

Spirit's Discontents: Nature Submerged by Art and Technology

Although both artefacts and organisms were ultimately conceived of as celestial imprints in matter, the semiotic forces of heaven reached matter in a more complicated way in the artefact than in the natural organisms. In the artefact, these forces had to make a detour, first, through the artisan's intellect (already a weakened version of celestial semiosis), and, second, through his hands – which, in a further distancing, had to collect dispersed materials and force them into shape with alien instruments. Compared with this circumstantial collecting, assembling and exterior shaping of the arts, natural creation appeared as more graceful, since it brought forth and bestowed its products spontaneously. After all, as mirrored in the etymological root of *natura* (related to *natus*, participle of the verb *nascor*, to be born), nature gives birth, whereas art merely constructs (the same relation characterizes the Greek *physis*).[25] Despite his belief in the possibility that nature could be perfected through the arts, Aristotle thus had to concede that 'the Good (Beautiful) is more fully present in the works of nature than in the works of Art'.[26]

This generative shortcoming of artefacts compared with nature could also be formulated in terms of violence. Because artefacts, unlike organisms, did not grow forth smoothly, arranging materials spontaneously, they forced Mother Earth, the former self-fertilizing super-organism still offering resistance to patriarchal ideology, into constellations – art – for which she had not wished. A strong symptom of this idea of violence against nature is the widespread topos of the fall from Paradise (Judaeo-Christian culture) or the sequence of World Ages, from the Golden Age to the Silver, Bronze and Iron Age (Graeco-Roman culture). Here, the development of civilization is construed as decadence, whether this be through ungraceful collecting and spreading of goods from unnaturally faraway places (commerce), scraping on Mother Earth's surface (agriculture), or worse, penetrating and exploiting the subterranean, her very womb, for minerals and metals (quarrying and mining).[27] Adding to this disgrace in the macro-handling of nature came the bodily pain associated with the activity necessary to produce all artefacts – work. Note, for instance, that the verb 'to work' and its Germanic variants, for instance, German *werken*, also had an earlier meaning of inflicting pain (German: *es werkt*). This disgrace associated with work was only deepened by the Graeco-Roman specialization of the artisan, for, since his product was destined for consumption by his superiors, rather than being a goal – *telos* – in itself (*praxis*), his work appeared as a servile activity serving his superiors' interests (*poiesis*).[28] This teleological displacement philosophically legitimized the ancient societal practice of enslaving artisans; for, according to a view formalized by Aristotle, were not artisans by nature destined to be living instruments serving their masters' needs?[29] In this relation, then, work not only inflicted violence on nature and pain on the workers' bodies; the workers themselves were converted into inferior instruments, prostheses for their masters. Thus in the ancient fluctuating attitude towards art – art as either the rape or the perfection of nature – we see an ambivalent relationship, with clear forebodings of modernity's dichotomy between cynical industrial exploitation and sentimental, aesthetic reanimation of nature; except that both a considerable part of the ancient instruments (the slaves) and ancient nature as a whole (Mother Earth) are more unambiguously alive than their modern counterparts.

Not surprisingly, all the ungraceful activities pertaining to the production of artefacts and upheld by 'necessary' violence added up to continuous civilizational discontents haunting advanced premodern cultures – a primitivistic trend striving for a pre-civilizational primordial nature without artefacts, be it Paradise or the Golden Age. In Paradise – a patriarchal raising of the earlier, subterranean womb to celestial heights – there was not yet, or no longer existed, a gap between spirit and matter, prototype and imprint, semiosis and act. This divine permeation not only made superfluous the domains of artefact, such as agriculture and, especially, metallurgy, but also hindered those natural organisms that succumbed to the post-Paradisiacal inertia of matter called death, the loosening of spirit's formative powers compensated for through nature's version of instrumental force: sex – that is, male penetration and fertilization of the female body. Just how ambivalent matters often were is seen again in Pliny's *Natural History*, which, despite its blessing of the collaboration between art and nature, nevertheless condemns human overexploitation of the earth and is looking back to a Golden Age–like state of things:

> We trace out all the fibres of the earth, and live above the hollows we have made in her, marvelling that occasionally she gapes open or begins to tremble – as if forsooth it were not possible that this may be an expression of the indignation of our holy parent! We penetrate her inner parts and seek for riches in the abode of the spirits of the departed, as though the part where we tread upon her were not sufficiently bounteous and fertile.[30]

In prehistory, and far into Egyptian and Mesopotamian cultures, technology comprised the entire procedure from the extraction of natural materials to their elaboration, and therefore was conceived of as an obvious and divinely sanctioned assistance to nature's processes. But technology was harder to legitimate in the elitist philosophies of classical antiquity and extensive parts of the Middle Ages and early modernity. Craft – reclassified from the ninth century into *artes mechanicae*, separating off from the *artes liberales* of antiquity – was now seen as a necessary evil, sustaining the whole of society, but in practice relegated to its lowest strata and as far as possible screened from the view of the elites because of its connection with pain (against nature and the body of the worker) and its engagement in matter at a far distance from spirit.[31]

This is not to say that the *artes mechanicae* did not sometimes receive praise from intellectual writers for their ability to help mankind or even incite wonder for their ingenious inventions: only that this happened in a contradictory and ambiguous manner, marked by the typical Western denigration of all things connected to the body and matter. If we look for later premodern ideologies that still wholeheartedly infuse technological processes with divine sanctification, we therefore tend to find them on the fringes of elitist society: among the artisans themselves, or among theoreticians, or amateurs, considered suspicious or heretical. In particular, a veneration of the whole metallurgical domain – right from the mining of the inner earth to the subsequent refinement and elaboration into artefacts – is represented by esoteric and secretive artisanal traditions that are often embraced under the umbrella of alchemy.[32] These traditions were continuously suspect, because they registered a presence of aliveness in material domains, which normally – that is, since the rise of the patriarchal hegemony – had been located at an unbridgeable distance from spirit and its organization of otherwise chaotic matter. Thereby the suspicion arose that these artisanal traditions still had not exorcized matter of its earlier forces of conception and did not decently reserve these forces to a transcendental beyond. The forces seemed thus to be operating in a space in which spirit was only a lighter form of a semiosis unfolding *within* nature, with its base in crude matter: in other words, a magical semiosis. Seeing earthly penetration and artistic elaboration of base matter as a fulfilment of nature's own intentions, completely devoid of problems – neither raping nature nor standing at a certain distance to her, transferring form via an alien intellect and its subservient instruments – was close to heretical.

Because of elitist scepticism about such a close artisanal collaboration with nature, that part of craft labelled 'mimetic arts' – the forerunners of modernity's aesthetic 'art' – also met with recurrent criticism in classical antiquity and the Middle Ages. The major criticism was the accusation that mimetic art was idolatrous: false or demonic manifestations of abilities to physically affect the surroundings. This scepticism thus denigrated no less than what we have proposed as the primordial function of art: to be a magical supplement to technology. In cultures which reserve both ordinary and extraordinary kinetic powers to a celestial Beyond and which restrict extraordinary kinetic powers to ever rarer miracles, things, and especially mimetic images, can no longer absorb in themselves any agency to physically move and affect. A worry that images are mistakenly believed to have such agency (a supernatural rather than supranatural performance) lies behind the iconoclastic movements of Hebrews, Muslims and Christians (including Protestants).[33] It can even be discerned in Plato's disappointment in the mimetic arts, resulting from his judgement of them by the same standards as used for functional objects and organisms, whose optical impression those arts allegedly imitate. Instead of demonizing images as idols – false gods with a poor or evil magical performance – Plato just denigrates them as *eidola,* false things with a poor ability to instruct about those phenomena whose mere sensual impression they imitate.[34] The mimetic arts thus come to be a category in themselves, to reach a quasi-autonomy, in antiquity, once their raison d'être is no longer either to supplement technology as magic artefacts or to incarnate the celestial divine as religious images – in both cases concentrating divine forces with the ability to physically affect – but rather to be merely an imitation of the optical, acoustic or otherwise sensual impression of things: in short, to move from performance to representation.

Art and Technology Acquire New Dignity

To gain a new dignity, no longer based on a now spurious magical performance, the mimetic arts in later premodern times thus had to form an alliance with that celestial Beyond from which Plato claimed they had been separated at two removes, imitating something (sensuous phenomena) that was already a weaker imprint of extrasensuous prototypes. This gap could be bridged through two strategies. In the first – hinted at by Plato, unfolded in greater detail in Aristotle's *Poetics* and having an expansive afterlife in Renaissance and neoclassicist cultures (e.g. Alberti, Vasari, Danti and Bellori) – the artist should improve upon the examples found in the natural world by, for instance, selecting among actual human body parts and restructuring these into an edifying whole. This would be a virtual version, then, of art as the perfection of that which nature intends but cannot carry through herself.[35] Like the apparently continuous form-giving which the celestial spheres exercised on passive terrestrial matter, this tradition is ultimately sculptural. In Hegel's clever reading, this tradition builds up ideal plastic volumes with clear contours and surfaces through which the soul of the figure, its formative principle, is fully expressed.[36] In the second strategy – the Neoplatonic heritage of the Middle Ages – the artist should as far as possible bypass sensuous phenomena and channel visions of the divine more directly into his image; this channelling also being possible through portrayal, however indirect, of holy personages and happenings, as these were thought to condense divinity in the mortal domain.[37] As Hegel also shows with his concept of the post-Classical Romantic era, this tradition is ultimately two-dimensional, turning the pictorial surface into a screen of revelation, of mediation of a reality not fully present in the range of outer material phenomena. Converted thus from a body to a screen filtering out material remains from the world represented, the image in this form obviously had to downplay magic, the performative alliance of sign and thing, and instead foreground what became the defining quality of modern art: the capacity to move the spectator on a purely imaginary level.

The draining of material qualities from the image occurred through a long and complicated exchange with the craft traditions, those of the ancient, common domain of *ars/techné*. Throughout the Middle Ages, and in the court cultures of the early modern period, from the 1400s to the 1700s, painters and sculptors were considered more or less sophisticated craftsmen, and regularly asked to perform 'lesser' tasks, such as banner-making, decoration and procession and stage design. Wishing to raise their status towards that of visionary intellectual and, ultimately, original author, both artists and their intellectual supporters appealed for the visual arts to be accepted among the ancient set of liberal arts, thus deepening and slightly shifting the break between *artes mechanicae* and *artes liberales* that was already under way in the Middle Ages.[38] Prior to the emergence of a formalized, systematic corpus of natural sciences in the late 1600s, a related pathway to a higher status of art was to directly blur the boundaries with emerging scientific observation and theory, injecting the crafts of the visual arts with the status of natural philosophy. As Claire Farago illustrates in her contribution to this volume, this is what the polymath Leonardo da Vinci strove for, targeting scholastic philosopher Thomas Aquinas's *scientia media*, a science grounded in theory and practice, *episteme* and *techné*. This strategy, handed down in Leonardo's corpus of writings, even proved useful for the Counter-Reformation, which had use for a legitimation of art based on the simple observation of God's creation, rather than on a virtuoso display of artifice, such as was evident in contemporary Mannerism. Also, in the new institution of the art academy, which emerged in Italy in the second half of the 1500s, an alliance between artistic and scientific principles was attempted and institutionalized, with the instruction of expanding scientific disciplines, such as geometry, arithmetic and anatomy. However, as the reader will note in Bjørn Okholm Skaarup's study, this alliance was marked by generous doses of wishful thinking, since what is truest to nature and cognitive theorizing about it is not necessarily the most beautiful, particularly not in relation to the classicizing decorum otherwise heralded by the art academies. Just as the pioneering art historian Giorgio Vasari had, in 1550, dismissed Paolo Uccello's obsessive inquiries into the laws of central perspective, so did the first overseer of the Florentine Accademia del disegno (established 1563); Vincenzo Borghini explicitly pleaded against 'excess knowledge' among the artists, who should 'do, and not reason' (*fare, e non ragionare*).

Northern European artists were less dominated by the constraints of Renaissance classicism, and often embraced less varnished renderings of natural observations, including explorations of anatomy. The early sixteenth-century wooden dolls made by the South German monogrammist I.P., explored by Markus Rath, are an extreme example of this. Jointed with strings, allowing their limbs to assume an infinite number of poses, these dolls transgress their simple function of being anatomical models for artists. Instead, they seem to be autonomous showpieces of artistic skill, presenting the artist as an *alter deus* who can not only imitate the appearance of human anatomy but also recreate and – through the numerous strange positions of the limbs – surpass the abilities of this anatomy. If the aliveness evoked by these dolls strikes the modern viewer as uncanny, in the Freudian sense of hovering in an ambiguous space between alive and dead,[39] this is no coincidence, for the 'natural' anatomy recreated in the dolls does not appear solely organic, but is to a large extent a piece of technology, a quasi-automaton. This joining of organic aliveness and dead mechanism marks a decisive step towards Descartes's explicit formulation of animal and human bodies as machines.[40] We find a more tortured fusion of dead and alive, one in which the mechanical aspect is primarily bound up with the dissecting gaze, in the pioneering work on anatomy, *De humani corporis fabrica* (1543), by the Flemish physician Andreas Vesalius.[41] In its accompanying woodcuts, made in the workshop of the Venetian painter Titian, a series of flayed bodies have arisen from the dead to perform a morbid, scientific striptease. Confronted with these exhibitionist zombies, one clearly senses that a sort of epistemological violence was necessary, in order to dissect and display the inner workings of the body (a corporeal architecture, turned inside out, not unlike the ancient ruins popping up in the background landscapes). Indeed, through the establishment of an anatomical science based on penetrative dissection (cf. 'anatomy', from the Greek *ana* = 'through' + *temnein* = 'to cut'), the very object of this science – the body – loses its divinely given vitality, and is transformed into a dead mechanism whose workings may, to a large degree, be described without recourse to the idea of a soul, and which, quite logically, presents itself as a moving corpse.

In the new natural sciences, all physical matter became the object of a new, absolute control, established in an alliance between mechanical instruments and an epistemological description based on ironclad laws. Francis Bacon, one of the founders of modern scientific method, designates the miner and the smith as the natural philosopher's ideal models: 'the one searching into the bowels of nature, the other shaping nature as on an anvil'.[42] In this exaltation of metallurgy, the ancient bad conscience about desecrating nature by exploiting her resources, as exemplified above in Pliny's *Natural History*, undergoes a complete volte-face, celebrating the enslavement of nature, comparable to the inquisition of witches during this period. According to Bacon, nature thus 'exhibits herself more clearly under the trials and vexations of art [i.e. mechanical art] than when left to herself'.[43]

Interestingly, the generation of knowledge through absolute dominance of nature – the origin of the previously mentioned 'cynicism' of modern technological culture – liquidates nature so thoroughly that she is turned into neutral, genderless matter, no longer calling for, or receptive to, sculpting by a heavenly spirit, or attraction by a Ptolemaic unmoved mover.[44] In the expanding, ultimately infinite world view of the later Middle Ages, the heliocentric universe presented by Copernicus in 1543 – the same year in which Vesalius dissected the microcosm of the human body – the ancient, dualist tension between celestial, motive spirit and terrestrial, passive matter loses its foundation in the observed universe, giving way to an infinite distribution of kinetic energy, *impetus*, with no intervention from God, who has seemingly long since abandoned his creation.[45] This is the mechanical Cartesian universe, whose mathematical laws were systematized by the progenitors of the new physical sciences, including Kepler, Galileo, Newton and Boyle, and the kinetic energy of which may be used by machines – automata superseding the animate force used in earlier handicrafts. Indeed, the prime metaphor for the modern world view – the universe as clockwork – itself indicates that the new natural sciences came into being in intimate dialogue with the expanding practice of making new mechanical instruments: technological devices that control, channel and distribute the forces of nature for human use, more efficiently than before, from mills and wheel ploughs to military and mining inventions to steam engines. There is seemingly no further need for the animation of matter from within (magical) or from without (divine), since machines and nature alike appear as constellations of neutral entities, each receptive to ever-new amounts of equally neutral and un-teleological kinetic energy.

Between Premodern Animation and High-Modern Disenchantment – and Beyond

The disenchanted, mechanistic world rose through a complicated give and take with the animate, organismic world inherited from the ancient and medieval cultures, and did not fully supplant it until the 1800s. Even more than the art academy, an early modern forum for maintaining the boundaries between nature, emerging natural science, technology and imitative art fluid until surprisingly late, was the Wunderkammer, or cabinet of curiosities, in which at least two of Monogrammist I.P.'s wooden dolls ended up. This heterogeneous progenitor of the more specialized Enlightenment institution of the museum combined a new culture of observant curiosity with the ancient idea of a fundamental creative drive in nature, the human agency of which was still merely a superstructure that imitated, competed with and in some ways perfected its natural base. In the Wunderkammer, which made many a prince proud of his possessions, mirroring the political macrocosmos in an epistemological microcosmos, all imaginable kinds of wondrous specimens were grouped together, crossing the boundaries between stunning *naturalia* (minerals, plants, animals) and amazing *artificialia* (handmade objects of any kind, including extraordinary pieces of decorative art, e.g. weapons, puppets, silver- and goldsmiths' work, jewellery and medals, ancient sculpture, painting and graphic art, not to mention the brand-new wonders of a rising technological culture: sophisticated mechanical instruments [*scientifica*] and automata).[46] A specialty of the Wunderkammer culture, one covered in Lisbet Tarp's contribution, were hybrid objects that combined nature's sketches with art and technology's fulfilments. Thus, around the year 1600, artists often used natural materials to enhance marble and alabaster images of hybrid mythological creatures, for example, a mermaid with breasts made from the naturally shaped bulges of an inserted pearl, or a small-scale sculpture of Daphne turning into a tree, the latter represented by an attached coral branch. According to Tarp, works such as these may be viewed as experiments conducted in parallel with the still-ongoing alchemical investigations, both sorts of activity perfecting nature's processes, using the principles of nature herself. Yet, these hybrid artworks also illustrate the gradual transformation to a modern world view, in which wondrous, natural products emerge as a result of pure accident, 'images made by chance', and in their incompleteness demand the projection of the artist's creative vision onto the passive material.

The expanding appreciation of nature's accidental happenings was informed, too, by what might be taken as the complementary position to artistic subjectivity: an objective desire to present and preserve natural phenomena in their naked and unadorned state. In fact, to both the modern artist and the modern scientist, the wonders of nature were increasingly revealed in ordinary phenomena, each of which appeared to be singular, irreducible and, therefore, paradoxically extraordinary. Nonetheless, whereas for the artist this extraordinariness inspired an even more singular subjective projection, for the scientist, it was only the variable, external appearance of a specimen, an example of an objective category in one of the new taxonomical systems that became the trademark of the new Enlightenment sciences, from botany and zoology to geology, chemistry and physics. With regard to the visual arts instrumentalized to preserve some of these numerous natural specimens in and outside the encyclopedic works of the scientists – miniatures, paintings, drawings, engravings – ideally, traces of human handicraft, projected subjectivity, should as much as possible make way for an objective gaze, one reactualizing in the disenchanted modern world the mythical *acheiropoietos*, the image made not by human hands.[47]

The ultimate fulfilment of the desire to objectively depict nature in its naked state is photography, a representational format that transfers the power of fixation from the hand of the artist to a combination of natural forces and automated technology. However, before the stage was reached at which light could be fixed with silver halide crystals, there were 'dress rehearsals' with various tracing and print techniques, one of the more remarkable of which was the direct, repeatable imprinting of plant species, studied in Alisa Luxenberg's contribution. Presented to the Spanish king in 1777 by the sculptor, and later art writer, Celedonio de Arce, this indexical technique promised to reduce labour and to bypass unreliable, subjective human handicraft – the more exquisite mimetic forms of which were now conceived of as 'art'. However, precisely because of this rising specialization of fine art, the Royal Art Academy of San Fernando in Madrid, to which de Arce in vain appealed to be a member, found the new technique's dry registration to be outside its area of competence and interest. Ironically, since botanists were accustomed to artistic, three-dimensional-looking, colourful renderings of plants, usually reflecting Platonic idealization, they too failed to appreciate this technique's flat, monochromatic imprints, often marked by imperfections, such as worm holes or broken leaf tips. Therefore this technology was soon neglected and forgotten. According to science historians Lorraine Daston and Peter Galison, we have to wait until the 1850s, and indeed the spread of photography, before the heritage of Platonic idealization in scientific illustration was thoroughly undermined, and a sense of objectivity, informed by optical precision, was institutionalized.[48]

An inexhaustible provider of singular occurrences in nature, and therefore a particularly challenging arena for subjective or objective representation, is the domain explored in Turner's *Rain, Steam and Speed*: the weather. As may be deduced from the dual connotations of the term 'atmosphere', describing both the objective air of the environment, and the subjective mood it creates, weather phenomena (sky, clouds, mist, wind, precipitation and their refraction of light) were crucial for landscapes – the genre that reanimates objective nature through our subjective projections on it. John Constable, Turner's great contemporary in British landscape painting, noted of the sky that, 'It will be difficult to name a class of Landscape, in which the sky is not the "*key note*", the *standard of "Scale"*, and the chief "*Organ of Sentiment*" … . The sky is the "*source of light*" in nature – and governs every thing [Constable's italics]'.[49] Interestingly, Constable's sensitivity to the mood-generating potential of the changing sky did not preclude his involvement with the emerging science of meteorology. As Hanna Johansson demonstrates in her essay, his many autonomous cloud studies from the 1820s, painted under the open sky, exist in an indeterminate position between objective study, complete with notes on weather conditions, and subjective response, culminating in a thick impasto and undecipherable forms that anticipate abstract expressionism.

Johansson takes Constable's cloud studies as a first step in the groundbreaking, historical transformation of seeing art, technology and nature as moving into a closer relationship with one another, as we approach our own time. Thus, Turner's above-mentioned desire to stain the canvas with the humidity he perhaps felt when transported through the mists by a locomotive, came closer to fulfilment when, in 1974, British filmmaker Chris Welsby adjusted his film camera to shifting weather conditions, literally letting rain drops hit and remain on the lens. It came even closer, when Danish-Icelandic artist Olafur Eliasson, in the 1990s, generated light-sensitive installations of real, if artificially produced, water droplets. Apparently, the price paid for art's embracing of a physically discernible atmosphere is that the 'weather' now presented is tainted in various ways, if not, as in Eliasson's case, directly generated by, technology. The getting closer to the 'real' in art presupposes a flowing together of artificial and natural, and of art and technology. A recreation of a whole weather environment was already attempted a hundred years earlier, when the physicist C.T.R. Wilson developed his famous cloud chamber. Elaborating on Peter Galison's findings, Kristine Nielsen notes in this volume that in this chamber, the border between exact sciences and aesthetic, even metaphysical, concerns was perforated. Despite its groundbreaking significance for our understanding of subatomic particles and the conditions of cloud formation, paving the way to Wilson's Nobel Prize in 1927, the cloud chamber was invented because the Scottish physicist desired to recreate the spectacular optical phenomena he had experienced on the summit of his native mountain, Ben Nevis. Wilson was seemingly guided by a partly idealized and nostalgic notion of the supposedly extraordinarily pure air existing on this highest of British summits, but, paradoxically, this hyperbolically informed purity was the very reason the cloud chamber worked as a scientific instrument.

The external atmospheres recreated in Wilson's and Eliasson's artificial interiors are both results of a certain Romanticist vision, caging bits of virgin nature in technological frameworks. As indicated in Caroline Jones's essay, this vision probably attained its most extreme modernist form when, in 1967, Hans Haacke expressed the explicit desire to make pieces that were 'absolutely independent of the viewer's mental participation', and eliminated human factors altogether. In the *System Art* shown at the Hayden Gallery at the Massachusetts Institute of Technology that year, the viewer was witness to the unfolding of natural and physical processes, seemingly without any appeal to his or her subjective participation: grass seeds germinating in a pile of dirt, silk chiffon flowing in ribbons and waves, 'rain' percolating through small holes in transparent acrylic plastic, and – again evoking Wilson and Eliasson – a whole 'weather' cycle of precipitation, from liquid to droplets to mist, unfolding inside a Plexiglas cube. Yet, the very fact that these instances of allegedly 'untouched' natural environments were staged inside technologically produced confines, and furthermore, that these confines themselves were presented inside the art institution, again presents us with a situation in which the borders between art, nature and technology are highly destabilized and negotiable. No wonder Haacke soon plunged into works explicitly embedded in social concerns, shortly after abandoning the interface to 'pure' nature still seen in his transformation of extremely polluted river water into clean, fishable water, in the *Rhine Water Purification Plant* (1972).

The idea of unspoilt pieces of nature caged in the frameworks of art (or technology) had already received an ironic treatment in Duchamp's glass ampoule, *Paris Air* (1919). The notion of 'Paris air' itself evokes the entire picturesque tradition that we saw represented in Turner's depiction of weather, and according to which, a slightly polluted atmosphere – here in the form of metropolitan air – evokes a mood, a feeling for a place. But Duchamp mocks this tradition through his overtly analytical stance, encapsulating a ridiculously small sample of air, thus draining it of all the usual sensuous and mood-evoking qualities of art, including visibility, and making it dependent on a semiotic designation. On a hugely amplified scale, the contemporary climatic situation that Søren Pold and Christian Ulrik Andersen analyze resembles Duchamp's disjunction between the atmosphere and the senses. Paradoxically, that technology's waste products have now penetrated nature's atmosphere to such an extent that the epistemological boundary between nature 'out there' and culture 'in here' has become thoroughly unstable, makes physical impressions (such as those Turner sought) a less reliable guide for acting in the globally warming climate. In the words of Pold and Andersen, the weather has become 'semioticized', since 'our immediate surroundings are now being mediated by complex visualisations, statistics and carbon quotas': a technological interface through which what you get (the weather) is bound up with what you expect (the climate). Thus, the rising genre of climate art acquires a role in making us more conscious of the choices involved in this interface. Hereby – so, at least, the utopia states – art provides a critical space for reflection, through which we may assume the roles of responsible actors in the technological-natural atmosphere; move from what Bruno Latour has termed 'matters of fact' (High-Modern Cartesian observation) to 'matters of concern' (postmodern participation).[50]

Modernist Technophilia, Avant-garde Subversion

The recent state of affairs, in which art becomes involved in the actual function of technology at the same time as the boundaries between technology and nature have become indeterminate, has evidently moved some steps away from the High-Modern situation – with nature functioning as a more or less passive background for the mutually contrasting enterprises of art and technology. In many ways, this was the distribution of roles that Heidegger mourned in his post-war meditation, 'The Question Concerning Technology' (1954), an essay that has itself moved a long way from the harmonious working together of art, nature and craft, which Heidegger (though in a spurious embedding of Nazism) presented in 'The Origin of the Work of Art' (1935–36).[51] Whereas in the earlier essay art is the happening of truth (for instance, the truth of tools or of hidden nature) through de-familiarization away from habit, in the later work, art has degenerated merely to something artistic and aesthetic, swallowed up in a sector of 'cultural creativity' (*Kulturschaffen*).[52] At the same time, modern technology is no longer the innocent craft born of the earth and its inhabitants postulated in the earlier essay, but has become a mere instrumentalization of physical science, which demands that nature present itself in a calculable, fixable manner as a system of information. In this applied version of theoretical knowledge, nature is transformed into what Heidegger calls *Bestand* (a reservoir of energy resources), which is permanently accessible to *Gestell* (modern technology as an inhuman 'scaffold', an insatiable apparatus for uncovering reality in its totality, as *Bestand*).[53] This complete separation of the functions of art and technology puts humankind in the utmost danger as it threatens to turn it too into mere *Bestand* for the industrial apparatus. To avoid this, Heidegger urges us to reclaim the essence of technics, pointing precisely to the ancient domain of *techné,* in which both practical and mimetic arts were pervaded with what in modern times has been reserved for art: a concern for beauty and poetry.[54] In both its sceptical description of an aesthetic art separated from a technology based on the calculation of physics, and in its hope for a renewed rapprochement between art and technology, inspired by the ancient, common domain of *techné,* Heidegger's analysis corresponds to our historiographic scheme, and should be seen as a crucial inspiration for it.

Heidegger's dream of a reawakened *techné* may be taken as the philosophical parallel to the well-established desire of certain parts of the art world to reintegrate artistic and technological practices, a desire which thrives as a shadow to modernity's mainstream separation of the two and cultivation of art in an ultra-non-utilitarian realm. This countertradition to Kant's vision of the artwork as its own *telos*, art for art's sake, is often, as in Heidegger, formulated in specific antagonism to industrialization and its mass production: so that the technology with which it is desired that art should fuse is once again a more primitive technology, typically reviving the medieval practice of collectively produced handicraft. So it is, for instance, in the English Pre-Raphaelite tradition from John Ruskin to William Morris, and also to a high degree in the ideals of *Gesamtkunstwerk* informing various versions of Art Nouveau, right up to their complication and purification in the German Bauhaus of the 1920s. Despite its name, evoking medieval collective handicraft, Bauhaus nevertheless represents a new embrace in the arts of industrial technology and mass production, something that typifies modernist functionalism as a whole. This embrace has as its emblem a universe of purified geometric forms, a distillation of that calculable domain which underpins the high-industrial production, and thus it marks a spreading of this domain away from its obvious area of implementation, machines. Firstly, it invades 'autonomous' art such as paintings and sculptures (which, however, also supplement geometry with its contrary pole, an abstract purification of that turbulent chaos which still in Turner includes naturalistic signifieds). Secondly, it marks utensils that, because of their status as prostheses for the human body, could in themselves be considered pieces of machinery, now aesthetically formed by functional design, the mass-produced version of handicraft. And thirdly, it is implemented in architecture, that shaping of the environment which has always had an insecure identity between high art and utilitarian craft, but which now ultimately turns into those 'machines for living' which Le Corbusier, a prime promoter of modernist functionalism, considers the ideal of building.[55] Because utensils can be considered easily detachable prostheses for the human body, they are less of an obvious threat to bend this body according to their own, possibly inhuman, interests. Architecture, however, which structures the whole environment into fixed spaces, is less easy to govern and therefore more susceptible to transformation into a hegemonic cage imprisoning the human being within preprogrammed norms. If functionalist buildings are to be considered 'machines for living', and those machines are interpreted as pieces of industrial technology marked by Heidegger's dialectics of *Bestand/Gestell*, then modernist architecture is a particularly illuminating example of Heidegger's most nightmarish vision: that humans themselves turn into a passive reservoir, *Bestand*, for an all-demanding industrial apparatus, *Gestell*. For as they are already inside the machines for living, humans are by definition already swallowed up by them, as by the machines in Chaplin's *Modern Times* (1936). In this perspective, critiques of modernist architecture as inhuman and technocratic, like Tom Wolfe's postmodernist classic *From Bauhaus to Our House* (1981), become particularly revealing.

Ironically, although Bauhaus and other modernist trends share with Heidegger a desire to fuse art and technology, to a Heideggerian sensibility they mostly fail, because the technology with which art is fused is the High-Modern technology that reduces material reality to cognitive calculability: that lawful domain which, according to Kant, is absolutely not part of the aesthetic realm. Calculability and lawfulness may seem not to instrumentalize human interests and intentions, as long as the laws are reserved for virgin nature, which, as we have seen, is portrayed by the modern natural sciences as emphatically devoid of purpose and therefore unresistant to the purposes for which 'she' is instrumentalized by humans. But again: as soon as those natural laws are implemented in human technology, absolute absence of purpose is converted into its opposite: absolute purpose. For the High-Modern ideal of technology is to be an instrument for human intentions, to act as extensions of the human body, smoothly fulfilling those intentions. The problem with architecture in relation to this dialectic between absolute non-purpose (mechanical nature, out there) and absolute purpose (the artefacts which humans make out of mechanical nature, in here) is that it is located in an unstable space between near and remote, in an in-between position or place. To refer to three of the roles for technology distinguished by the philosopher of technology, Don Ihde, architecture is both too distant to fulfil the role of mere prosthesis for the human body (*embodiment*) and too encompassing to allow the body to operate upon it as a quasi-autonomous object (*alterity*); but at the same time it is too close to appear as a securely distanced *background* for this body.[56] So when ultra-purposeful technology, based on nature as calculable entities, colonizes this in-between zone and turns it into a machine for living, the roles between master and servant are inverted: what ought to be an instrument for humans – technology – becomes an uncontrollable monster, *Gestell*, swallowing up humans as part of its *Bestand*.

So even if architecture and design are parts of the artistic domain 'born' with a functional rather than merely contemplative dimension, this heritage is no guarantee that Heidegger's dream of a coming together of art and technology will turn out to be a happy meeting. In fact the likelihood of unhappiness increases as artistic interests are formulated increasingly in terms of the functional goals of technology – as happens in Adorno and Horkheimer's *Dialectics of Enlightenment* (1944), in which the machinery of capitalism has absorbed large parts of the arts into a whole Culture Industry. Besides satisfying capitalist consumption, this industry of easily digestible goods makes humans forget about a rationality that no longer serves to liberate them from wild nature, but rather masters them in the form of a perverted 'second nature' comparable to Heidegger's *Gestell*.[57] Despite this expanding movement of technology, perhaps the truly scary synthesis is arrived at when technology meets and merges with a movement in the opposite direction: a (just as colonizing) aesthetization of everything. Thus in Hal Foster's *Design as Crime* (2002), the Heideggerian *Gestell* is reformulated as an industry of total design, in which all artefacts – 'not only architectural projects and art exhibitions but everything from jeans to genes' – are formed to serve the capitalistic circuit of consumption most efficiently.[58]

What then might a happy – or at least a less problematic – meeting of art and technology look like in the culture after 1900? Although we have traced the ambition to merge art and technology in both modernist functionalism and various popular, mass-produced art forms, according to Peter Bürger's famous thesis, it was first and foremost the historical avant-gardes who wanted to interfere in real life and thus transgress the Kantian notion of art as disinterested in the purposes of its surroundings. However, although engagement with technology does seem to be a continuous feature of avant-garde movements, we must note that in innumerable cases technology seems to have become alienated from smooth functionality. It has been portrayed as demonstratively non-utilitarian (as in the industrially produced and geometric forms of minimalism), worn out (as in the garbage of *Nouveau Réalisme*), ironic (as in Dada collages which combine machine parts in pictorial form), abstract (as in the paintings and sculptures of concrete modernism, modernism thereby transcending smooth functionality when remaining autonomous), playful (as in the everyday tools of Fluxus), reflective (as in readymades) and even directly destructive (as in the exploding machines of Tinguely). Thus life-interfering modernism and avant-garde art can perhaps be said to part ways in relation to technology's functionality: modernism wishing to facilitate it, avant-garde art wanting to interrupt it by appropriating existing technology in extraordinary ways or dreaming of a technology not yet seen. At the negating end of this spectrum we find avant-garde ideologies such as *pataphysics*, a proto-Dada fake science inverting physics' ambition of lawfulness into a concern for the extraordinary and singular.[59] On a more constructive note, the post-war Situationists, for instance, desired to break down the barrier between art and technology, play and work, and to create in the borderland a playful technology beyond utilitarian concerns.[60] The same could be said of Experiments in Art and Technology (E.A.T.), a collaborative enterprise between artists and engineers, including Robert Rauschenberg and Billy Klüver, launched in 1966 with an innovative series of performances in New York City (*9 Evenings: Theatre and Engineering*).[61]

However, that there still exists a 'decorum of subversion' pertaining to art appropriating technology can be felt in the hesitance with which art that explores technology explicitly is received in the art institution. This kind of art – foregrounding digital media, mechanical kinetics, electric light, interaction or biotechnology – is to a large extent ghettoized in special technologically oriented institutions and festivals like ZKM in Karlsruhe, InterCommunication Center (ICC) in Tokyo and Ars Electronica in Linz.[62] On the other hand, art museums and galleries maintain a certain technophobic resistance to new media, with video art remaining their outermost frontier. The 'decorum of subversion' aside, a contributory factor in the art institution's scepticism is sheer confusion about how to tackle kinetics and interaction in art. Kinetics has become accepted in video art (because it is part of basically static installations) and to a certain degree in performances with human bodies, but when sculptures really begin to artificially move and interact with their surroundings, as in technologically based art, the art institution is especially alert as to the artistic quality. Doesn't this art risk dropping the avant-garde's obligatory bite and degenerating into gadgets from the experience economy – toys for grown-ups (too much bodily fun) or didactic presentations of new technologies (too much cognitive fun)? Thus although modern art has in a way been flooded with technology for some time, the gates between the two domains can probably only be declared safely open when the principles of free creativity are no longer threatened by capitalist rationality camouflaging as sensuous (but senseless) spectacle.

In any case, whatever the exact method used, avant-garde art typically alienates technology from its already-known utilitarian functions. This artistic alienation corresponds quite closely with Heidegger's observation in the mid-1930s that an artwork recovers things, including utilitarian tools, from the oblivion of habit through making them uncanny in relation to their normal settings and uses.[63] In *Being and Time* (1927) this realm of habitual use was termed 'ready-at-hand' (*zuhanden*), a subconscious prolonging of the actions of the human body. In everyday life this state is only interrupted when equipment gets broken, exhibits a functional lack, or demonstrates a friction in its use so that the user becomes conscious of it and is led towards (although not fully arriving at) a state of distanced observation, the so-called present-at-hand (*vorhanden*). Fully realized – that is, not having its point of origin in the ready-at-hand but beginning from a distanced observation – the present-at-hand can be identified with the above-mentioned modern subject/object relationship, that oppositional stance which Heidegger considers the deeply problematic heritage of modern Western culture and which culminates in technology as *Gestell*.[64]

In its treatment of technology, then, avant-garde art artificially stages a withdrawal from the ready-at-hand to an almost (but not fully) realized present-at-hand, in a displacement that is both contemplative and non-contemplative, both aesthetic and non-aesthetic, generating a positive friction towards a potentially dangerous non-reflected use of technology. The prototypes of this practice would be Duchamp's readymades, with their disrupted everyday technologies: the upside-down, and therefore non-utilitarian bicycle wheel simultaneously sabotaging the stool's function (*Bicycle Wheel*, 1913), the toppled urinal implicitly returning bodily waste (*Fountain*, 1917), the snow shovel's potential for crippling you deferred by instead exhibiting it (*In Advance of the Broken Arm*, 1915).[65] If we factor in here the Heideggerian scepticism towards modern industrial technology, we could perhaps even hear Duchamp ask if the readymades were ever fully ready-at-hand, since having an identity as designed and mass-produced items coming out of an industry based on that present-at-hand-ness, that distanced calculation, which Heidegger despises, they could again be potential representatives of a totalizing *Gestell*, a machinery enslaving the human, rather than smoothly prolonging his or her actions in ready-at-hand relationships. Whatever the exact relationship being questioned by the readymades, it is at least interrupted through their reflective aura.

Although the readymades are symptomatic of a new engagement with technology in art – one that breaks with the technophobic tendency of the Romantic *L'art pour l'art* heritage – in their wish to interrupt technology in movement and function rather than surpass it they look rather distant from their prehistoric, magical forerunners. Yet magical artefact and readymade do meet in a fusion of functional tool and semiosis, generating power through an aura of signification. In fact, this performative dimension – created in the readymade by displacement from normal use – often gives the avant-gardes' appropriated pieces of technology a character of fetish, of neo-primitive artefact animated by some hidden force. In some cases the fetishism invoked transforms an aura of fetish already possessed by the piece of technology via its status as a (whole or fragmented) commodity circulating on the capitalist market.[66] In any case, the point is that even though the readymade does not behave exactly like a magical artefact, allegedly exerting physical influence on bodies or things, it does share the magical property of extraordinary agency. By transforming both the potential for ready-at-hand-ness (the tool as enslaved by the human body) and *Gestell* (the human body as enslaved by technology), the readymade creates an intensified alterity, a relation to technology, which Ihde exactly compares to anthropomorphism and of which he considers the animistic idol a forerunner.[67] This, then, is not the subject/object relationship of distanced epistemology (*vorhanden*) but something akin to Emmanuel Levinas's otherness, the ethical quality of the infinite difference of another human or living being.

Among the utopian possibilities lying behind so much avant-garde alienation of technology, an important one could thus be that the interruption of habitual use with its over-strict teleological thinking and practice (the dichotomy of master/slave, *Gestell*/embodiment) and the generation of a new alterity establish the potential for humans to truly interact with technology: that is, to act in an ensemble whose constituents each have their own agency. Indeed, this utopia seems to be further explored in different kinds of actionism and performance in which bodies act together with all kinds of artefacts and raw matter, from Dadaism to Vienna Actionism, from happenings to feminist performances. Analytical frameworks for such ensembles could be theories in which the human subject is decentred in relation to the co-actors – whether animate or inanimate, organic or artificial – with which it moves: from Peter-Paul Verbeek's idea of technology and the human agent as a new thirdness, generating an internal and ad hoc subject/object relation, to Bruno Latour's actor-network-theory (ANT), fully relinquishing the subject/object relationship, to Deleuze and Guattari's radically antihumanistic notion of the assemblage, the Body without Organs, switched on to the rhizomatic network.[68]

Reentering Nature

What now of nature, in this performative field that includes a plethora of heterogeneous agents? First of all, nature loses its status as the object of distanced contemplation, whether cognitive or aesthetic, in which modernity strove to posit it. Indeed, in becoming co-player in the same space as culture, nature can no longer perform the role of the one pole in the dichotomy nature/culture, and thus ultimately loses its meaning as a defining concept.

In visual art, this loss is manifested most obviously in a crisis of the genre which Turner explored so eloquently in *Rain, Steam and Speed*, and which could be seen as a hallmark of nineteenth-century painting: landscape as panoramic representation. To be sure, as Robert Rosenblum was perhaps the first to indicate,[69] much of the stuff of which landscape is made – the sublime, with its monotonous vastness and disorganized natural matter – lives on in abstraction – namely, in that chaotic tendency which, together with its above-mentioned supplementary side, geometry, purifies representation from unambiguously identifiable figurations or referents, notably organic bodies. Abstraction, then, does not signify an exclusive seeing away from any physical world, the 'downcast eyes' of Martin Jay;[70] rather, the eyes are directed towards hitherto unclaimed areas of reality, be they 'inner' sensual and bodily realities or 'outer' realities without unambiguous referents, both parts being more accessible through the ever-expanding micro and macro prostheses of optics. However, through this dissolving of known referents, familiar distances based on perspectival foreshortenings also break down, and with them the dualism of subject and object which is visually founded in placing a detached observer, a subject, at the apex of a perspectival cone whose sides lead to an infinity of objects against which the subject measures itself.[71]

As this infinity of objects amounts to nature, the crisis of the subject/object dualism is closely bound up with that of the culture/nature dualism (or the technology/nature dualism). 'Culture' here is to be understood as the subject's outer scaffolding, the near material surroundings that establish subjectivity through their being processed according to the needs of the subject, which they simultaneously build up. In this generation of subjectivity, the notion of landscape is a double-edged sword. On the one hand, it ensures that the subject can be profiled against an unprocessed area (nature) outside the subject's outer scaffolding (the obviously processed surroundings of the city). To take pictorial art as a case in point, in the fifteenth century, the era in which the very notion of landscape is introduced, a new prospect of cultivated nature appears that has not been seen since ancient Egyptian and Neolithic image culture: a grey zone interspersed with fields, hedges, fences and roads, which expands in the landscape imagery of the following centuries.[72] To secure a stable, even intensified culture/nature dualism, then, these technological interventions should act in accordance with Ritter's principle of sentimentality: landscape as aesthetic consolation for the sins of industry. In other words, they should cast off their obvious associations of utility and appear as if reabsorbed into nature's bosom – which happens in such frictions with nature as erosion, ruination, patina, overgrowth, rustic irregularity, phenomena that can all be summarized in the notion of 'the picturesque', a sort of sublime light.[73] However, this primitivizing absorption of culture into nature, the 'scaping' part of landscape, reaches critical saturation with the Industrial Revolution. Industry's newcomers – factories, railways, telegraph lines and an expanding, less strictly delimited metropolis, not to speak of later additions like power plants, motorways and aeroplanes – are not so easy to naturalize in the picturesque fashion. And so either the artists have to dream away the more recent inventions of technology (as happens in different versions of Romanticism) or else the 'scaping', a general tendency of all perspectival space, becomes an interlacing in which technology has just as prominent a part as nature in the panorama. This de-sentimentalized convergence of modern technology, its laborious production processes and nature, is most directly represented in nineteenth-century Realism and its twentieth-century followers, notably Socialist Realism and other totalitarian art movements.[74] A more distanced representation of this convergence, in which perceptual mediation and its pictorial equivalents in brushwork are foregrounded, is found in Impressionism. Thus, with its shimmering vision of a train rushing forth in the middle of a sublime fog composed of both nature's and the machine's exhalations, Turner's *Rain, Steam and Speed* could be placed precisely at the turning point of these later developments, mixing far-flung Romanticism (the image of unbridled nature) with both harsh realism (the image of technology in control of nature) and distanced proto-Impressionism (the image of the shimmering atmosphere as the stuff of unbridled artistic licence).

So, in whatever form we meet the remnants of 'nature' in twentieth-century art, it is in a clearly de-sentimentalized form in which nature no longer stands out as a clearly recognizable other stabilizing culture and the subject. As we discussed previously, this was the case even in late Romantic offshoots, such as Haacke's nature cages, or Eliasson's misty environments, all of which let their otherness bloom inside a double armature of technology and art. In Jacob Wamberg's interview in this volume, Eliasson elaborates on the way in which these explicit, techno-artistic frameworks allow the spectator to co-author the artistic experience. In this artificial setting, he or she may enter a space between the singular and the plural, subjective and collective, that interestingly evokes the experience of physically moving about in a natural landscape. In both situations, the contemplation of landscape as merely a spectacle, a two-dimensional panoramic view, is broken. Through the rediscovery of a relation to nature for which our bodies are evolutionistically equipped, it is Eliasson's hope that we can interact more responsibly in the urban space, from which we have become even more alienated than from nature. Thus, Eliasson wishes for two seemingly contrasting routes, constructivism and phenomenology, to join in a new feedback loop: making us more conscious of the ways we see and construct the world (an agenda for which postmodern science and technology are often better equipped than art, according to Eliasson) *and* connecting us with the physical and social environment through a comprehensive engagement of our bodies.

The sophisticated staging of nature in the work of artists such as Eliasson means that landscape as a lens unproblematically presenting nature as sublime or picturesque has become 'displaced' to popular culture, particularly tourism, the ideals of which, regarding local nature, are precisely that it can be framed as a landscape panorama, unspoiled by modern technology (including the technology transporting the tourist to the desired sites). However, in representations by modernist and avant-garde art, if landscape is not abstracted from unambiguous content (cf. abstract expressionism), it is laden with either ironic or otherwise alienating markers (e.g. Naivism or Surrealism), or interspersed with technological traces to such a degree that its sentimental, High-Modern aura disappears (cf. the *terrains vagues* of photography's New Topographics). Technology may also be foregrounded by the media frameworks themselves, as is the case with the images of nature by Rodney Graham, Roni Horn and Diana Thater, analyzed in Hans Dickel's essay. Thus, imaging technology (film, video and photography) is presented as fundamentally other than nature – foregrounding itself as technology and media that (like Eliasson's scaffoldings) transcend the Neo-Romantic gaze of the tourist and culture industries. Nonetheless, they also probe nature more deeply than is possible with the naked human eye, revealing the extent to which our present perception of nature is made possible through technology. This is underscored when Rodney Graham takes 75 blurry Polaroid snapshots at night, using the camera flash to find his way through a forest (1976), or when Roni Horn visualizes the River Thames (2000) through a fixed camera at the bottom of a boat, capturing images of the movement of the water that the unaided eye could never catch.

With twentieth- and twenty-first-century art, things become less strained if we remove the panoramic frame, or the expectation of it, and look at nature as it appears in a more graspable material form. The most obvious example of such physical presence of nature – indeed, perhaps nature's last artistic domain as 'wild', its area of revenge – would be Earth Art, with its interventions in and displacements of raw stony matter. Earth Art could be conceived of as a collapsed sublime, sharing the sublime's concern for the disorganized forms of inorganic nature, even the monotony of geometry. But unlike the distanced landscape vistas of the sublime, where vapour is a marker of the void necessary to emancipate the contemplating subject from the threat of nature (culminating in the work of Turner and the Impressionists), Earth Art removes the void and desublimates the vapour, returning it and the ethereal subject whose reflection it is to the tellurian load, a concentration of the clay of Genesis, from which the subject sought to distance itself – evaporate – in the civilizing process. Yet, apart from Earth Art, physical remnants of 'nature', whether as inorganic or organic raw matter, chaotic patterns or temporal coincidences, will typically slip into post-1900 artwork with a less homogeneous alliance to genre, and a less clear Romantic recollection, interrupting the works' aura of being products of detached spiritual forces, in various ambiguous ways – consider Joseph Beuys's fat, Piero Manzoni's excrement or Anselm Kiefer's ash and lead; or Duchamp's standard stoppages, formed through coincidental meeting with the ground. In such miniaturized and embedded appearance, the actual agents behind or within the raw matter or temporal pattern involved – from stones to paints and pigments, from eggs to bodily fluids, from coincidences to arbitrary rules – become increasingly difficult to disentangle. Where does nature stop and technology, or indeed art, take over?

Whatever the form of nature in these examples, it is not organized in the closed assemblages of organic bodies. And indeed, in both representational and physical art forms, such bodies, from plants to animals to humans, are difficult to absorb smoothly into avant-garde contexts. For Peter Bürger a defining characteristic of avant-garde art was, already on a structural level, that it interrupts the idea of the artwork as an organic entity in which all ingredients are closely interwoven, having instead the machinic assemblage, the montage, as its prototype.[75] It does seem that if organic forms are absorbed into an avant-garde work, they must be posed in ironic, alienated or contorted ways, ranging from questioning the body's relations to its spatial surroundings (e.g. Bruce Nauman), to physical transformations (bioart) or mutilations (e.g. Vienna Actionism), to different sorts of imaginary disfigurements. The last strategy, for instance, is seen in Surrealism, where bodies are always open, unstable and subversively recombined, never closed or unchanging, as in neoclassical art forms and their kitschy descendants in totalitarian art, with its heroic, nude bodies. Nevertheless, Surrealism does not break so completely with the past as suggested by Bürger, since its disturbing effect emerges precisely because its strange assemblages pose as monstrous unities, erasing the joints that are exposed in 'canonic' avant-garde montages. Indeed, in their essay, Camilla Skovbjerg Paldam and Maria Fabricius Hansen refer to an organic avant-garde, revealing a genealogy reaching back to the Italian Renaissance, and Mannerist grotesques. At that time, however, the monstrous imagery of strangely mixed parts of plants, animals, humans and even artefacts was displaced to the periphery of the visual culture, appearing either in the borders of history paintings or in decorative frames, whereas in Surrealism they occupy the centre stage of the work of art. Moreover, although the grotesque bodies of Surrealism are still 'marvellous', in a sense recalling the culture of the Wunderkammer, the forces of nature they evoke are more explicitly chaotic and dangerous, subversively breaking the Renaissance utopia of civilized, mimetic art as the superstructure to nature's sketches. By unleashing those forces from his or her conscious control, the Surrealist strives to channel the surplus of virtual images into the realm of the real, thereby re-evoking the original ambition of art: to magically transform the environment through extraordinary semiotic acts.

An example of a hybrid creature walking more or less directly, as through a historical wormhole, from the Wunderkammer into Surrealism, indeed, into contemporary art and military experiments, is the figure of the insect, discussed in Marion Endt-Jones's essay. With its metallic and shimmering colours and blind, instinctual movements, the insect seems to be a natural-born cyborg, a creature hovering ambivalently between the animal and mineral kingdoms, and between life and automation. This classificatory evasion not only made insects ideal for virtuoso recreations in silver or gems in the Wunderkammer, or for disturbing representations in Surrealism. Today, living insects are at the forefront of U.S. military experiments with remote-controlled cyborgs; for example, experiments involving inserting electronic microcomponents into dragonflies and moths at their pupal stage.

When art enters the domain of combining living organisms with artificial components, or wholly altering living materials through the intervention of technology, it becomes subject to the same kinds of challenges as those mentioned previously in relation to 'technological art', of which it is in fact an intrinsic part. On the one hand, this genre – often labelled 'bioart' – seems literally to incarnate the avant-garde dream of being part of, and shaping life; on the other, it is so concerned with technological procedures that here too, the artistic aspect seems threatened by cognitive didactics and a puerile techno-fetishization. One response to this challenge, pursued in Pernille Leth-Espensen's essay, is to foreground bioart's critical potential. For instance, when the artist Anne Niemetz uses an atomic force microscope as microphone and transforms the sounds she captures from cells into music, this raises intriguing questions as to the mediating qualities of scientific instruments. Where exactly in the indeterminate space between microscopic nature and macroscopic technology are the sounds produced? Integrating philosopher of technology Don Ihde's observation that technologies give all scientific representations a hermeneutic dimension (acknowledged or not), Leth-Espensen identifies bioart as a relevant forum for reflecting on technological mediation.

As we already saw with Eduardo Kac's green rabbit, it is in bioart that the triangle of art/technology/nature becomes most obviously compromised, and therefore that some of the basic borderlines of modernity, the era of the emancipated human subject, are challenged. With bioart one is presented with palpable examples of the cyborg, the figure fusing organism and machine that has become paradigmatic for the idea of the posthuman. Based on the premise that we are about to take control of our own evolutionary destiny through technological interventions in the human body, including its genetic programme, the posthuman indeed presents itself as a possible framework for the epistemological crossovers explored in this book.[76] As Donna Haraway has noted in her 'Manifesto for Cyborgs', the cyborg dissolves a whole row of dualisms on which Western culture has been founded: not only the more specific artefact/organism but also human/animal, culture/nature, civilized/primitive, subject/object and self/otherness. If we include art/technology in this array, our triangle of art/technology/nature could be wholly absorbed by cyborgian hybridization.[77]

How art may be more specifically inscribed in such a posthuman convergence of technology and nature is suggested in Adorno's posthumously published *Aesthetic Theory* (1970), particularly in its upgrading in avant-garde art of the otherwise obsolete concept of natural beauty. Adorno here makes clear that natural beauty no longer has anything to do with picturesque naturalistic illusion, but on the contrary, as already demonstrated in Impressionist landscapes, shows an alliance with technological elements. With a sensitivity approaching that of Earth Art, his contemporary, Adorno remarks, for instance, that alpine moraines and slag heaps may in themselves resemble dumping grounds for industrial waste.[78] Furthermore, though still dependent on a human subject's reception, this natural beauty transgresses the subject and gives voice to an objective uncertainty or non-identity lying beneath the subject: 'Art wants with human means to realize the speech of the non-human'. This non-human speech, contorting bourgeois labour, commodities and hegemony, points towards something more: a surplus or dignity of nature which seems to lie ahead in time.[79] Thus Adorno points futuristically towards a Heideggerian meeting-point at which art is reconciled with both technology and non-human nature. Indeed, by first letting art show the surplus for which nature strives, and, second, including technological elements in this surplus, Adorno presents a posthuman version of the Aristotelian idea of technical art as a fulfilment of that which nature strives for, but cannot realize with its own means.

But is the speech of the non-human – nature – realized through human means nothing more than a wishful futuristic dream alluded to in art? Certainly not. So far, in our discussion of the postmodern triangle of art/technology/nature, we have discussed technology and nature as objects primarily intended for artistic *re*presentations, or as entities that intermingle physically in artistic *pre*sentations, but not as forces that themselves determine representational media. Yet in the new media emerging since the early nineteenth century – photographic, and later also acoustic – representations are no longer exclusively traces of the artisan's or artist's handiwork, as was the case in all previous media (at least as understood in a disenchanted retrospective view). From photography and phonograph to film, tape recorder and television, and on to their common multimedia convergence, the computer, a collaboration between nature and technology emerges, in which traces or indexes from the surrounding world – in the broadest sense: nature – are stored automatically in technological media and may be reproduced and manipulated infinitely through further technological means. This absorption of natural imprints in technological matrices not only breaks down the nature/technology dichotomy; it also seriously challenges the art/technology dichotomy. For if images and sounds are produced automatically, of what competence does artistic activity consist?

To really understand the mechanisms of this media revolution, one should conceive of it in bodily terms, as an augmented, quasi-organic nature emerging from within culture. If human artefacts are associated with bodily functions and understood as prostheses for such functions, then the above-mentioned High-Modern divide between art and technology can be reformulated as the result of a Cartesian split between mind and body projecting itself further onto our fabricated surroundings – that is, as a sharp divide between media and machines. Media, including art, are thus prostheses for the senses and brain operations (virtuality); machines for bodily motor and metabolic functions (physicality). The property we observed with regard to High-Modern non-magical art – that it cannot move anything except the mind – is thus easily extended to media in general. On the other hand, when Marshall McLuhan described the electric age – the period that followed what he termed mechanic modernity – as an exteriorization of the human central nervous system,[80] that could be understood, in surprisingly literal terms, as a putting together, a corporeal re-coordination, of those formerly separated functions of media and machines, the virtual and the physical domains. Thus the repertoire of indexical media – visually, for instance, registering first the frozen past (photography), then the animated past (film) and finally the animated present (television) – could be described as an artificial recreation of the human sense apparatus, including its memory storage. The expanding computer manipulation of indexical media after World War II could similarly be compared to the brain's processing of those sense data: from recalling memories, to selecting between sense impressions with regard to action in the present, to creatively recombining sense impressions. And finally, computers' being connected to one another and actively manipulating machines in an ever more ubiquitous network since the 1990s could be compared to the way neurons give signals to motor and metabolic functions in our bodies. In sum, our recent technologies have artificially recreated a whole feedback loop of sense impressions, processing and proactive reaction.

The High-Modern split between virtual and machinic, including the divide between art and technology, was perhaps the price for the project of human autonomy in the widest sense, generating a free subject moving around in an infinite extracorporeal world of objects. The postmodern, if not posthuman, coming together of virtual and machinic, on the other hand, indicates a compromising of human autonomy: that our bodies and brains are being immersed and interacting with a world which has certain intracorporeal qualities. Thus the postmodern interface is no longer truly an extracorporeal medium, hovering in splendid cerebral isolation and unable to move anything physical. It is a medium with depth, a penetrated representation, which may lead directly to physical action through electrical signals – a semiotic ability otherwise exclusive to our brain images and their triggering of movements through intracorporeal nerves. By exteriorizing our nerve system in the ubiquitous network of computers, we thus transfer to the surrounding world abilities of dense communication between signs and things which, in the recent past at least, were known only on the insides of our bodies.

For, of course, if we expand the time scale, this fusing of semiotic and physical dimensions once again acquires a neo-primitive quality, which reminds us of art's original magical function, its supplemental role to technology. However, whereas the transferral of semiotic powers to the physical world – nature – seemingly amounted to wishful thinking in preliterary cultures, it is concretely realized in our computerized world. On the other hand, now this happens not in art but in technology. So is there any supplementary function left for art? At this point we should probably reconsult the avant-garde tradition of fusing sign and physicality alluded to above, the critical performative tendency from readymades to actions. But in order to transcend the obvious caricature – technology wanting to facilitate the action of semiotically empowered machines, avant-garde art wanting to interrupt it – less dichotomous thinking is needed. This is the thinking of heterogeneous assemblages, swarms decentring the human subject and including mutations of what used to be categorized as nature, technology and art.

A Remark on Organization

To illuminate our triad of art, technology and nature as diversely as possible, the 14 essays in this volume are organized into three sections that combine a thematic and a chronological approach. In the first section, 'Assistance/Interruption: Art and Technology Interlacing with Nature', all the essays concern fifteenth- and sixteenth-century art(efacts), but most also draw parallels with twentieth-century Surrealism and contemporary art, disclosing connections, differences and similarities in hybrid works across time. The second section, 'Hermeneutics: Art and Technology Representing Nature', is devoted to differing representational regimes, starting with the aim of transparent representation in art and science, and their integration in the fifteenth and sixteenth centuries, moving on to an account of new representational techniques (such as botanical prints and photography) in the eighteenth and nineteenth centuries, which challenge the former ideas of objective representation, and ending with a discussion of the hermeneutic interpretation of twentieth-century scientific technology critically investigated by contemporary art. In the third and final section, 'Localization/Exposure: Art and Technology Revealing Paradigms of Nature', all the essays concern nineteenth- and twentieth-century art and science, demonstrating an increasing interrelation among art, nature and technology, from nineteenth-century artistic and scientific cloud studies to mechanical constructions of nature and weather in contemporary art.

We hope this collection of essays and its extensive introductory framework will provide the reader with a fresh and eye-opening view of the relations among art, technology and nature. However, knowing that the volume's premises may be controversial, we chose to also include a critical discussion of the entire undertaking. We are pleased that James Elkins, a scholar known for his incisive, at times abrasive, opinions, accepted the invitation – even if this meant entering the line of fire.

Notes

1 Daniel Lord Smail, *On Deep History and the Brain* (Berkeley, Los Angeles and London: University of California Press, 2008); David Christian, *Maps of Time: An Introduction to Big History* (Berkeley, Los Angeles and London: University of California Press, 2004); Kevin Kelly, *What Technology Wants* (New York: Viking, 2010).

2 Pliny the Elder, *Natural History*, trans. H. Rackham (vols 1–5 and 9), W.H.S. Jones (vols 6–8) and D.E. Eichholz (vol. 10), Loeb Classical Library (Cambridge, MA: Harvard University Press; London: William Heinemann, 1947–80) (10 vols).

3 Eduardo Kac, *Telepresence and Bio Art: Networking Humans, Rabbits, and Robots* (Ann Arbor: University of Michigan Press, 2005), pp. 264–67.

4 According to a somewhat shaky account by Lady Richmond, Turner had seen a train approaching in a tempest, while gazing from the window of either another train or a horse-drawn coach. See John Gage, *Turner: Rain, Steam and Speed* (London: Allen Lane, Penguin Press, 1972), pp. 16–19 and 85, and William S. Rodner, *J.M.W. Turner: Romantic Painter of Industrial Revolution* (Berkeley, Los Angeles and London: University of California Press, 1997), pp. 155–56.

5 'Landschaft. Zur Funktion des Ästhetischen in der modernen Gesellschaft' [1962], in Joachim Ritter, *Subjektivität: Sechs Aufsätze* (Frankfurt am Main: Suhrkamp, 1974), pp. 153–54.

6 See especially Friedrich Schiller, 'On Naive and Sentimental Poetry' [1795], trans. Julius A. Elias, in *German Aesthetic and Literary Criticism: Winckelmann, Lessing, Hamann, Herder, Schiller, Goethe*, ed. H.B. Nisbet (Cambridge: Cambridge University Press, 1985).

7 Immanuel Kant, *Kritik Der Reinen Vernunft* (Frankfurt am Main: Suhrkamp, 1974 [1788]), and Immanuel Kant, *Kritik Der Urteilskraft* (Frankfurt am Main: Suhrkamp, 1992 [1790]).

8 John D. Barrow and Frank J. Tipler, *The Anthropic Cosmological Principle* (Oxford and New York: Oxford University Press, 1986), pp. 51–60.

9 Kant, *Kritik Der Urteilskraft*, § 44, p. 240.

10 Ibid., § 45, p. 240.

11 For the connection of art and magic agency, see James George Frazer, *The Golden Bough: A Study in Comparative Religion* (Edinburgh: Canongate, 2004 [1922]), pp. 12–14; David Freedberg, *The Power of Images: Studies in the History and Theory of Response* (Chicago: University of Chicago Press, 1989), pp. 270–77; Marcel Mauss, *A General Theory of Magic*, trans. Robert Brain (London and Boston: Routledge and Kegan Paul, 1972 [1950]), p. 63; Alfred Gell, *Art and Agency: An Anthropological Theory* (Oxford: Clarendon Press, 1998), pp. 96–154.

12 Ellen Dissanayake, 'What Art Is and What Art Does: An Overview of Contemporary Evolutionary Hypotheses', in *Evolutionary and Neurocognitive Approaches to Arts, Aesthetics and Creativity*, ed. Colin Martindale, Paul Locher and Vladimir M. Petrov (Amityville, NY: Baywood, 2007), pp. 1–14.

13 Alondra Yvette Oubré, *Instinct and Revelation: Reflections on the Origins of Numinous Perception* (London and New York: Taylor and Francis, 1997).

14 Jacob Wamberg, 'Toward a (Re)constructed Endosemiotics? Art, Magic, and Augmented Reality', in *Throughout: Art and Culture Emerging with Ubiquitous*

Computing, ed. Ulrik Ekman (Cambridge, MA: MIT Press, 2012), pp. 976–1002; on the underworld as the first beyond, see Erich Neumann, *Ursprungsgeschichte des Bewusstseins* (Zurich: Rascher, 1949); idem, *The Great Mother: An Analysis of the Archetype*, trans. R. Manheim (Princeton: Princeton University Press, 1963 [1955]); and Anne Baring and Jules Cashford, *The Myth of the Goddess: Evolution of an Image* (New York: Viking Arkana, 1991).

15 Jacques Lacan, 'The Mirror Stage as Formative of the I as Revealed in Psychoanalytic Experience' [1949], in *Écrits: A Selection*, trans. Alan Sheridan (New York and London: W.W. Norton, 1977), p. 4.

16 Jean Clottes and David Lewis-Williamw, *The Shamans of Prehistory: Trance and Magic in the Painted Caves*, trans. Sophie Hawkes (New York: Harry N. Abrams, 1998 [1996]), pp. 12–17; David Lewis-Williams, *The Mind in the Cave: Consciousness and the Origins of Art* (London: Thames and Hudson, 2004), pp. 208–14; Jacob Wamberg, *Landscape as World Picture: Tracing Cultural Evolution in Images* (Aarhus, Denmark: Aarhus University Press, 2009 [2005]), vol. 1, p. 108.

17 J.L. Austin, J.O. Urmson and Marina Sbisà, *How to Do Things with Words: The William James Lectures Delivered at Harvard University in 1955* (Oxford: Oxford University Press, 1962).

18 Larry Shiner, *The Invention of Art: A Cultural History* (Chicago: University of Chicago Press, 2001).

19 Martin Heidegger, *Die Technik Und Die Kehre* (Pfullingen: Neske, 1962), p. 12.

20 Aristotle, *Poetics*, ed. and trans. Stephen Halliwell, Loeb Classical Library (Cambridge, MA: Harvard University Press, London: William Heinemann, 1995), 1454b.

21 Neumann, *Ursprungsgeschichte*, p. 252; Max Horkheimer and Theodor W. Adorno, *Dialektik Der Aufklärung: Philosophische Fragmente* (Frankfurt am Main: Fischer Taschenbuch Verlag, 1971 [1944]), p. 14.

22 Wamberg, *Landscape*, vol. 1, pp. 166–70; Stephen Toulmin and June Goodfield, *The Fabric of the Heavens* (London: Hutchinson, 1960), pp. 65–67.

23 Aristotle, *Generation of Animals*, trans. A.L. Peck, Loeb Classical Library (Cambridge, MA: Harvard University Press, London: William Heinemann, 1953), 1, 22, 730b.

24 Pliny, *Natural History*, books 33–37.

25 Arthur O. Lovejoy and George Boas, *Primitivism and Related Ideas in Antiquity* (Baltimore: Johns Hopkins University Press, 1935), p. 103ff.

26 Aristotle, *Parts of Animals*, trans. A.L. Peck, Loeb Classical Library (Cambridge, MA: Harvard University Press, London: William Heinemann, 1961), 1, 1, 639b.

27 Lovejoy and Boas, *Primitivism*.

28 Jean-Pierre Vernant, 'Arbeit und Natur in der griechischen Antike', in *Seminar: Die Entstehung von Klassengesellschaften* (Frankfurt am Main: Suhrkamp, 1973), pp. 246–70.

29 Aristotle, *Politics*, trans. H. Rackham, Loeb Classical Library (Cambridge, MA: Harvard University Press, London: William Heinemann, 1990 [1944]), 1, 1253b and 1256b 8.

30 Pliny, *Natural History*, 33, 1, 1.

31 Pamela H. Smith, *The Body of the Artisan: Art and Experience in the Scientific Revolution* (Chicago: University of Chicago Press, 2004), esp. pp. 7–8 and 147; for the mechanical arts in the Late Middle Ages and Renaissance, see David Summers, *The Judgement of Sense: Renaissance Naturalism and the Rise of Aesthetics* (Cambridge: Cambridge University Press, 1987), pp. 241–65.

32 Mircea Eliade, *The Forge and the Crucible*, trans. S. Corrin (Chicago and London: Chicago University Press, 1962 [1956]).

33 For the Hebrew and early Christian contexts, see Moshe Barasch, *Icon: Studies in the History of an Idea* (New York: New York University Press, 1995).

34 Plato, *Republic*, trans. Paul Shorey, in *The Collected Dialogues of Plato including the Letters*, ed. Edith Hamilton and Huntington Cairns, Bollingen series LXXI (Princeton: Princeton University Press, 1989 [1938]), 597c–598b.

35 Erwin Panofsky, *Idea: A Concept in Art Theory*, trans. Joseph J.S. Peake (Columbia: University of Southern California Press, 1968). In ibid., Giovanni Pietro Bellori, 'The Idea of the Painter, Sculptor and Architect, Superior to Nature by Selection from Natural Beauties'.

36 Georg Wilhelm Friedrich Hegel, *Vorlesungen Über Die Ästhetik* (Frankfurt am Main: Suhrkamp, 1986 [1835]).

37 Umberto Eco, *Art and Beauty in the Middle Ages* (New Haven: Yale University Press, 1986).

38 Summers, *The Judgment of Sense*, p. 259.

39 Sigmund Freud, 'The Uncanny' [1919], in *The Uncanny*, trans. David McLintock; with an introduction by Hugh Haughton (London: Penguin, 2003), pp. 121–62.

40 René Descartes, 'Discourse on the Method of Rightly Conducting the Reason and Seeking Truth in the Sciences' [1637], trans. John Veitch, in *The Rationalists* (New York: Doubleday, 1960), section V, p. 80.

41 J.B. de C.M. Saunders and Charles D. O'Malley, *The Illustrations from the Works of Andreas Vesalius of Brussels: with annotations and translations, a discussion of the plates and their background, authorship and influence, and a biographical sketch of Vesalius* (Cleveland and New York: Dover, 1950).

42 Cited in Carolyn Merchant, *The Death of Nature: Women, Ecology, and the Scientific Revolution* (New York: Harper and Row, 1980), p. 171.

43 Ibid.

44 On the deanimation of nature in early modernity, also see Hans Jonas, *The Phenomenon of Life: Toward a Philosophical Biology* (Chicago: University of Chicago Press, 1982), pp. 72–74.

45 Toulmin and Goodfield, *The Fabric of the Heavens*, pp. 221–23.

46 Horst Bredekamp, *Antikensehnsucht und Maschinenglauben: Die Geschichte der Kunstkammer und die Zukunft der Kunstgeschichte* (Berlin: Verlag Klaus Wagenbach, 1993), pp. 33–35.

47 Joseph Leo Koerner, *The Moment of Self-Portraiture in German Renaissance Art* (Chicago and London: University of Chicago Press, 1993), pp. 80–126.

48 Lorraine Daston and Peter Galison, 'The Image of Objectivity', *Representations* 40 (Autumn 1992): pp. 81–128.

49 Cited in Leslie Parris and Ian Fleming-Williams, *Constable* (London: Tate Gallery, 1991), pp. 228–29.

50 Bruno Latour, 'Why Has Critique Run Out of Steam? From Matters of Fact to Matters of Concern', *Critical Inquiry* 30/2 (Winter 2004): pp. 225–48.

51 Martin Heidegger, *Der Ursprung Des Kunstwerkes*, ed. Hans-Georg Gadamer (Stuttgart: Philipp Reclam, 1970 [1935–36]).

52 Heidegger, *Die Technik und die Kehre*, p. 34.

53 Ibid., pp. 13–23.

54 Ibid., pp. 34–36.

55 Le Corbusier, *Towards a New Architecture* (London: Architectural Press, 1960 [1927, 1923]), p. 10.

56 Don Ihde, *Technology and the Lifeworld: From Garden to Earth* (Bloomington: Indiana University Press, 1990), p. 72ff.

57 Horkheimer and Adorno, *Dialektik Der Aufklärung*.

58 Hal Foster, *Design and Crime (and Other Diatribes)* (London: Verso, 2002), pp. 17–19.

59 Raymond Fleury, Paul Gayot and Thieri Foulc, *Les Très Riches Heures du Collège de 'Pataphysique* (Paris: Fayard, 2000).

60 Stewart Home (ed.), *What Is Situationism? A Reader* (Edinburgh: AK Press, 1996).

61 Billy Klüver, Julie Martin and Barbara Rose (eds), *Pavilion: Experiments in Art and Technology* (New York: E.P. Dutton, 1972).

62 Stephen Wilson, *Information Arts: Intersections of Art, Science, and Technology*. (Cambridge, MA: MIT Press, 2002), pp. 851–60.

63 Heidegger, *Ursprung*.

64 Martin Heidegger, *Sein und Zeit* (Tübingen: Max Niemeyer, 2001 [1927]), § 16.

65 To our knowledge, it is only in Amelia Jones's brilliant *Irrational Modernism: A Neurasthenic History of New York Dada* (Cambridge, MA, and London: MIT Press, 2004) that the readymades are analyzed in that perspective of interrupted technology – and still without Heidegger. For a broader discussion of Duchamp's relation to contemporary technology, see Linda Dalrymple Henderson, *Duchamp in Context: Science and Technology in the Large Glass and Related Works* (Princeton: Princeton University Press, 1998).

66 For the Marxian notion of the capitalist commodity fetish, see, for instance, William Pietz, 'Fetish', in Richard Shiff and Robert S. Nelson, *Critical Terms for Art History* (Chicago: University of Chicago Press, 1996), pp. 201–2.

67 Ihde, *Technology*, pp. 97–108.

68 Peter-Paul Verbeek, *What Things Do: Philosophical Reflections on Technology, Agency, and Design* (University Park: Pennsylvania State University Press, 2005), pp. 129–30; Bruno Latour, *Reassembling the Social: An Introduction to Actor-Network-Theory* (Oxford: Oxford University Press, 2007); Gilles Deleuze and Félix Guattari, *A Thousand Plateaus: Capitalism and Schizophrenia*, trans. Brian Massumi (London: Continuum, 1987 [1980]).

69 Robert Rosenblum, *Modern Painting and the Northern Romantic Tradition: Friedrich to Rothko* (London: Thames and Hudson, 1975).

70 Martin Jay, *Downcast Eyes: The Denigration of Vision in Twentieth-Century French Thought* (Berkeley: University of California Press, 1994).

71 Erwin Panofsky, *Perspective as Symbolic Form* (New York: Zone Books, 1991 [1927]).

72 Wamberg, *Landscape*, passim.

73 Christopher Hussey, *The Picturesque: Studies in a Point of View* (London: F. Cass, 1967 [1927]).

74 Jacob Wamberg, 'Wounded Working Heroes: Seeing Millet and Van Gogh through the Cleft Lens of Totalitarianism (Adding Reflections from Kiefer and Baseliz)', in *Totalitarian Art and Modernity*, ed. Mikkel Bolt Rasmussen and Jacob Wamberg (Aarhus, Denmark: Aarhus University Press, 2010), pp. 26–104.

75 Peter Bürger, *Theory of the Avant-Garde* (Minneapolis: University of Minnesota Press, 1984 [1980, 1974]), p. 78.

76 For more details, see Jacob Wamberg, 'Dehumanizing Danto and Fukuyama: Toward a Post-Hegelian Role for Art in Evolution', in *The Posthuman Condition: Ethics, Aesthetics and Politics of Biotechnological Challenges*, ed. Kasper Lippert-Rasmussen, Mads Rosendahl Thomsen and Jacob Wamberg (Aarhus: Aarhus University Press, 2012), pp. 141–54.

77 Donna J. Haraway, *Simians, Cyborgs and Women: The Reinvention of Nature* (London: Free Association, 1991).

78 Theodor Adorno, Gretel Adorno and Rolf Tiedemann, *Ästhetische Theorie* (Frankfurt am Main: Suhrkamp, 1970), pp. 106–7.

79 Adorno, *Ästhetische Theorie*, pp. 111–21, citation p. 121.

80 Marshall McLuhan, *Understanding Media: The Extensions of Man* (New York, Toronto and London: McGraw–Hill, 1964), pp. 3 and 247.

PART I

Assistance/Interruption: Art and Technology Interlacing with Nature

1

Creatio ex ligno: The Characteristics of Wooden Renaissance Dolls[1]

Markus Rath

> Was ist also Natur in der Kunstmäßigkeit? Autonomie in der Technik?
> Sie ist die reine Zusammenstimmung des inneren Wesens mit der Form,
> eine Regel, die von dem Dinge selbst zugleich befolgt und gegeben ist.
> —Friedrich Schiller, Letter to Gottfried Körner, February 23rd 1793

Monogrammist IP's Autonomous Figurines

In sixteenth-century sculpture, six-jointed dolls not larger than a hand form a distinct and unique group. The boxwood figurines were probably made by the so-called Monogrammist I.P., an anonymous woodcarver active in the region between Nuremberg, Salzburg and Passau in the first third of the sixteenth century. Three reliefs signed with the initials IP were the starting point for reconstructing his œuvre. Due to similarities in character and anatomical details between the jointed dolls and the figures in these reliefs, the former have been convincingly attributed to this artist.[2]

The representation of the dolls' bodies testifies to the substantial developments in the study of human anatomy at the time and astonishes the viewer by dint of the interplay between the various parts. Indeed, their highly complex technical construction leads to a striking articulation, while the precise representation of human anatomy adds to the impression of autonomy. The classic discussion of the term 'autonomy' by Immanuel Kant (*Kritik der Urteilskraft*, 1790) and Friedrich Schiller (*Kallias–Letters*, 1793) refers to an aesthetic, more or less independent and even 'useless' conception of the objet d'art, not directly linked to a clear social purpose, whether scientific, religious or judiciary.[3] In addition to this complex, this essay highlights the role of technical refinement (i.e. hypermotility, smaller size, increasingly elaborate details) that leads to the autonomy of the object and challenges its traditional purpose.[4] The example of the old jointed dolls uncovers different elements of artificial autonomy: it will be argued that the growing sophistication of artistic techniques gives rise to a more individual status of the artefact. In doing so, the following factors will be considered: the technical methods of carving; the origin and the status of the jointed doll as *the* artificial model of the human body; and finally the intellectual and artistic context of exploration and creation of the period in which these figures were created, as exemplified by Albrecht Dürer (1471–1528), one the most important artists of the time. Finally, a new reading of the dolls will be presented. These are not children's dolls, erotic toys for adults or models for the artist. The essay will reveal that these objects were created as 'natural artefacts', which emphasized the status of the artist (and later that of the collector) as *alter deus* and served as ideal human models for obtaining natural knowledge.

Today, one of the jointed dolls is on permanent display at the Berlin Bode Museum. It can be regarded as the earliest example of the group (Colour plate 2).[5] The joints of the single phalanges of the fingers and the toes may be moved by means of hinge joints, all of the other limbs by finely turned ball joints. A large spherical joint is fitted between the higher and the lower body. The extremities are formed very precisely; every detail of the finger- and the foot-bones, including the fingernails, is carved and even the lines of the palm seem to be represented. On the head, the artist placed a woven bonnet and he finely elaborated the round breasts and the curly pubic hair to represent the gender-specific features of the female body. The whole body consists of 57 individual components, each of them meticulously adjusted to the others. Despite its small size, the composition of the multipartite figurine results in a coherent, lively form.

The five other specimens of the group include a pair consisting of a young male and a female figure in the Tiroler Landesmuseum Ferdinandeum in Innsbruck, and three individual dolls in Hamburg, Leipzig and Madrid.[6] All of them have been made in a similar manner, and equal attention has been paid to individual details. The female doll in the Grassi Museum in Leipzig is taller and has a bigger stomach than the figure in Berlin, and the expression of the face is more serious. Moreover, the artist placed a floral wreath on her bonnet, a detail similar to the more elaborate headdress of the male jointed doll in the Museum für Kunst und Gewerbe in Hamburg, which wears a laurel wreath.[7] These additional details suggest the wish to represent different types of human beings – a desire confirmed by the obvious age variety of the dolls.[8] Finally, in contrast to the five others, the jointed male doll in the Museo del Prado in Madrid has a movable jawbone that enables him to perform movements of speaking and eating.[9] The latter figure is not as well preserved as the others, and its anatomical details are obscured by a layer of dye.[10] Three other specimens of this type disappeared in 1945, a pair and a single jointed doll known only through written sources.[11]

As indicated, all six preserved figurines represent different ages and types of human beings. All may adopt an infinite number of poses. Each figure has an air of vivacity, oscillating between human behaviour and, because of the possibility to take on countless positions, an uncanny, superhuman ability. Both these characteristics are crucial components of an autonomy that is based on two principles: skilful mimesis (*imitatio*) combined with the desire to overcome the principles observed in nature (*superatio*) through innovative form and technique (*techné*).[12] Respecting these principles, each of these jointed dolls has become an animated being that seems to be able to interact with the beholder.

The Art of Motility

The history of self-operating artefacts such as automata harks back to Greek mythology. The ancient myths and, in particular, the sources of the Alexandrian school (first period c. 300–330 BC), describe a vast range of artificial specimens, such as singing birds, highly mobile statues or musical fountains.[13] From this period onwards, the wish to construct 'acting' machines, named androids if they take on a human form, draws a continuous line in the history of artist-engineers until today.[14] For artists of the Renaissance such as Leonardo da Vinci, the creative act always included virtual engineering by exploration of hydraulically, mechanically or physiologically based (re)constructions.[15]

By creating an automaton, the engineer invents an artwork that is autonomous to some extent, insofar as it is able to move by itself; for example, it sings or writes (that is, after the mechanism has been started). However, even the most sophisticated automata have a limited number of predefined functions which can be repeated but from which the object cannot escape. The individual set of skills is predefined by its mechanical structure. Thus, this process not only implies a multiplicity but also a restriction of capacities. The jointed dolls of the Monogrammist I.P. cannot be classified as automata; as shown above, they are not able to move (speak, eat, etc.) by themselves. Nevertheless, they have another kind of autonomy, embedded within the potentialities of their structure and use. Like other sculptures which, after they have been finished by their creator, become autonomous as 'finished' *Gestalt* (this is the case with all sculptures, even in case of the *non-finito*),[16] the jointed dolls are also the final result of a creative process with specific properties – defined proportions, weight and style – and therefore can be considered as autonomous as other figurative objects. However, in contrast to sculptures made of one single piece, the figures may be arranged and rearranged ad infinitum by the human hand. Therefore, these figures cross the borders of the genre of sculpture. Every pose is an intermediate state, but it has the same autonomous quality of a sculpture made out of one piece. This potential for changes generates transitory forms. Thus, the dolls are neither automata nor sculptures, but belong to a third class of autonomous artwork.[17]

The agility and mobility of the dolls are achieved by a refined system of strings, hidden inside the wooden bodies. This internal secret can be revealed by modern methods of CT-based investigation: three-dimensional

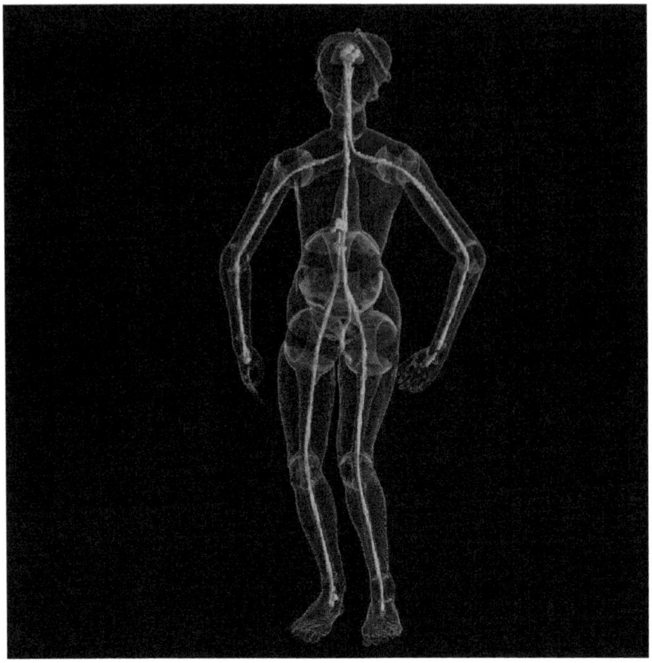

1.1 K. Bente, A. Franz, Ch. Jürgens, computed micro-tomographical studies and animations of wooden Dutch dolls (Monogrammist I.P., sixteenth century). Grassi Museum for Applied Art, Leipzig

microphotographs (3D–μCT) of the doll in the Grassi Museum in Leipzig show that its technical construction is as advanced as its mobility is impressive (Figure 1.1).[18] The individual components are connected by this intricate system of internal strings. Arms and legs as well as the ball joints have funnel-shaped holes to ensure that the strings may smoothly slip inside them. Fixed by locking handles in hands and feet, all of the well-fitted components are threaded like pearls on a necklace using four hidden cords made of catgut. These elements are led through the extremities and finally tied together in a knot in the head, which gives the construction the appearance of following the structure of the central nervous system. A small opening in the doll's bonnet enables the last production step of putting the tension on the strings and fixing them in the head. Even if this system of cords is hidden from sight, it can be seen as an extension of the figures' imitative character from outside appearances (i.e. correct anatomical forms) to internal structure. Thanks to their delicate and unique carving technique and complex inner mechanics, the movable figurines seem to possess an animated human body.

The invention of this string-balanced organization of the turned ball joints represents a crucial factor for the development of the dolls' moving system. Essential for this progress has been the art of turning. This art is a special form of wood – or ivory – carving by means of a machine that produces rotational-symmetric forms.[19] From the sixteenth to the eighteenth century, the art of turning was not a simple craft like any other, but that of the emperor.[20] As Klaus Maurice has shown, the art of turning 'represented "learning through play" with the aim of extending the scope of human organs with the aid of the machine. ... A prince learned how to form raw material according to plan, to regulate a sequence of predetermined steps. Through play he could identify with the idea of translating mechanical causality into the governance of society'.[21] The example of the finely carved and turned objects made by the noble artist reveals, beyond the transitory form, the second and more traditional semantic background of the term 'autonomy' as it is used here: as a product of the highest finesse, without an immediate function. The artificial object is thus linked to the philosophical aspect of creation itself.

In the case of Renaissance dolls, turned movable elements obtained by a dynamic production process and individually carved components complement one another. Instead of creating an antagonism between carved and turned elements, the combination of these two kinds of technical craftsmanship leads to a higher degree of 'naturalness', natural mimesis respecting form *and* mobility. These figures reflect the desire to imitate, and even surpass the human ability of motion (*superatio*), thanks to the hypermotility created by hinge and ball joints in combination with the string system. By virtue of this harmonic interplay of all limbs, mobility and the artist's aim to create natural human forms, the figurines appear as entities that represent more than the sum of their parts.

Invention of a 'Living' Model – The Lay Figure

Sources for this kind of movable, human-shaped sculpture may be found in the articulated dolls of Ancient Greece, of which numerous examples survive.[22] Moreover, in the Middle Ages, both small and life-size jointed figures were used in ecclesiastic contexts.[23] However, it was only in the wake of the artistic and technical developments of the Quattrocento that the sculptural archetype was invented: the lay figure as the artist's model.

In the Renaissance, drawing practice changed drastically. While some innovative drawings were used and passed on as precious patterns in the medieval workshop, the draughtsman of the Renaissance sought to re-enact creation by searching for an equivalent of the 'divine' *disegno*.[24] A production of new forms was directly linked to artistic capacities of creation. As a result of this new conception of drawing, the use of three-dimensional models became essential: plastic models made of clay, plaster, wax or wood followed the nobilitation of drawing as the incarnation of the artistic plan.[25]

In his *Vite*, Giorgio Vasari often uses the term *modello* to indicate different kinds of modifiable models used to try out various motifs. His high esteem for this kind of plastic model (one that is flexible during the process of invention) and its role in the practice of drawing are conveyed in his 'Introduzione' to the *Vite* of 1568. Here he not only emphasizes the importance of the drawn model but also that of the three-dimensional model in the process of artistic creation: 'the said perfection of judgment is in no way more necessary for sculptors than for painters, it being sufficient for the former to execute good models in wax, clay, or something else, even as the latter make their drawings on corresponding materials or on cartoons'.[26]

It may be argued that, as a result of its flexible nature, the manikin could function as such a *modello buono* as well. In the *Vita* of Fra Bartolomeo (1472–1517), Vasari describes the lay figure as a *modello di legno*, a life-size wooden sculpture which could be arranged in any position and used for drapery studies. In his description, Vasari already refers to the quasi-natural status of the manikin: the *artificial* model was created to help Fra Bartolomeo, who preferred to paint his figures *after life* (*tenere le cose vive innanzi*).[27] According to Vasari, it was Garofalo (1481–1559) who used this new technology for the first time in Lombardy. By arranging and rearranging the lay figure in different poses, he could make complicated drawings such as drapery studies.[28] Furthermore, he used both lay figures and clay models in different positions to study light and shadow.[29]

In fact, more than a hundred years before Vasari's biographies, the lay figure was already known as an important device for artists. The very first description of a movable manikin can be found in the third book of Antonio Averlino Filarete's (c. 1400–69) treatise on architecture of 1464: 'Provide yourself with a small wooden figure with articulated arms, legs and neck; then make a linen cloth, and with that clothe it, as you like, as if it was a living being'.[30] This first reference to jointed dolls insists on their hybrid nature: the inanimate manikin should be arranged as if it were alive – *come se fussino d'uno vivo*. An impetus of animation can be found, or so it seems, in the movable human form itself.

One of the crucial areas in which the lay figure was employed was the study of the clothed, or rather draped, human figure. The study of draped figures was of great importance, since arguably the most important precondition for successfully conveying natural forms in artificial ones – that is, with visual forms – was the so-called *rilievo*. The term *rilievo* was coined by Cennino Cennini (c. 1370–1440) in his *Libro dell'arte*, written at the end of the fourteenth century and considered to be one of the first treatises on art. Cennini gives instructions for the 'right' use of light and shadow, whereby the highlighted areas are described as *rilievo*. He calls the process of creating this effect *incarnazione*.[31] Only by respecting the physicality of the three-dimensional model could a corporal effect be achieved. As the German art historian Julius von Schlosser already pointed out, models made of wood, clay or plaster were used to obtain that very 'clear bodily effect',[32] as indicated by the term *rilievo*.[33] The invention of mobile models like the manikin seems to be closely associated with this artistic demand of the Quattrocento. In addition to the possibility of fixing a bodily experience, that is, a physical form onto a surface, the lay figure was also used for the invention of highly dynamic scenes and complex postures. Thus, the lay figure was an effective model between motion and immobility.

Of course, one can hardly imagine these early lay figures resembling the highly developed figurines of Monogrammist I.P. With regard to the size of these models, we know that the wooden *modelli* of Fra Bartolomeo were life-size, as described by Vasari, or the length of one *braccio* (arm-length), as revealed by an inventory drawn up after the death of the artist.[34] Nevertheless, the use and reception of lay figures shows a crucial aspect that was to be present in the much smaller and more elaborate dolls as well: the impetus of animation.

Dürer's Dolls

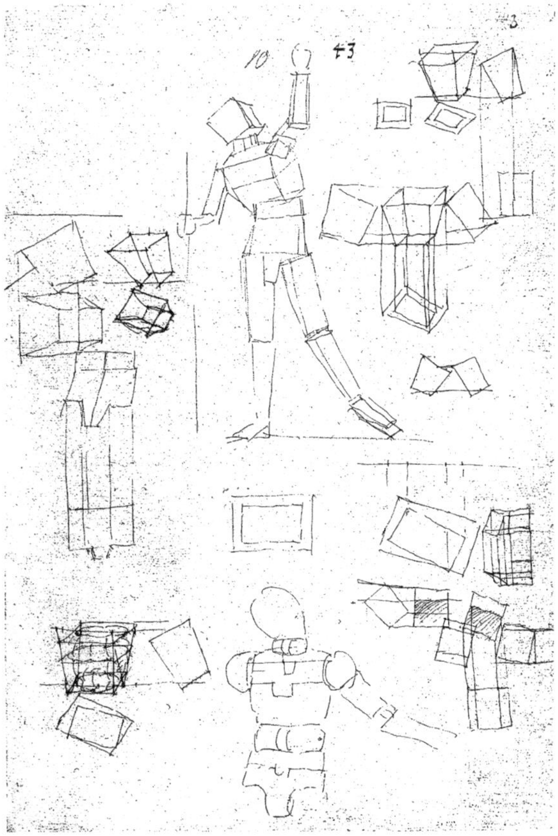

When Vasari wrote his biographies in the mid-sixteenth century, Nuremberg had already long established itself as one of the most important artistic centres north of the Alps. Furthermore, the city had been a stronghold of *Dockenmacher* (doll makers), for centuries.[35] It may be assumed that the artist and art theorist Albrecht Dürer brought the Italian invention of the lay figure to his atelier in Nuremberg and from there into the studios of northern artists.[36] Having encountered them during his travels to Italy, he may have imported a specimen to Germany, most likely in 1505.[37] The context of artistic exploration and creation in the period when Monogrammist I.P. created his figurines finds its prototype in the person of Dürer.

One of Dürer's sheets with sketches may illustrate this relation. On folio 143b of the Dürer codex, several line drawings of two- and three-dimensional geometrical figures are placed at the right- and the left-hand sides of the sheet (Figure 1.2).[38] A complex and dynamic figure composed of cubes is placed at the top of the central axis of the page. It stands in the classical *contrapposto*, bending forward with its left arm raised and holding a stick in the right hand. A coherent body arises only from the combination of the lightly sketched cuboids. This body not only appears to be human but also appears to be moving. In the lower part of the folio, however, the mechanical upper torso of a lay figure is drawn. Despite the absence of hatchings and the summary drawing style, the shape and structure of the jointed doll emerges clearly. Because of a wide range of different joints – hinged joints next to ball joints in the shoulders – this model appears to be very similar to modern kinds of wooden models, as they are used today.

1.2 Albrecht Dürer, pen drawing (undated), folio 143b of the Dresden Dürer-codex. Source: Heribert Schulz (ed.), *Geometrie der Figur. Luca Cambiaso und die moderne Kunst* (Osnabrück, 2007), p. 26, figure 1

Dürer, who used the lay figure in a similar manner as his Italian colleagues, also employed the wooden model for his empirical studies, especially for the *Vier Bücher von menschlicher Proportion (Four Books of Human Proportion)*.[39] The first three of these books illustrates the multifaceted characteristics of the human shape as a conglomerate of natural and geometrical paradigms, whereas the fourth book gives a description of the human figure in motion. By the use of parallel projection, non-dynamic corpora may be set into action.

Here Dürer invented the well-known method of 'cube-proceeding' (*Kubenverfahren*)[40] by transforming organic forms into geometrical ones. As folio 143b shows, the lay figure had made an important contribution to the development of this new method: oscillating between human and artificial, static and mobile, fixed and flexible, the technical development of this model led to its establishment as an instrument of artistic scholarship. The lay figure represented more than a passive form: for Dürer, it was a medium of thought.

The more mechanical type of lay figure that is sketched in the lower part of the Dürer sheet is best compared not to the lay figures of Monogrammist I.P. but to a jointed doll from the Copenhagen 'Kunstkammer' (Colour plate 3).[41] The figurine represents a male body, more slender than muscular, probably dating back to the late sixteenth century. It is of the same height as the dolls of Monogrammist I.P. and made of the same material, but differs in shape and motion technology. While, as previously shown, the components of Monogrammist I.P.'s jointed dolls are fixed by an internal system of strings, the Copenhagen doll is held together by a combination of hinges and ball joints that are fixed by screws. Consequently, the combinatorial principles of construction and motility are immediately visible, as in the case of Dürer's sketched model. Furthermore, the head and the hands are finely shaped, but without veins or muscles, and they lack the moving toes and fingers. The hands are fixed in a single position, the left one outstretched, the right one clenched.

The Copenhagen doll was clearly created to be arranged in the main poses of the human body. Although its creator must have been a master of woodwork, the figure seems less individualized than the figurines of Monogrammist I.P. At the same time, the intrinsic proportionality of both kinds of Renaissance dolls shows the influence of Dürer's 'natural' principles of measurement and could therefore be of interest as geometrical or mathematical models.[42] This is reflected by the location of this specimen: since at least 1674, this lay figure has been kept in 'Det mathematische Kammer' of the royal curiosity cabinet.[43] In general, the 'Kunstkammer' collection of the sixteenth century contained, in addition to mineral, vegetable and animal items (*naturalia*), various artworks, including wooden statuettes,[44] and extraordinary pieces of decorative arts (*artificialia*) as well as technically sophisticated instruments of all types (*scientifica*). The historically based system 'nature-antiquity-artefact-machine'[45] may be considered as the ideal classification of any collection since the middle of the sixteenth century.[46] The jointed Renaissance dolls, especially the ones of Monogrammist I.P., incorporate the fundamental parameters of such collections.[47]

The Sophisticated Doll as 'Natural Artefact'

The jointed dolls of the Monogrammist I.P. were created in a climate of artistic research and interest in natural principles. It is very likely that this master of woodcraft was familiar with Dürer's empirical studies, already widely disseminated before the publication of the *Four Books on Human Proportion*. Due to the quasi-autonomous status of the dolls, it has remained difficult to understand the purpose of their creation and use. Firstly, the nakedness of the dolls has led to the assumption that they were meant as erotic toys.[48] Yet, one can argue that, even at the time, more was needed than a mere mimetic representation of the nude body to portray an erotic subject matter.[49] Their faces, for example, hardly show the kind of seductive facial expression one would expect for such a subject.[50] Nor can we assume that they functioned simply as models for the artist. Their fragility and scale speak against this, as well as the characteristic, individualistic details of the anatomy, and the multiple, agile means of torsion enabled by the hidden motion technology. These dolls are too small and too precious to be related to the simple, human-shaped models mentioned by Vasari. As indicated, precious lay figures represented collectors' items in the 'Kunstkammer'. Given their size, the Renaissance dolls of Monogrammist I.P. seem primarily made to be 'manipulated'. Taking them in his hands, the collector could study human anatomy and proportion, and give the small figurines any desired position. He was both the owner and the originator of the human form, an *alter deus*.

Certainly, the early lay figure described by Filarete already had the capacity to be arranged as if it was alive – at the very least, it could be used for very vivid compositions. However, with the creation of small, rearrangeable dolls that could be held in the hands, a new relationship between model, owner and knowledge emerged. By reducing the size of the dolls and at the same time increasing their motility and artistic refinement, a new level of sophistication was reached that accounts for their independent status. Bridging the gaps between art, technology and nature, I.P. created a group of individual statuettes which, due to their endless possibilities of rearrangement, size and skilfulness, escape the categories of self-operating tools to become more autonomous. Moreover, these works of art may be said to be the crowning achievement of movable, miniature sculpture. The latter became a prevalent artistic genre in the sixteenth century that was characterized by a complex perception not reduced to the sense of sight. Thus, these artefacts can be seen as visual-haptic models of understanding and aesthetic experience.

Notes

1 Many thanks to Lee Chichester, Joris van Gastel, Jaya Remond, Lisbet Tarp and Francesca Whitlum-Cooper for their helpful remarks.

2 Jutta Reisinger-Weber, *Der Monogrammist IP und sein Umkreis* (Passau: Dietmar Klinger Verlag, 2007), pp. 111–21, with further references.

3 Cf. Jürgen Habermas, *Theorie des kommunikativen Handelns* (Frankfurt am Main: Suhrkamp, 1981) p. 229ff.

4 For the history of 'artistic autonomy', cf. Andrea Esser (ed.), *Autonomie der Kunst? Zur Aktualität von Kants Ästhetik* (Berlin: Akademie Verlag, 1995); Ralph Ubl, 'Autonomie', in *Metzler Lexikon Kunstwissenschaft*, ed. Ulrich Pfisterer (Stuttgart and Weimar: Metzler, 2011), pp. 45–49,with further references. For the moral implications of aesthetic autonomy, cf. Norbert Meuter, 'Ästhetische Autonomie. Einige Gedanken über Kunst und Moral', in *Bodies in Action and Symbolic Forms. Zwei Seiten der Verkörperungstheorie*, ed. Horst Bredekamp, Marion Lauschke and Alex Arteaga (Berlin: Akademie Verlag, 2012), pp. 219–36.

5 The figurine measures 216 millimeters in height and has a weight of 78.8 grams. Within the group of IP a development of complexity and details may be observed. The figure in the Bode Museum seems to be the most typified so that it could be chronologically seen as the first. Cf. Reisinger-Weber, *Der Monogrammist IP*, p. 113ff.

6 Ibid., pp. 111–21.

7 Christian Theuerkauff, 'Gliederpuppe', in *Der Mensch um 1500. Werke aus Kirchen und Kunstkammern* (Berlin: Staatliche Museen Preussischer Kulturbesitz, 1977), pp. 166–71, p. 170.

8 The pair in Innsbruck shows two young adults (maybe about 20 years old), the dolls in Hamburg and Leipzig (about 25–30 years old) seem to be a bit older that the figure in Berlin (20–25), whereas the male figurine in Madrid is the oldes (about 40–50).

9 Reisinger-Weber, *Monogrammist IP*, p. 120ff.

10 Francisco Javier notes a 'leve pintura rosada'. Francisco Javier Sánchez Cantón, 'Un maniquí del siglo XVI', in *Archivo Español de Arte* XXV (1952): pp. 101–9, p. 103.

11 With the pair once kept in the Berlin 'Akademie der Künste' it seems that the anatomically most developed exemplars are lost today. Cf. Ingrid Hägele, Gudrun Schmidt, Gudrun Schneider (eds), *Kriegsverluste der Preußischen Akademie der Künste: Kunstsammlung und Archiv* (Berlin: Akademie der Künste, 2005), p. 45, no. 114 and no. 115. Concerning the third one, cf. Arpad Weixlgärtner, 'Von der Gliederpuppe', in *Göteborgs Konstmuseum Årstryck* (1954): pp. 37–71, p. 48; Arpad Weixlgärtner, *Dürer und Grünewald. Ein Versuch, die beiden Künstler zusammen – in ihren Besonderheiten, ihrem Gegenspiel, ihrer Zeitgebundenheit – zu verstehen* (Göteborg: Elanders Boktryckeri, 1949), p. 102ff.

12 Götz Pochat, 'Imitatio und Superatio in der bildenden Kunst', in *Imitatio. Von der Produktivität künstlerischer Anspielungen und Mißverständnisse*, ed. Paul Naredi-Rainer (Berlin: Reimer, 2001), pp. 11–48.

13 For example: *The Pneumatics of Hero of Alexandria*, ed. Marie Boas Hall (facsimile of London, 1851, London, 1971), http://www.history.rochester.edu/steam/hero/, accessed November 21, 2011.

14 Alfred Chapuis and Edouard Gélis, *Le Monde des Automates: Etude Historique et Technique* (Geneva: Slatkine, 1984). Klaus Völker (ed.), *Künstliche Menschen. Dichtungen und Dokumente über Golems, Homunculi, lebende Statuen und Androiden* (Frankfurt am Main: Suhrkamp, 1994 [1971]); Rudolf Drux, *Menschen aus Menschenhand. Zur Geschichte der Androiden. Texte von Homer bis Asimov* (Stuttgart: Metzler, 1988). Bernhard J. Dotzler, Peter Gendolla and Jörgen Schäfer, *MaschinenMenschen. Eine Bibliographie* (Frankfurt am Main and New York: P. Lang, 1992). Pia Müller-Tamm and Katharina Sykora (eds), *Puppen Körper Automaten – Phantasmen der Moderne* (Cologne: Oktagon, 1999); Gisela Febel, 'Neue Forschungsliteratur zum Thema "Menschenkonstruktionen"', in *Menschenkonstruktionen. Künstliche Menschen in Literatur, Film, Theater und Kunst des 19. und 20. Jahrhunderts*, ed. Gisela Febel and Cerstin Bauer-Funke (Göttingen: Wallstein Verlag, 2004), pp. 249–71.

15 Luca Garai, *Gli automata di Leonardo* (Bologna: Bononia University Press, 2007).

16 Cf. Horst Bredekamp, *Theorie des Bildakts, Frankfurter Adorno-Vorlesungen 2007* (Berlin: Suhrkamp, 2010).

17 A related form of hybrid artificial figure may be seen in the marionettes that, as Heinrich von Kleist writes, 'possess an infinite consciousness'. Heinrich von Kleist, *Über das Marionettentheater*, in Heinrich von Kleist, *Briefe und Werke*, ed. Siegfried Streller and Peter Goldammer (Berlin and Weimar: Aufbau Verlag, 1978), vol. 3, pp. 473–80, p. 480.

18 Special thanks to Klaus Bente and Christian Jürgens for this contribution. K. Bente, A. Franz and Ch. Jürgens: computed microtomographical studies and animations of wooden Dutch Dolls (Master IP, sixteenth century) of the Grassi Museum for Applied Art (Leipzig) – personal communication (2011).

19 For the multifarious implications of the art of woodcarving at that time, cf. Michael Baxendall, *The Limewood Sculptors of Renaissance Germany* (New Haven: Yale University Press, 1980).

20 Klaus Maurice, 'The Princely Art of Turning: On the Aesthetic Significance of the Natural Sciences and Technology', in *Gedrehte Kostbarkeiten/Turned Treasuries*, ed. Georg Laue (Munich: Kunstkammer Georg Laue, 2004), pp. 17–27.

21 Ibid., p. 23.

22 Some older examples existed in ancient Egypt, but the doll appears in the late geometrical period for the first time as an artefact of great social importance. Cf. Kate McK. Elderkin, 'Jointed Dolls in Antiquity', *American Journal of Archaeology* 34/4 (1930): pp. 455–79. Mirjam Andres, *Die Antikensammlung. Griechische, Römische, Altorientalische Puppen und Verwandtes* (Hanau-Wilhelmsbad: Hessisches Puppenmuseum, 2000).

23 Gesine Tauber and Johannes Tauber, 'Mittelalterliche Kruzifixe mit schwenkbaren Armen. Ein Beitrag zur Verwendung von Bildwerken in der Liturgie', *Zeitschrift des deutschen Vereins für Kunstwissenschaft* 23 (1969): pp. 79–121. Johannes Tripps, *Das handelnde Bildwerk in der Gotik. Forschung zu den Bedeutungsschichten und der Funktion des Kirchengebäudes und seiner Ausstattung in der Hoch- und Spätgotik* (Berlin: Mann Verlag, 1998). Valeria E. Genovese, *Statue vestite e snodate. Un percorso* (Pisa: Edizioni della Normale, 2010).

24 For the concept of *disegno*, cf. Heinrich Th. Schulze Altcappenberg and Michael Thimann (eds), *Disegno. Der Zeichner im Bild der Frühen Neuzeit* (Berlin and Munich: Deutscher Kunstverlag, 2007). Wolfgang Kemp, 'Disegno. Beiträge

zur Geschichte des Begriffs zwischen 1547 und 1607', *Marburger Jahrbuch für Kunstwissenschaft* 19 (1974): pp. 219–40.

25 Laurie Fusco, 'The Use of Sculptural Models by Painters in Fifteenth-Century Italy', *Art Bulletin* 64/2 (1982): pp. 103–20. Johannes Myssok, *Bildhauerische Konzeption und plastisches Modell in der Renaissance* (Münster: Rhema, 1999).

26 'ladetta perfezzione di giudizio non è punto più necessaria alli scultori che a' pittori, bastando a quelli condurre i modelli buoni di cera, di terra o d'altro, come a questi i loro disegni in simili materie pure o ne' cartoni'. Giorgio Vasari, *Le vite de' più eccellenti pittori, scultori e architettori nelle redazioni del 1550 e 1568*, ed. Paola Barocchi, Rosanna Bettarini, 6 vols (Florence: Sansoni, 1966–87), vol. 1, p. 22.

27 'Aveva openione Fra' Bartolomeo, quando lavorava, tenere le cose vive innanzi, e per poter ritrar panni et arme et altre simil cose fece fare un modello di legno grande quanto il vivo che si snodava nelle congenture, e quello vestiva con panni naturali; dove egli fece di bellissime cose, potendo egli a beneplacito suo tenerle ferme fino che egli avesse condotto l'opera sua a perfezzione: il quale modello, così intarlato e guasto come è, è apresso di noi per memoria sua'. Vasari, *Le vite*, vol. 4, p. 101. Italics by the author.

28 'Ma egli è ben vero che, in facendo quest'opera, fece Benvenuto quello che insin allora non era mai stato usato in Lombardia, cioè fece modelli di terra per veder meglio l'ombre et i lumi, e si servì d'un modello di figura fatto di legname, gangherato in modo che si snodava per tutte le bande, et il quale accomodava a suo modo con panni adosso et in varie attitudini'. Vasari, *Le vite*, vol. 5, p. 412.

29 Cf. Gianni Carlo Sciolla, 'Il disegno dal manichino in legno e dal modello in terra e cera nella tradizione dal Quattrocento al Seicento. Appunti per una ricerca', in *Conservazione dei materiali librari archivistici e grafici*, ed. Marina Regni and Piera Giovanna Tordella (Turin: U. Allemandi, 1996), pp. 209–26.

30 'Fa' d'a[v]ere una figuretta di legniame, che sia disnodata le braccia e le ghambe e ancora il collo; e poi fa' una [v]esta di panno di lino, e con quello habito, che ti piace, come se fussino d'uno [v]ivo'. Antonio Averlino Filarete, *Trattato di architettura*, in *Antonio Averlino Filaretes Tractat über die Baukunst nebst seinen Büchern von der Zeichenkunst und den Bauten der Medici*, ed. Wolfgang von Oettingen (Vienna: Graeser, 1890), p. 654ff, my translation.

31 Christiane Kruse, 'Fleisch werden – Fleisch malen: Malerei als "incarnazione". Mediale Verfahren des Bildwerdens im Libro dell'Arte von Cennino Cennini', *Zeitschrift für Kunstgeschichte* 63/3 (2000): pp. 305–25.

32 Julius von Schlosser, 'Aus der Bildnerwerkstatt der Renaissance', *Jahrbuch der kunsthistorischen Sammlungen des Allerhöchsten Kaiserhauses* XXXI (Wien, Leipzig, 1913/14): pp. 67–135.

33 Leon Battista Alberti stipulated the use of the 'right' *rilievo* that was even more important than the *disegno* itself: 'E forse più sarà utile exercitarsi al rilievo che al disegno'. Leon Battista Alberti, *Della pittura*, ed. Luigi Mallé (Florence: Sansoni, 1950), s. 109. Cf. Markus Rath, 'Die Haptik der Bilder. *Rilievo* als Verkörperungsstrategie der Malerei', in *Das haptische Bild. Körperhafte Bilderfahrung in der Neuzeit*, ed. Markus Rath, Jörg Trempler and Iris Wenderholm (Berlin: Akademie Verlag, 2013), pp. 3–29.

34 'cioè uno modello di legno quanto el naturale, cioè una figura, e ancora uno altro modello circa d'un braccio ghangherato'. Vincenzo Marchese, *Memorie dei piú insigni pittori, scultori e architetti* (Florence: A. Parenti, 1846), vol. 2, p. 608.

35 Cf. Manfred Bachmann and Claus Hansmann, *Das große Puppenbuch* (Leipzig: Edition Leipzig, 1971), p. 42. Concerning the term 'Docken', cf. Ignaz Zingerle, *Das deutsche Kinderspiel im Mittelalter* (Innsbruck: Wagner, 1873), pp. 132–35.

36 Weixlgärtner, *Gliederpuppe*, p. 40.

37 Arpad Weixlgärtner, 'Dürer und die Gliederpuppe', in *Beiträge zur Kunstgeschichte. Franz Wickhoff gewidmet* (Vienna, 1903), pp. 80–90, p. 87.

38 Gone nearly unnoticed for 400 years, the Dresden sketchbook was first published in 1905. Cf. Robert Bruck, *Das Skizzenbuch von Albrecht Dürer in der königlichen öffentlichen Bibliothek zu Dresden* (Strasbourg: Heitz, 1905).

39 Hans Rupprich, *Albrecht Dürer. Schriftlicher Nachlass*, 3 vols (Berlin: Deutscher Verein für Kunstwissenschaft, 1956–69), vol. 3, p. 122. Cf. Jürgen Fredel, *Ideale Maße und Proportionen. Der konstruierte Körper*, in *Die Beredsamkeit des Leibes. Zur Körpersprache in der Kunst*, ed. Ilsebill Barta Fliedl and Christoph Geissmar (Salzburg and Vienna: Residenz, 1992), pp. 11–42, p. 33. Greatest thanks to Elisabeth Weymann who let me participate in her latest research on the relation of Dürer's proportional studies and the proportions of the dolls of the Monogrammist I.P. (which seem to be directly related to the printed 'guidelines').

40 Fredel, 'Ideale Maße', p. 32. There are some early forms of this technique articulating the human body with cubes, i.e. in the Dresden sheet fol. 166v, c. 1515.

41 Bente Gundestrup, *Det kongelige danske Kunstkammer 1737 / The Royal Danish Kunstkammer 1737*, 2 vols (Copenhagen: Nationalmuseet/Arnold Busck, 1991–95), vol. 1, p. 317.

42 The Berlin jointed doll is closely related to Dürer's study of a woman with the height of eight lengths of the head: Albrecht Dürer, 'Seiten und Vorderansicht der Frau von 8 Kopflängen (Typ 1 B)', 22.2 x 14.3 cm; cf. Albrecht Dürer, *Vier Bücher von menschlicher Proportion*, 1, fol. B5v, in *Albrecht Dürer. Das druckgraphische Werk*, ed. Rainer Schoch, Matthias Mende and Anna Scherbaum (Munich: Prestel, 2001–2004), p. 342.

43 On folio 29a, two small lay figures are noted: '2 Smaa Leemend'. Gundestrup, *Kunstkammer*, p. 317.

44 As well as in Copenhagen, the 'Praunsche Kabinett' in Nuremberg documents a lay figure in the inventory of 1616: 'ein hülzes gliedmenlein, anderthalb schuch hoch auch in einem kestlein' (a little wooden lay figure, about 1.5 feet high, in a small box), *Die Kunstsammlung des Paulus Praun. Die Inventare von 1616 und 1719*, ed. Katrin Achilles-Syndram (Nuremberg: Stadtrat, 1994), p. 148, no. 382. For wooden statuettes cf. also Jörg Rasmussen, *Deutsche Kleinplastik der Renaissance und des Barock* (Hamburg: Museum für Kunst und Gewerbe Hamburg, 1975), p. 9. Concerning the system of such collections, cf. Horst Bredekamp, *Antikensehnsucht und Maschinenglauben. Die Geschichte der Kunstkammer und die Zukunft der Kunstgeschichte* (Berlin: Verlag Klaus Wagenbach, 1993), p. 35.

45 Bredekamp, *Antikensehnsucht*, p. 33.

46 Ibid. As Bredekamp shows, Samuel Quiccheberg first presented this kind of ideal collection in his *Inscriptiones vel tituli theatri amplissimi* in 1565. Cf. ibid., pp. 33–39.

47 Due to their small-scale size they have even become more precious. Cf. Dagmar Eichberger, *Leben mit Kunst. Wirken durch Kunst. Sammelwesen und Hofkunst unter Margarete von Österreich, Regentin der Niederlande* (Turnhout: Brepols, 2002), p. 410.

48 Ernst Friedrich Bange, *Die Kleinplastik der deutschen Renaissance in Holz und Stein* (Leipzig: Schmidt and Günther, 1928), p. 49. Ernst Friedrich Bange, *Die Bildwerke in Holz, Stein und Ton: Kleinplastik* (Berlin and Leipzig: De Gruyter, 1930), p. 41, Weixlgärtner, 'Gliederpuppe', p. 54. Theuerkauff, 'Gliederpuppe', p. 170. For an alternative currency cf. Rasmussen, *Kleinplastik*, p. 85, and Carlos Obergruber-Boerner, 'Menschenpaare', in *Nackt. Die Ästhetik der Blöße*, ed. Wilhelm Hornbostel and Nils Jockel (Munich: Prestel, 2002), pp. 35–44, p. 40.

49 Robert W. Scribner, 'Vom Sakralbild zur sinnlichen Schau. Sinnliche Wahrnehmung und das Visuelle bei der Objektivierung des Frauenkörpers in Deutschland im 16. Jahrhundert', in *Gepeinigt, begehrt, vergessen: Symbolik und Sozialbezug des Körpers im späten Mittelalter und in der frühen Neuzeit*, ed. Klaus Schreiner and Norbert Schnitzler (Munich: Wilhelm Fink Verlag, 1992), pp. 309–30.

50 In the face of the jointed doll in Hamburg, Rasmussen observes, 'scharf-blickende … Augen und de[n] Ausdruck kühler Beherrschtheit' (a strict … glance and the expression of frigid self-control). Rasmussen, *Kleinplastik*, p. 85, cat. 6.

2

Genesis of Images: Intersections of Art and Alchemy in Early Modern Europe

Lisbet Tarp

Introduction

Alchemy is a disputed subject in relation to art, and it has even been stated that alchemy did not have a significant influence on painting.[1] Nevertheless, scholars have been recognizing inspiration from alchemy in a wide variety of motifs in early modern art.[2] In this essay, the intersections are regarded as structural correspondences in the line of thought of artistic and alchemical experiments rather than analyzed at the iconographical level. Philosophy and theories of nature were intricate parts of workshop practices of art and alchemy, and the manufacturing process was made visible in the final appearance of the works of art presented here.[3]

During the premodern era, alchemy and art were connected as natural philosophy was the common source of both. Natural philosophy was not a uniform science, but a great variety of ideas about the world's structure and processes including inherited antique knowledge, workshop practice, and the Christian framework. The creation of artificial objects – *ars* in Latin – was dependent on these ideas and procedures, and the results reflected the hierarchy of the world, properties of matter and actions of nature. This essay focuses on decorative art objects that not only rely on natural philosophy but rather take it a step further by using and testing ideas of natural philosophy in a way similar to alchemy. Both artist and alchemist interrogated and twisted materials in order to merge natural and artificial creation. One of the major pursuits of alchemy was to control and transgress the border between natural and artificial creation, and this pursuit is also evident in the decorative arts. In this way the workshop experiments with materials of alchemy, and decorative arts reveal an intertwined corpus of knowledge both manual and intellectual.[4]

For centuries the relation between natural and artificial has been negotiated in different fields of production – medicine, engineering, gardening, art and, as mentioned above, alchemy.[5] A recurring topic in this negotiation has been the imitation of nature. Aristotle defined two ways of imitating nature: one is to copy the shapes and forms produced by nature; another is to imitate the inherent processes of nature.[6] Early modern alchemists aspired to practice the latter; it may also be argued that some artists in this period likewise practiced such imitation and understood their art as a process of collaboration with nature. The imitation of processes of nature was related to the quest to go beyond mere representation and bring into the man-made object some kind of natural growth and genuineness. This can also be said about alchemy and its ultimate goal of creating real gold and not simply materials that imitated its appearance. As this essay shows, there are works of decorative art from the sixteenth and seventeenth centuries in Europe that appear to absorb inspiration from the same pool of ideas and to parallel the methods of alchemy; these artists experimented with properties of matter in order to use nature's forces and magnify the illusion with material presence.

The decorative art objects presented here feature pearls, marble, alabaster and other types of stone bearing patterns with image-like qualities. Viewed as part of an experimental culture that tried to connect to and exploit the powers of nature, they will be analyzed as 'test cases' for the collaboration between nature and art.

Alchemy and Art: Practiced Theory

William Newman states that alchemy historically originated in the workshop practices of decorative arts in the antique period, and furthermore, that the practitioners of decorative arts and alchemy were united in their desire to replicate rather than represent nature.[7] Despite rejections of alchemy by artists in the Renaissance period (for instance, Leonardo da Vinci and Bernard Palissy)[8] the period's workshop activities of manufacturing art, drugs and pigments can be difficult to separate from alchemical production.[9] To be 'made by alchemy' could simply mean to be made artificially.[10] In defining 'alchemy', the scholastic philosopher Roger Bacon wrote in the thirteenth century of 'theoretical alchemy, which theorizes about all inanimate things and about the whole generation of things from the elements. There is also an operative and practical alchemy, which teaches how to make precious metals and pigments, and many other things better and more plentifully than they are made by nature'.[11] A common characteristic of many of the artistic and alchemical processes was the importance of the material's purity and the ingredients' composition including a perfection strived for in both the process and the result. Historian Pamela H. Smith writes, 'alchemy and artisanry possessed a common essence: the ennobling of matter through manual work'.[12] Impurities had to be removed and proportions measured to establish the right conditions in which to alter matter, preparing it to turn into a work of art. This had both a practical side and a philosophical one; today it is crucial to isolate an experiment if you want to create a specific product or result and prevent side effects, but in the early modern experiment or manufacturing process materials would not be isolated but intentionally linked to and influenced by external forces. Qualities and measures were viewed as related to cosmic orders and divine proportions that required synchronization. Alchemy was the experimental science par excellence, and experiments as acts of 'earthly astrology' were supposed to be connected to the forces of the cosmos in order to make them act upon the substances, stimulating the 'inner desires' of the materials.[13] For instance, if an impurity had a quality related to the element of fire, it could be extracted by exposing the contaminated material to heat or another material with this quality; the impurities would then seek and join their right element.[14] The materials' properties and their interwoven connections to cosmic influences were fundamental to the logic of the experiments, as the conservator Spike Bucklow has shown.[15] Among other natural philosophical ideas, the causes of planets and theories of humours and elements were used to explain the choice of ingredients, treatments and processes, and in this sense the alchemists put concepts into action and paved the way for experimental science.[16] A famous example is the influential doctor and natural philosopher Paracelsus (1493–1541), who based his doctrine of signatures on similar assumptions about interconnections in the world, and like alchemists before him applied his ideas about nature in actual experiments.

With this description of alchemy in mind, the process of refining immature images found in nature as practiced in the decorative arts could be viewed as part of the urge to apply as well as scrutinize and question the ideas of nature, for instance, the concept of *natura pictrix* – the image-making nature. In the workshop practice, materials were investigated and boundaries between nature and art were negotiated, making the workshop a place fundamental to the development of the modern scientific method.[17] The maker of decorative arts may be seen to have had similar ambitions as the alchemist: to experiment with nature and engage with issues such as the genesis of an image or the role of the maker. In this way, decorative arts not only brought different kinds of images to trial but also displayed skill and knowledge about the world and the philosophical questions at stake.

Associated Images

In the treasure collection at Rosenborg Castle in Copenhagen, there are several pieces with pearls depicting body parts of humans or animals, and one of these is a pendant depicting a crane with a pearl incorporated into its body (Colour plates 4a and 4b).[18] The bulky and irregular surface of the pearl represents the shiny plumage on the body and the upper part of a leg, where the foot is attached. The pearl's 'potential image' is made visible and framed by the iconography of the embellishment, and the pearl and additions together form an exquisite bird. The artist finished what nature had begun.

Visually, the pearls are both hidden and revealed in the compositions. A play with materiality versus image is staged. Most often these kinds of pieces show mermaids, birds, dolphins, sea monsters and the like, but it soon becomes equally apparent that the pearls are alien elements, shaped by nature and not by art.[19] These pieces work like the *multistable* images in which one can see two images in one picture, but not at the same time.[20] In the case of three-dimensional objects, however, the focus shifts between viewing the material components and the representation that unifies the composition. The work of art calls attention to its simultaneous roles as illusionistic image and material thing.

The use of alabaster and marble has resulted in pieces in which the veins and colours of the stones were thought to resemble clouds, water, mountains or cityscapes. Examples of this are a well-known double-sided painting on an alabaster slab in Philipp Hainhofer's Augsburg Art Cabinet from the first half of the seventeenth century, now at Uppsala University in Sweden,[21] and paintings on stone slabs by Hans von Aachen (1552–1615) and Filippo Napoletano (1589–1629).[22] The artists enhanced the lines, shapes or colours of the latent, natural image in order to complete the work of art. Although such pieces exhibit finely elaborated passages and parallels between the artistic contributions of nature and artists, they also reveal that the transitions between art and nature were not seamlessly made. The viewer's action of locating points of transition where surfaces are connected or paint meets stone colour catalyzes reflection on how to perceive and understand the constitution of the work of art: what did nature create and how did the artist contribute? The underlying question is: what is the genesis of images?

Artists' insistence on exposing their works' hybridity suggests that their aim was not primarily to dissolve boundaries between art and nature but rather to investigate and negotiate each category's limitations and possible transgression. The stating of the formula that nature can produce images can also be interpreted as a statement about the 'naturalness' of the production of art, thereby contributing to the status of art as such. The idea that images are inherent in raw materials makes it possible for the artist to claim that art is an extension of nature, and that artistic and natural processes are mutually analogous creative acts dependent on divine forces.[23]

In some objects this line of thought is taken a step further as there is a tendency to favour not only natural objects that may be embellished but also materials that contain references to natural creative acts. This is the case in the small-scale sculptures that show Daphne as she turns into a tree, represented by an attached coral.[24] The coral's resemblance to branches is used, but coral was also considered a mysterious material, capable of radical change that had a mythological origin as the petrified blood from Medusa's head.[25] Furthermore, coral was considered to be a zoophyte – that is to say, it avoids clear ontological identification – due to its resemblance to both mineral and plant, depending on the conditions to which it is exposed.[26] By using coral in art, it seems as though natural transformative powers were added to the depictions of stories of Daphne or Actaeon's metamorphosis.

Even though the pieces did not blend a natural and an artificial image like the jewelry with pearls described above, the use of coral contributes to the theme of transformation on a material level.[27] Daphne's depicted metamorphosis is mirrored in the coral's inherent quality of change. In this way, the composition can also be seen as portraying the process of matter being (re-)shaped, as occurs in the process of creating an image. Likewise, a pearl mirrors the process of art in a similar fashion by being a grain of sand nourished by an oyster; it represents the refinement of common materials into exquisite works of art.

Metamorphosis and stages of change were, indeed, common themes for works of art in early modern Europe, as were decay and distortion caused by optical effects. Such themes can be seen in the flourishing grotesques in Italian palaces, the use of motifs from Ovid's stories and the attention to the viewer's physical position and experience, challenging both the visual apparatus and recognition of the world's phenomena.[28] In the meandering of the grotesques, a bow turns into an animal head, a vase or a flower; paintings and sculptures depict transformations, for example, from a human into a stag or a tree, lingering at the moments of ontological shifts. And accordingly, the fascination with change is also evident in the collecting and use of materials, such as the mysteriously created fossils and corals that motivated debates about ontological categories and the petrifying powers of the earth.[29] The artistic experiments with infusing natural powers into images are also evident in the use of marble or alabaster slabs in depictions of miraculous events like the Annunciation, the Resurrection or the Assumption of St Mary Magdalene.[30] The divine forces appear on the surface of the stone in the shape of clouds, light and whirling vapours, and this kind of amorphous, naturally emerging image mirrors the direct intervention of God represented in the narrative. For instance, in a Florentine picture of the Resurrection, Christ is painted on a shining, cloudy spot of veined alabaster that represents the divine light (Figure 2.1).[31] In this way, Christ is shown as if he were emerging from the stone, and the 'light' creates a bridge between heaven and earth as well as between natural and artificial creation. Again, the natural image infuses the work of art with transformative powers supposedly initiated by God and as such reflects its own origin and creation. In this way the topoi of transformation and transgression are legitimized by the religious subjects with supernatural events. It seems reasonable that pieces of decorative art with secular themes and traces of natural processes were also connected to the capacity of the works' natural components to suggest a divine origin.

Also in question here is the material components' ability to subtly introduce another level of meaning, possibly making the viewer reflect on the appearance of the work of art and its constitution by displaying its materiality as such along with the signs of its supposedly divine making.

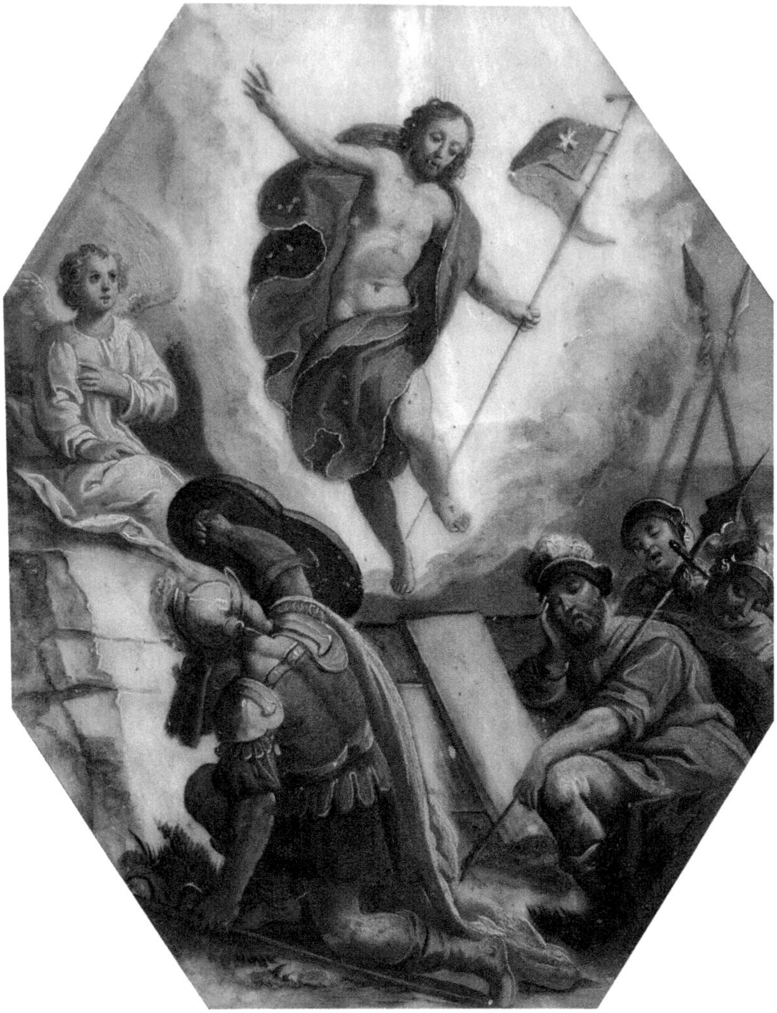

2.1 *The Resurrection* (late sixteenth century–beginning of seventeenth century), painted alabaster, inv. Oggetti d'Arte 1911, n. 1446. Galleria Palatina, Florence. Photo credit: Galleria Palatina

The Spark of Life: When Matter Is Shaped

To collect and insert 'natural agents' into works of art is similar to the alchemical procedure of adding 'seeds' to substances.[32] The laboratory setup, especially the vessel used for experiments, was an attempt to duplicate the womb of the earth, a place for accelerating the generation of natural counterparts, as was articulated by the influential medieval scholar Albertus Magnus.[33] The most famous examples of alchemy are efforts to transmute base metals into gold and to create a copy of a human being, the homunculus.[34] But there were also numerous drugs and pigments produced in the alchemical laboratory in the quest to copy and perfect natural substances.[35] The alchemist tried to imitate and control the cycle of unification, putrefaction, purification and spiritualization to stimulate, simulate and perform natural creation. In those processes the quest to find and add 'seeds' – sparks of life – was crucial for catalyzing the transformation process and turning the mixture into the desired result.[36] More than any other human activity, alchemy encompassed an ambition to alter and improve on nature, which garnered criticism from those who believed that its practitioners aspired to attain godlike abilities.[37] Nevertheless, God's creation from nothing, ex nihilo, was radically different from the creative acts of nature and human beings. In the defence of alchemy, it was argued in the late Middle Ages that the alchemist could *not* create the spark of life but only had limited access to manipulating the substances at hand.[38] For instance, in the preface to the lastingly influential 1487 witch-hunting manual *Malleus Maleficarum,* it was asserted that human beings were only able to modify and adapt nature's products like metals, wood and stone, but not transmute the species.[39] Nature, on the other hand, was capable of reducing an existing substance to prime matter and giving it a new substantial form.[40] Alchemy balanced on a razor's edge by intending to imitate nature's processes and indirectly simulate God's creative acts; one of its aims was to improve nature and complete her unfulfilled intentions; another was to challenge and cross the ontological borders determined by nature and God.

By representing imaginary monsters and hybrid figures like the grotesques,[41] art also challenged the limits of acceptable creations. For those who were offended, suitable critique was found in the writings of Horace, which were read and paraphrased during the Renaissance. In the decorative arts, depictions of monsters flourished on household accessories (ink stands, cutlery, vessels, cabinet decorations, door handles, and the like) as well as on jewellery and armour.[42] Whereas the patterns of *pietra paesina* – a polychrome marble type – tend to be used to represent seascapes or landscapes (trees, cities, mountain profiles and the like) into which various creatures, ships, figures and other details were painted, the three-dimensional natural components like pearls, corals, fossils, shells and ostrich eggs more frequently represent parts of animals, monsters and hybrid creatures.[43] Such motifs were rejected by the Council of Trent in 1563 as inappropriate subjects due to their unnaturalness and supposed lack of a counterpart in the real world.[44] Nevertheless, reports of existing monsters represent a grey zone, and the emergence of monstrous things in nature was interpreted in various ways in the early modern period, as monsters generally moved from being understood as portentous signs to mistakes of nature due to bad conditions or influences.[45] A topic still in need of study is whether the many reports and depictions of monsters during the Counter-Reformation can be interpreted as a way of trying to circumvent restrictions on artistic licence.

The Renaissance humanist Benedetto Varchi considered Aristotle's four causes – the Efficient Cause, the Formal Cause, the Material Cause and the Telic Cause – in his treatise on creation and more specifically monsters.[46] The four interdependent causes were thought to determine the production of things in the earthly sphere. The Telic Cause was explained by Varchi as the 'intent' of the creator – this being Nature/God – that is, the purpose inherent in all created things.[47] This intent could be fulfilled or not, according to disturbances in the stages of development. In alchemy and the decorative arts, the idea of an inherent cause seems to be similarly interpreted; the immature natural image or semi-noble metal was treated as an object of unrealized intent, representing the point of departure for artists and alchemists, who 'assisted' nature in completing the work begun; the 'cause' of nature became fulfilled through the human extension of the process.[48] To associate the human act with God's grand masterpiece in the works of art can be interpreted as a way of stating that the natural order was not depraved or violated by the manual interventions. In the words of Albertus Magnus in his book on minerals, 'skilful alchemists proceed as skilful physicians do ... by strengthening nature, they aid the power of nature, directing it so as to bring about natural health'. The alchemist is here interpreted as working in line with nature, and it is concluded that 'nature performs the work, and not art'.[49]

The act of trying to connect art to natural forces may also be related to the old wish to make works of art come alive, often exemplified by the story of Pygmalion, who falls in love with his own sculpture, finally brought to life by Venus.[50] Likewise, the famous ancient story of the artists' contest between Zeuxis and Parrhasius illustrates the desire to bring illusions into reality. The birds pecked at the grapes painted by Zeuxis, who then tried to draw aside the curtain painted by Parrhasius; Zeuxis deceived the birds, but Parrhasius deceived a fellow artist.[51] Whereas the Pygmalion sculpture underwent a complete transformation, the works of art under discussion here could only indicate and play on the potential slippage between imitation and material reality. Inserting bits of transformed natural objects like pearls and stones with latent images, however, did provide a way to make works of art become more 'real' and allude to parallels between art and nature. Zeuxis's discovery of the illusion made the curtain retreat into the realm of imitations, but the painting had interacted with the viewer for a moment as if the depiction were real. Especially in the sixteenth century, the deception of the senses was a typical game in art, gardens and architecture, and the oscillation between viewing the illusion and breaking it is a key feature of the decorative works of art described here as well.

Concluding Remarks

A wide range of exquisite, strange and beautiful materials embodied the heterogeneous group of decorative arts in the early modern period. The objects were displayed in collections, households, on the body and at banquets. The materials were shaped into secular as well as Christian and mythological motifs, and occasionally the material reflected or even interfered with the subject depicted, as in the case of Daphne with coral branches.

The significance of the materials used in art in the early modern period is still a subject in which further research is needed. Another aspect of the play with materiality in decorative arts was the influence from the tradition of relics, folklore medicine and magic. This is evident in the objects of coral that children wore for protection, or when drinking vessels were equipped with snake or dragon's teeth (fossil shark's teeth) to de-poison the contents.[52] Again, natural philosophy created correspondences between magical practice, pharmacology and the making of beautiful artefacts.

The ways materials were put into play tells a story about the different concept of art at the time. Art-making was but one among other manual arts – from shoemaking to alchemy or painting – but the fundamental act of creation was performed by nature, and ultimately by God. From this point of view, the works of art that featured the threshold between art and nature interrogated the hierarchy of making, usurping or claiming to usurp some of the primary powers of creating. Nevertheless, the artists' efforts to physically test and examine the limits of collaborations between nature and art also exposed the limitations of nature's art as they may have seemed crude and less refined compared to the imagery by the artist. Eventually, this might have contributed to the decline of interest in the idea of nature making art.

The (in)famous peculiar symbolism and philosophy of alchemy survived its experimental aspect, thus twisting our picture of early modern alchemy. The increasingly esoteric and sectarian development of alchemy during the eclipse of the geocentric world picture contributed to the obscurity of an art form originally tied to both artistic and 'scientific' experimentation.[53] Through time, the concept of nature's art and alchemy's refined experiments lost their framework, thus letting the two arts drift apart.

Notes

1 For instance, James Elkins and Didier Kahn's discussion in *Konsthistorisk Tidskrift* starting with Elkins's article 'On the Unimportance of Alchemy in Western Painting', no. 1–2 (1992): pp. 21–26, continued in no. 64 (1995): pp. 47–53. Lawrence M. Principe and William R. Newman, 'Some Problems with the Historiography of Alchemy', in *Secrets of Nature*, ed. Newman and Grafton (Cambridge, MA, and London: MIT Press, 2006 [2001]), pp. 385–431.

2 Jacob Wamberg (ed.), *Art and Alchemy* (Copenhagen: Museum Tusculanum, 2006).

3 This article is based on a paper presented at the conference 'The Artwork between Technology and Nature', National Gallery of Denmark, January 22, 2010.

4 Pamela H. Smith has developed an interesting concept of 'artisanal epistemology': Pamela H. Smith, *The Body of the Artisan: Art and Experience in the Scientific Revolution* (Chicago: University of Chicago Press, 2004).

5 Bernadette Bensaude-Vincent and William R. Newman (eds), *The Artificial and the Natural: An Evolving Polarity* (Cambridge, MA, and London: MIT Press, 2007), the introduction.

6 William Newman, *Promethean Ambitions* (Chicago and London: University of Chicago Press, 2005), chapter 5.

7 William Newman, *Promethean Ambitions*, p. 116.

8 Ibid. p. 118ff.

9 See the writings of Cennino Cennini (1370–1440) or Benvenuto Cellini (1500–71).

10 'For the alchemist is the baker in baking bread, the vintner in making the wine, the weaver in weaving cloth. Thus, whatever arises out of nature for human use is brought to that condition ordained by nature by an alchemist'. Andrew Weeks (ed. and trans.), *Paracelsus (Theophrastus Bombastus von Hohenheim 1493–1541): Essential Theoretical Writings* (Leiden and Boston: Brill, 2008), p. 211. Quoted in another version in Smith, *The Body of the Artisan*, pp. 141–42.

11 William Newman, *Promethean Ambition*, p. 118. Robert Halleux, *Les textes alchimiques* (Turnhout: Brepols, 1979), p. 43.

12 Pamela H. Smith, *The Body of the Artisan*, p. 144.

13 Spike Bucklow, *The Alchemy of Paint: Art, Science and Secrets from the Middle Ages* (London: Marion Boyars Publishers, 2009), p. 51 and p. 109ff.

14 See Bucklow on Cennini's experiment with lapis lazuli, p. 57ff.

15 Ibid.

16 Pamela H. Smith, *The Body of the Artisan*, p. 17ff.

17 Ibid.

18 Jørgen Hein, *The Treasure Collection at Rosenborg Castle*, vols 1–3 (Copenhagen: Museum Tusculanum Press, 2009), p. 188.

19 See Yvonne Hackenbroch, *Renaissance Jewellery* (London: Sotheby Parke Bernet Publications, 1979), and Marco Chiarini and Cristina Acidini Luchinat, *Bizzarrie di pietre dipinte dalle collezioni dei Medici* (Florence: Silvana Editoriale, 2000).

20 W.J.T. Mitchell, *Picture Theory: Essays on Verbal and Visual Representation* (Chicago: University of Chicago Press, 1995). Jurgis Baltrušaitis also calls this type of image a 'Vexierbild'. Baltrušaitis, *Anamorphic Art* (Cambridge: Chadwyck-Healey, 1977 [1969]), p. 11ff.

21 http://www.gustavianum.uu.se/en/node152, accessed March 17, 2013.

22 Maria Sframeli (ed.), *I gioielli dell'Elettrice Palatina al Museo degli Argenti* (Florence: Centro Di, 1988).

23 Fredrika Jacobs, *The Living Image* (Cambridge: Cambridge University Press, 2005), chapter 2. Michelangelo famously interpreted the idea of the inherent image in natural forms in art and poetry.

24 See Wenzel Jamnitzer's *Daphne*, 1550, at Musée national de la Renaissance, and Abraham Jamnitzer's *Daphne*, late sixteenth century, at Staatlicher Kunstsammlungen Dresden.

25 Michael Cole, *Cellini and the Principles of Sculpture* (New York: Cambridge University Press, 2006), chapter 2.

26 Paula Findlen, 'Jokes of Nature and Jokes of Knowledge: The Playfulness of Scientific Discourse in Early Modern Europe', *Renaissance Quarterly* 43/2 (Summer 1990): p. 303.

27 Ibid.

28 About the viewer's experience, see Jacobs, *The Living Image*, p. 11, and Baltrušaitis, *Anamorphic Art*.

29 See also Palissy's considerations on fossils: Bernard Palissy, *The Admirable Discourses*, trans. Aurèle la Rocque (Urbana: University of Illinois Press, 1957). Maria Fabricius Hansen, 'Maniera and the Grotesque', in Wolfgang Braungart (ed.), *Manier und Manierismus* (Tübingen: Max Niemeyer Verlag, 2000).

30 Marco Chiarini and Cristina Acidini Luchinat, *Bizzarrie di pietre dipinte*, p. 15 and n. 79.

31 Ibid. p. 76.

32 William Newman, 'Art, Nature, Alchemy, and Demons: The Case of the *Malleus Maleficarum* and Its Medieval Sources', in *The Artificial and the Natural*, ed. Bensaude-Vincent and Newman, chapter 5, p. 120, and Newman, *Promethean Ambitions*, p. 51ff.

33 Albertus Magnus, *Book of Minerals* (Oxford: Clarendon Press, 1967), III, i, 9, pp. 178–79.

34 Base metals were thought to be immature stages of gold. Newman, *Promethean Ambitions*, p. 75 and chapter 4.

35 Spike Bucklow, *The Alchemy of Paint*.

36 Richard Kieckhefer, *Magic in the Middle Ages* (Cambridge: Cambridge University Press, 2010 [1989]), p. 13.

37 William Newman, *Promethean Ambitions*, p. 43ff, and Newman, 'Art, Nature, Alchemy, and Demons', in *The Artificial and the Natural*, ed. Bensaude-Vincent and Newman.

38 William Newman, *Promethean Ambition*, p. 48ff.

39 William Newman, 'Art, Nature, Alchemy, and Demons', in *The Artificial and the Natural*, ed. Bensaude-Vincent and Newman, p. 111.

40 Dennis Des Chene, 'Forms of Art in Jesuit Aristotelianism (with a Coda on Descartes)', in *The Artificial and the Natural*, ed. Bensaude-Vincent and Newman, p. 137.

41 See the essay by Maria Fabricius Hansen and Camilla Skovbjerg Paldam in this volume.

42 Luke Syson and Dora Thornton, *Objects of Virtue: Art in Renaissance Italy* (London: British Museum Press, 2001), and Marina Belozerskaya, *Luxury Arts of the Renaissance* (London: Thames and Hudson, 2005).

43 See Hackenbroch, *Renaissance Jewellery*, Chiarini and Luchinat, *Bizzarrie di pietre dipinte*; and Sframeli and Hackenbroch, *I gioielli dell'Elettrice Palatina*.

44 Robert Klein and Henri Zerner, *Italian Art 1500–1600: Sources and Documents* (Englewood Cliffs, NJ: Prentice-Hall, 1966), pp. 119–22.

45 Fredrika Jacobs, *The Living Image*, p. 161ff, Laura Lunger Knoppers and Joan B. Landes (eds), *Monstrous Bodies/Political Monstrosities in Early Modern Europe* (Ithaca and London: Cornell University Press, 2004), and Lorraine Daston and Katharine Park, *Wonders and the Order of Nature, 1150–1750* (New York: Zone Books, 2001 [1998]).

46 Benedetto Varchi, *Lezione sopra la generazione de' Mostri*, 1548, quoted in Jacobs, *The Living Image*, p. 161ff, and note 83, p. 157.

47 Fredrika Jacobs, *The Living Image*, p. 161 and p. 11.

48 David Summers, *The Judgment of Sense* (New York: Cambridge University Press, 1987), p. 269.

49 Albertus Magnus, *Book of Minerals* (III, i, 9), p. 178.

50 For elaboration on the subject, see Fredrika Jacobs, *The Living Image*, p. 49, and Stephen J. Campbell '"Fare una Cosa Morta Parer Viva": Michelangelo, Rosso, and the (Un)Divinity of Art', *Art Bulletin*, 84/2 (December 2002): pp. 596–620.

51 Fredrika Jacobs, *The Living Image*, p. 186, and Newman, *Promethean Ambitions*, p. 12.

52 Wilfried Seipel, *Alle Wunder dieser Welt: Die kostbarsten Kunstwerke aus der Sammlung Erzherzog Ferdinands II (1529–1595)* (Vienna: Kunsthistorischer Museum Wien, 2001), p. 48ff.

53 Principe and Newman, 'Some Problems with the Historiography of Alchemy', pp. 385–431.

3

Living Jewels, Creepy Crawlers and Robobugs: Insects in the Wunderkammer, Surrealism and Contemporary Art

Marion Endt-Jones

Paradoxes of Classification

A *BBC News* headline published in March 2006 stated, 'Pentagon Plans Cyber-Insect Army'.[1] The article introduced a new scheme developed by the American Defence Advanced Research Projects Agency (DARPA), which was aimed at creating remotely controlled insect-machine hybrids that could one day be used for surveillance purposes or for search-and-rescue missions – to send transmissions and to 'sniff out' explosives. The idea was to take advantage of the insects' ability to regenerate body parts and reposition internal organs during the process of metamorphosis; if electronic micro-components were implanted into dragonflies and moths at their pupa stage, scientists at the U.S. military agency argued, they would gradually fuse with the insects' neural or muscular tissues, resulting in a tightly integrated animal-machine interface at the fully grown stage. What sounded like wild speculation at the time – entomologists at London's Natural History Museum dismissed the proposal as 'ludicrous' and 'squarely in the realm of fiction'[2] – has since made a significant step forward: in January 2009, a team of researchers from the University of California, Berkeley, and University of Michigan, Ann Arbor, presented the prototype of a remote-controlled flower beetle with implanted electrodes and a radio receiver on its back at the International Conference on Micro Electro Mechanical Systems in Italy. Although this U.S. government-funded biotechnological hybrid is still quite far from being operational – to be used for search-and-rescue missions, for example, the beetle would need to carry a sufficiently small camera and heat sensor – DARPA's insect cyborg created for electronic warfare may leap from science fiction to fact before long.[3]

As biologist and feminist theorist Donna Haraway famously argued in her 'Cyborg Manifesto' of 1985, our recent culture's token of hybridity, the cyborg, is a creature 'simultaneously animal and machine, who populate[s] worlds ambiguously natural and crafted'.[4] According to Haraway, cyborgs such as the U.S. Army's wirelessly controlled beetle, are characterized by a breakdown of several boundaries: they breach well-established binary opposites such as human versus animal, organism versus machine, autonomy versus automaton, physical versus non-physical, and social reality versus fiction, and thus pose a threat to certain entrenched, naturalizing and totalizing ideologies of identity, gender, sexuality, class, race and nation. Due to their hybrid, boundary-crossing and transgressive nature, cyborgs become in Haraway's analysis 'promising and dangerous monsters'[5] with the political potential to challenge and subvert Western, male-dominated dualisms, ideologies and power systems.

This essay traces the legacy of the cyborg to both Surrealism and an early modern 'age of curiosity'[6] in which the meanings of objects and the boundaries between different fields of knowledge and modes of creative production were flexible and permeable rather than set in stone. Recent scholarship on Surrealism has focused on the group's reactions to the fast-paced progress in science and technology which, at the end of the nineteenth and beginning of the twentieth centuries, provoked enthusiasm and anxiety in equal measure, and to the political developments leading to the two world wars.[7] Departing from and going beyond these observations, I argue that the Surrealists recognized and employed the transgressive, subversive possibilities of hybridity as epitomized by the cyborg more readily and more effectively than any other avant-garde movement. The transgression of boundaries is a notion that lies at the very core of the Surrealist marvellous. It is a concept the members of the group embraced and practiced in art and life, launching an all-encompassing, forceful attack on petit-bourgeois values related to family, country, religion, work and honour.[8]

Reading the Surrealist marvellous historically rather than in a psychoanalytical context,[9] I propose that it is deeply rooted in sixteenth- and seventeenth-century natural philosophy and pre-scientific discourse as manifest, at the time, in the ubiquitous collections of curiosities, the cabinets of curiosities or Wunderkammern.[10] Defining the marvellous in the *Manifesto of Surrealism* (1924) as associated with fear, the attraction of the unusual, chance and the taste for things extravagant,[11] and describing the related concept of convulsive beauty as 'veiled-erotic, fixed-explosive, magic-circumstantial' in *Mad Love* (1937),[12] André Breton highlights the ambivalent, border-crossing nature of the marvellous. His discussion of objects like coral, crystal, an anthropomorphic mandrake root and a rubber statue that exudes a dark liquid upon being scratched[13] taps into an early modern sensitivity to curious and wondrous phenomena that straddle the boundaries of the three kingdoms of nature: the animate and inanimate, the organic and inorganic, and nature and representation. The Surrealist marvellous can thus be traced historically to what historian of the natural sciences Paula Findlen has termed 'paradoxes of classification'.[14] I further argue – suggesting a trajectory stretching from the Wunderkammer to Surrealism to contemporary art – that boundary-transcending concepts and practices related to both the early modern and Surrealist marvellous resurface in contemporary artistic creativity. The main question I am concerned with is whether contemporary artists who draw on these precedents of the marvellous embrace a purely aestheticizing strand of this legacy, or whether the radical, political impact of earlier practices is retained in their works. Do they keep the promise of subversion that Haraway saw in the animal-machine hybrid of our age of high technology? An object that, I believe, encapsulates different aspects of transgression – and especially transgression of the boundaries between art, nature and technology – is the insect with its characteristics and connotations of metamorphosis, of mimicry and of instinct-driven behaviour drawing it close to the automaton and the machine. Tellingly, for Georges Bataille, reflecting on the eye in a dictionary entry published in 1929 in the journal *Documents*, for which he acted as the chief editor, the fear of insects is among the most singular and developed of human horrors, equalled only by an inexplicable fear of the eye.[15] Our relationship with insects is always ambivalent, oscillating between fascination and fear: insects are extremely difficult to domesticate in an intellectual sense.[16] By nature unsettling, the insect becomes at different moments in cultural history a potent symbol of transgression. Thus representations of, and various forms of artistic engagement with, the insect serve in this essay as a case study of intersections between art, technology and nature in three distinctive but interconnected moments of history: the sixteenth- and seventeenth-century Wunderkammer, Surrealism, and contemporary art and visual culture.

Reclassification of Life

In the seventeenth century, collectors of curiosities treasured insects primarily for their diversity and microscopic size, but also for their scintillating appearance. Their multicoloured, shimmering carapaces granted them a status similar to that of delicately crafted, precious jewellery. That insects included in Wunderkammer collections held a double association with the kingdom of nature on the one hand, and the sphere of art, craft and decoration on the other, is exemplified by a box which the Nuremberg goldsmith Wenzel Jamnitzer crafted for Archduke Ferdinand II of Habsburg's collection at Ambras Castle in Tyrol.[17] The small silver casket, dating from around 1560–70, was originally intended as a container for writing tools. Representative of the naturalistic *style rustique* introduced by Bernard Palissy, it is ornamented with silver life casts from natural objects, which are placed on the lid within 10 neatly separated compartments.[18] The orderly arrangement of the creatures – among them a grasshopper, a moth, a stag beetle and a rhinoceros beetle – into sections calls to mind the specimen boxes of a *naturalia* collection. Complementing and contrasting the rather tidy representation on the lid, even more lifelike insects, arthropods, molluscs and amphibians are crawling up the sides of the casket, seemingly working their way through a thicket of grasses and flowers.

The ambiguity of this piece demonstrates that many Wunderkammer objects testified to the inextricable link between art and nature, often stretching the boundaries between those realms.[19] 'These pleasing paradoxes, at once art and nature', Daston writes, 'aroused wonder most intensely by blurring the line between nature and human artisan. The early modern *Wunderkammer* was the showcase for the aesthetic of the marvels of art and nature, mingled at several levels: juxtaposition, fusion, and imitation'.[20] As for the writing casket, nature finds its way into art and art takes the guise of nature, playing out the ultimate rivalry between God, the first and superior creator, and the artist and artisan's craftsmanship. Aided by the latest discoveries in science and technology – and further encouraged by the invention of instruments such as the microscope at the beginning of the seventeenth century – supposedly 'inferior' creatures such as insects were deemed noble enough to decorate exquisite objects designed from precious materials, thus substituting purely ornamental items (precious stones or gold patterns) for scientifically accurate imitations of natural specimens.

The Surrealists make this early modern ambiguity of what constitutes art and what belongs to nature, technology and science, carried out at the levels of juxtaposition, fusion and imitation, their own. Breton and Jacqueline Lamba's jointly created object *Le Petit Mimétique* (Colour plate 5) for instance,[21] takes the confusion between art and nature to the extreme: is the pale dried mantis, pinned to a stack of leaves or flaking tree bark and placed in a glass-covered box, a work of art or an entomologist's specimen box? The combination of cellophane-lined box, specimen and pin clearly points to a scientific collection, to a specimen box of the kind Breton frequently purchased at Emile Deyrolle's taxidermy shop in 46 rue du Bac in Paris, which was and still is particularly famous for its well-stocked entomology section.[22] It calls to mind the beetle, cicada, moth and butterfly cases on the wall of Breton's studio in 42 rue Fontaine, which are part of his famous collection of artworks, tribal objects, natural specimens, *objets trouvés* and curiosities,[23] but which he presumably also used as 'raw material' for assembling some of his poem-objects, most notably the silkworms in *A Torn Stocking* (1941) and the iridescent blue-green chafer beetles in *Pan-(h)oply for Elisa (Day and Night)* of 1953.[24] Not unlike *Le Petit Mimétique*, both poem-objects draw revolutionary energy from their hovering on the border between poetry, desire and scientific classification;[25] they may be termed artworks, scientific specimens, natural objects, containers or souvenirs of love and longing – but ultimately, by refusing to fit any such category, they come close to achieving the kind of 'reclassification of life' Antonin Artaud called for in the movement's early years.[26]

Another device dating from the late sixteenth century, devoted to the display of insects, illustrates that there were more boundaries at stake in the Wunderkammer than art and nature: collectors were equally fascinated by objects which oscillated between life and death, or movement and stillness.

The so-called *Schüttelkasten* (box to be shaken), also kept by Archduke Ferdinand II at Ambras Castle, contains papier mâché replicas of beetles, turtles and snails, which are camouflaged to resemble the greenish-brown mix of moss, shells and plaster at the bottom of the case that mimics forest soil.[27] Whenever the collector-owner picked up and shook the shallow box, the bugs, whose heads and limbs are flexibly jointed to their bodies with pieces of iron wire, suddenly seemed to awaken to life, twitching and squirming as if creeping and crawling across forest ground.

The desire to imitate life by inspiring movement in seemingly immobile objects culminated in the fashion of the automaton. Automata are self-moving mechanical devices that became staple articles in cabinets of curiosities from the beginning of the seventeenth century onwards.[28] It is rather intriguing to picture August the Strong, Elector of Saxony, who founded the Green Vault in Dresden in 1723, playing tricks on visitors to his treasure chamber by activating the mechanism of his spider automaton.[29] Any unsuspecting observer must have been truly horror-struck when the previously harmless-looking creature, with its 2.6 centimetres in length about as big as the palm of an adult's hand, unexpectedly began to move, since the brazen device, created by watchmaker to the court Tobias Reichel around 1604, quite persuasively imitated a natural specimen.

Arguably, insects make for particularly convincing automata, because the oscillation between the animate and inanimate is inherent in their behaviour. Thus the uncanny ability of some species to grow severed body parts back or to survive in segments, points to their mechanized nature, and the question whether insects were driven by instinct or by intelligence puzzled entomologists and philosophers particularly since the second half of the nineteenth century – among them Jean-Henri Fabre and Henri Bergson, on whose writings the Surrealists drew extensively. The question whether insects are driven by instinct or guided by intelligence underlies the majority of Fabre's experiments, which are recorded in his 10-volume *Souvenirs entomologiques,* published between 1879 and 1907; frequently, the insects' instinct-driven behaviour led the self-taught naturalist to compare them to automata and machines.[30]

The Surrealists' concern with automatism and unconscious, instinctual drives found its most powerful expression in their fascination with the praying mantis. They recognized that the insect offered the most impressive demonstration of instinct in the animal kingdom, and that this quality was evident to an exceptional degree in the praying mantis. The mantis's cannibalistic nuptial habits exemplified for the Surrealists the struggle between the two Freudian primordial drives of love and death, Eros and Thanatos.[31]

For sociologist and 'polymath' Roger Caillois, who was closely involved with the Surrealist movement in the early 1930s and who made the praying mantis the primary object of his research in two interdisciplinary articles published in the Surrealist journal *Minotaure,*[32] the psychological force of the praying mantis stems in particular from the dual nature of its anthropomorphism: the 'fearsome insect' is associated with the 'savage desires'[33] of both love and violence owing to the fact that the female mantis has been found to decapitate and devour the male during the act of copulation. The female's insatiable mechanical drive towards reproduction and nourishment represented for the Surrealists the violence of instinctual behaviour; and the male mantis's mechanical impulse to copulate, ignoring mortal danger, provided evidence for the dominance of the sexual instinct and pleasure principle over the instinct for self-preservation, thus testifying to a violent, delirious and self-abandoning strand running through all of nature. The insects' inevitable, mechanical desire to attack and consume, whatever the odds for survival, revealed the natural world as governed by merciless forces, with potentially subversive and alarming consequences, threatening as they were to identity and order.

What Caillois pinpoints in particular, then, is the identification with the insect's *mechanical* sexual instinct. By likening the female to 'a machine with highly advanced parts, which can operate automatically' and, in the same passage, comparing it to an 'automaton' and 'a female android',[34] whilst anthropomorphizing it all along, Caillois constructs a very modern human-animal and human-machine hybrid – or cyborg – condensed into the compelling image of the praying mantis. The ultimate evidence of the insect's machine-like existence is that it can carry out certain actions in a decapitated state, as Caillois emphasizes:

> Above and beyond its joined rigidity, which recalls a coat of armor or an automaton, it is a fact that there are very few reactions the mantis cannot perform in a decapitated state – that is, without any center of representation or of voluntary activity. In this condition, it can walk; regain its balance; sever a threatened limb; assume the spectral stance; engage in mating; lay eggs; build an ootheca; and (this is truly frightening) lapse into feigned rigor mortis in the face of danger or when the peripheral nervous system is stimulated. I am deliberately expressing myself in a roundabout way as it is so difficult, I think, both for language to express and for the mind to grasp that the mantis, when dead, should be capable of simulating death.[35]

The attempt to describe the mantis's ability to simulate death in a decapitated state brings Caillois to the edge of reason and threatens to unravel the logic of his argument. At the same time, however, the fact that he struggles to get his head around it and to express the inexpressible, elegantly proves his point: the praying mantis, with its instinct-driven, mechanical behaviour undermining all reasoning, is the ultimate symbol of transgression, challenging description, representation and categorization.

Caillois's interdisciplinary research into the praying mantis had profound repercussions for the group's perception of automatism. The implications of his argument were all the more powerful for focusing on hybridity and transgression in *real* insects rather than the mythological beasts that haunted the Surrealists' imagination (the Minotaur, the Sphinx, Melusine and Acephalus), and for emphasizing their uncanny anthropomorphism.[36] While in the early phase of Surrealism, during the 1920s, the circle around Breton had almost unreservedly embraced automatism in the form of automatic writing and drawing, as well as techniques such as collage and frottage, in order to overcome the restrictions that reason and conscious thought (supposedly) inflicted on creativity and imagination, Caillois's writings exposed a more alarming aspect of automatism: the unconditional reliance on instinct and unconscious drives stipulated by automatic techniques implied a vertiginous involution back to a more primitive, mechanized stage of life. The sense that hard-won evolutionary and civilizational autonomy threatened to become replaced by senseless and deathly automation reinforced the doubts that Salvador Dalí and others had begun to cast upon the concept of automatism and strengthened Caillois's case for providing Surrealism with a more scientific foundation. Caillois's research suggested that automatism, instead of liberating the artist and writer's creativity and imagination, might really result in a complete shutdown of creative activity and development. Asked in an interview if the possibility to be wrong was in fact a possibility to make progress, Caillois responded: 'This seems obvious to me! And every time there is automatism involved, there is a shutdown, there is a loss of conscience, and a loss of the possibility to make a step forward'.[37]

Epistemological Anarchism

We have seen that the Surrealists embraced a pre-Enlightenment notion of the marvellous characterized by a breakdown of distinctions between art, nature and technology, thus engaging with an early modern episteme in which the meanings of objects and boundaries between different spheres of knowing were constantly being negotiated – appropriating and subverting it to go against the grain of dominant political and cultural currents of their time. Both the aestheticization of the insect and the foregrounding of its uncanny, transgressive qualities, situating it between art and nature, instinct and reason, and living organism and automaton/machine, resonate in contemporary uses and representations of insects in art and culture: in biomimetic insect robots used for warfare as much as in the works of artists like Hubert Duprat and Cornelia Hesse-Honegger. They belong to a strand of contemporary artistic production which suggests that art, nature and technology are dovetailed with rather than separate from one another and that the Surrealists' revolutionary project of upsetting linear, straightforward taxonomies informed by Enlightenment rationalism and separation deeply resonates in the postmodern era, which is characterized by transition and uncertainty.

Hubert Duprat's caddis-fly larvae clearly chime with the transgressive, boundary-crossing qualities of the insect evident in both early modern and Surrealist discourse and practice. Since the early 1980s, the artist, who lives and works in the south of France, has been using a species of aquatic fly whose larva, the caddis-worm, produces a case from plant and mineral material that it collects from its immediate surroundings, gluing the bits together with a silky secretion it gives off. Duprat simulates the insect's habitat in aquariums, but at the same time manipulates it by replacing the natural materials – pebbles, pieces of driftwood, snail shells and so forth – with gold, pearls, coral and gemstones. As a result of the process, Duprat's larvae manufacture cases that resemble jewellery.[38]

Duprat is known for partly handing over responsibility for his work to collaborators, questioning the status of authorship, creativity and the expert. But here he goes even further by transferring the accomplishment of the artwork to the insect, thus introducing an element of chance and instability to the process. The caddis-worm becomes the actual artist, leaving us to wonder whether we are looking at an object of nature or an artefact – exactly as the visitor to a cabinet of curiosities or a viewer standing in front of Breton and Lamba's *Le Petit Mimétique* did. But more boundaries than that between art and nature and between art and science are being negotiated: the creatures and their creations are also suspended between the aesthetic, or beautiful, and the abject; and between a mimetic impulse to use inconspicuous natural material for camouflage and the ostentatious pull towards the flashy, precious materials provided by the artist – a dilemma which evokes Caillois's anti-utilitarian argument for an inherently mechanical, aesthetic and luxuriant force in nature, driving organisms to delirium and self-abandonment. But at the same time, reflecting further on ideas related to Darwinian evolution, removing the insects from their natural habitat and introducing an element of 'noise' and perturbation, also highlights the insects' adaptability, their capability of surviving pollution and other disruptions caused by the human species.

The Swiss artist Cornelia Hesse-Honegger similarly operates from a position of 'epistemological anarchism'.[39] Her work exists at a point of intersection between classifications, foregrounding the insect's aesthetic qualities while at the same time highlighting its uncanny, transgressive allure. Hesse-Honegger, who trained as a scientific illustrator, produces delicate watercolour paintings of insects, which, only upon careful scrutiny, reveal slight mutations and aberrations (Colour plate 6). The images, which at first sight look like exquisite, ornamental close-ups in the style of Poul Beckman's *Living Jewels*,[40] turn out to be disturbing graphic records of deformed and irregular specimens showing tumours, bulges, missing segments or unnatural protuberances. Hesse-Honegger's illustrations are carefully copied from specimens she collected starting in 1987, shortly after the Chernobyl disaster, from sites around nuclear power stations all over Europe. She believes that the bugs' anomalies result from the increasing impact of low-level radioactive radiation which steadily and quietly contaminates the environment, largely unnoticed by the unsuspecting general public.[41] She instantly earned scorn from the scientific establishment, who branded her as a sensation-seeking amateur, but she has since made a point of describing her methodology as scientific and publishing her research in recognized, peer-reviewed scientific journals,[42] as well as exhibiting in art galleries and science museums alike.[43]

Hesse-Honegger's use of insects, a class of animals that is tremendously resilient, invasive and successful in colonizing every possible corner of our planet, makes her demonstration of the creeping danger of nuclear technology all the more powerful. Haraway argued in her 'Cyborg Manifesto' that the two defining characteristics of modern machines are miniaturization and invisibility: '[o]ur best machines are made of sunshine; they are all light and clean because they are nothing but signals, electromagnetic waves, a section of a spectrum'.[44] Hesse-Honegger's genetically modified, painted insects, which are works of art, but at the same time products of a nature potentially compromised by modern technology, make the ecologically and ethically questionable effects of this invisible, radiation-based technology visible. Beautiful and grotesque at the same time, they point to the fact that, for all the promises of progress and salvation that high technology might hold, there is always a darker force lurking in the recesses of mechanized life. Thus Duprat's precious caddis-worms and Hesse-Honegger's mutant insects partake of the aestheticizing and playful legacy of the Surrealist marvellous informed by early modern ambiguities, but they also, and maybe more importantly, retain its more radical, political impact, negotiating issues related to biopolitics, the promises and threats of biotechnology, and the challenges of climate change and conservation.

Acknowledgements

A first version of this essay was presented at the AAH Conference at Manchester Metropolitan University in 2009. I would like to thank the organizers of the session on 'Art, Nature and Technology', Jacob Wamberg, Camilla Skovbjerg Paldam and Maria Fabricius Hansen, as well as the other speakers and members of the audience, for their questions and comments.

Notes

1 Gary Kitchener, 'Pentagon Plans Cyber-Insect Army', *BBC News*, March 16, 2006, accessed February 20, 2013, http://news.bbc.co.uk/1/hi/world/americas/4808342.stm.

2 Ibid.

3 For a synopsis of the current state of research into remote-controlled animals, see Emily Anthes, 'The Race to Create 'insect-cyborgs'', *Observer*, February 17, 2013, accessed February 20, 2013, http://www.guardian.co.uk/science/2013/feb/17/race-to-create-insect-cyborgs.

4 Donna Haraway, 'A Cyborg Manifesto: Science, Technology and Socialist Feminism in the Late Twentieth Century', in *Simians, Cyborgs and Women: The Reinvention of Nature* (New York: Routledge, 1991), p. 149.

5 Ibid., p. 179.

6 Krzysztof Pomian calls the period in the sixteenth and seventeenth centuries that was characterized by voyages of discovery and collections of curiosities the 'age of curiosity', in *Collectors and Curiosities: Paris and Venice, 1500–1800*, trans. Elizabeth Wiles-Portier (Cambridge: Polity, 1990), p. 45. It has also been termed 'culture of curiosities', in Neil Kenny, *The Uses of Curiosity in Early Modern France and Germany* (Oxford: Oxford University Press, 2004), p. 19, and the 'age of the marvelous', in Joy Kenseth (ed.), *The Age of the Marvelous* (Hanover, NH: Hood Museum of Art/Dartmouth College, 1991).

7 Recent, relevant book-length studies include Gavin Parkinson, *Surrealism, Art, and Modern Science: Relativity, Quantum Mechanics, Epistemology* (New Haven and London: Yale University Press, 2008); Steven Harris, *Surrealist Art and Thought in the 1930s: Art, Politics, and the Psyche* (Cambridge: Cambridge University Press, 2004); Johanna Malt, *Obscure Objects of Desire: Surrealism, Fetishism and Politics* (Oxford: Oxford University Press, 2004); and Raymond Spiteri and Donald LaCoss (eds), *Surrealism, Politics and Culture* (Aldershot and Burlington: Ashgate, 2003).

8 For a compelling exploration of Surrealism and hybridity, see Donna Roberts, 'Surrealism and Hybrid (Psycho-) Morphologies', *Henry Moore Institute Online Papers and Proceedings*, March 19, 2008, http://www.henry-moore.org/docs/donna_roberts_surrealist_hybrids_0.pdf.

9 For studies exploring the relationship of the Surrealist marvellous with psychoanalytic discourse and the Freudian uncanny, see especially Hal Foster, *Compulsive Beauty* (Cambridge, MA, and London: MIT Press, 1993); Briony Fer, 'Surrealism, Myth and Psychoanalysis', in Briony Fer, David Batchelor and Paul Wood, *Realism, Rationalism, Surrealism: Art Between the Wars* (New Haven and London: Yale University Press, 1993), pp. 170–249; Rosalind E. Krauss, 'The Photographic Conditions of Surrealism', *October* 19 (Winter 1981): pp. 3–34; and David Lomas, *The Haunted Self: Surrealism, Psychoanalysis, Subjectivity* (New Haven and London: Yale University Press, 2000).

10 For a more detailed elaboration of this argument, see Marion Endt-Jones, 'Between *Wunderkammer* and Shop Window: Surrealist *Naturalia* Cabinets', in *Sculpture and the Vitrine*, ed. John Welchman (Aldershot and Burlington: Ashgate, 2013).

11 André Breton, *Manifestoes of Surrealism*, trans. Richard Seaver and Helen R. Lane (Ann Arbor: University of Michigan Press, 1972), p. 16.

12 André Breton, *Mad Love*, trans. Mary Ann Caws (Lincoln, NB, and London: University of Nebraska Press, 1987), p. 19.

13 Ibid., pp. 10–13, 16.

14 Paula Findlen, 'Jokes of Nature and Jokes of Knowledge: The Playfulness of Scientific Discourse in Early Modern Europe', *Renaissance Quarterly* 2 (Summer 1990): p. 302.

15 Georges Bataille, 'Œil', *Documents* 4 (September 1929): p. 216.

16 For an examination of this idea from different perspectives, see the following contributions to the growing field of cultural entomology: Eric C. Brown (ed.),

Insect Poetics (Minneapolis and London: University of Minnesota Press, 2006); Hugh Raffles, *Insectopedia* (New York: Vintage, 2010); J.F.M. Clark, *Bugs and the Victorians* (New Haven and London: Yale University Press, 2009); *Cabinet* 25: *Insects* (Spring 2007); *Antennae* 3: *Insect Poetics* (Autumn 2007, 2 vols); and *Antennae* 11: *Insecta* (Autumn 2009).

17 Kunsthistorisches Museum, Vienna, inventory no. KK_1155, http://bilddatenbank.khm.at/viewArtefact?id=87354, accessed February 20, 2013.

18 For literature relating to and illustrations of this object, see Ernst Kris, 'Der Stil "Rustique". Die Verwendung des Naturabgusses bei Wenzel Jamnitzer und Bernard Palissy', *Jahrbuch der Kunsthistorischen Sammlungen* 1 (1926): pp. 137–208; Yves Cambefort, 'A Sacred Insect on the Margins: Emblematic Beetles in the Renaissance', in *Insect Poetics*, ed. Eric C. Brown, p. 214; Patrick Mauriès, *Cabinets of Curiosities* (London: Thames and Hudson, 2002), p. 103; and Janice Neri, *The Insect and the Image: Visualizing Nature in Early Modern Europe, 1500–1700* (Minneapolis and London: University of Minnesota Press, 2011), pp. 95–99.

19 Mark A. Meadow analyzes Jamnitzer's writing box as a 'copious object' representing different spheres of knowledge: craftsmanship, the natural world, writing tools and scholarship, physical and conceptual containment and ordering, and technological research and development. See Mark A. Meadow, 'Quiccheberg and the Copious Object: Wenzel Jamnitzer's Silver Writing Box', in *The Lure of the Object*, ed. Stephen Melville (New Haven and London: Yale University Press, 2005), pp. 39–58.

20 Lorraine Daston, 'Nature by Design', in *Picturing Science, Producing Art*, ed. Caroline A. Jones and Peter Galison (New York and London: Routledge, 1998), p. 238.

21 *Le Petit Mimétique* was displayed in the *Surrealist Exhibition of Objects* at Galerie Charles Ratton in 1936 and reproduced in the accompanying issue of *Cahiers d'Art*, 2/1–2 (1936): p. 59.

22 See Claude d'Anthenaise, Louis-Albert de Broglie and Patrick Mauriès, *Nature fragile: le cabinet Deyrolle* (Paris: Beaux Arts Editions, 2008). While no surviving records prove the provenance of Breton's specimen boxes from Deyrolle, he did own a wooden butterfly-rearing cage 'purchased at Deyrolle, rue du Bac'. See online archive for André Breton's collection, http://www.andrebreton.fr/fr/item/?GCOI=56600100705310, accessed February 20, 2013.

23 Examples of such specimen boxes owned by Breton include a box of cicadas (ibid., http://www.andrebreton.fr/fr/item/?GCOI=56600100145140); a collection of moths (ibid., http://www.andrebreton.fr/fr/item/?GCOI=56600100267480); and a box filled with butterflies, beetles and other small animals (ibid., http://www.andrebreton.fr/fr/item/?GCOI=56600100460000; all accessed February 20, 2013). A black-and-white photograph dating from c. 1939 shows Breton sitting in an armchair in his rue Fontaine apartment, surrounded by five entomological specimen boxes mounted on the wall. See *André Breton: la beauté convulsive*, ed. Agnès Angliviel de La Beaumelle and Isabelle Monod-Fontaine (Paris: Editions du Centre Pompidou, 1991), p. 73.

24 See Jean-Michel Goutier (ed.), *André Breton, Je vois, j'imagine: poèmes-objets* (Paris: Gallimard, 1991), pp. 34–35, 46–47.

25 See Malt, *Obscure Objects of Desire*, p. 170, and Endt-Jones, 'Between Wunderkammer and Shop Window'.

26 Antonin Artaud, 'L'Activité du Bureau de Recherches Surréalistes', in *Bureau de recherches surréalistes: Cahier de la permanence octobre 1924–avril 1925*, ed. Paule Thévenin (Paris: Gallimard, 1988), p. 130.

27 Kunsthistorisches Museum, Vienna, inventory no. KK_5432, http://bilddatenbank.khm.at/viewArtefact?id=91439, accessed February 20, 2013.

28 See, for example, Horst Bredekamp, *The Lure of Antiquity and the Cult of the Machine: The Kunstkammer and the Evolution of Nature, Art and Technology*, trans. Allison Brown (Princeton: Markus Wiener Publishers, 1997), and Mauriès, *Cabinets of Curiosities*, pp. 116–21.

29 Tobias Reichel, *Automat in Form einer Spinne*, c. 1604, Green Vault, Dresden, inventory no. VI 7 qq, http://skd-online-collection.skd.museum/en/contents/show?id=118405, accessed February 20, 2013.

30 Examples include '[t]he insect, a sort of jointed automaton', Jean-Henri Fabre, *The Glow-Worm and Other Beetles* (Miami: University Press of the Pacific, 2002), p. 260; 'the insect's rigid tools, its broad, toothed forearms, whose spasmodic movements remind us of the stiff gestures of an automaton', ibid., p. 81; 'his long legs move with awkward jerks, as though driven by some mechanism within his belly', Edwin Way Teale (ed.), *The Insect World of J. Henri Fabre*, trans. Alexander Teixeira de Mattos (Boston: Beacon Press, 1991), p. 95; and 'the Cricket's egg is a little marvel of mechanism … the Cricket pops out like a Jack-in-the-box', Teale, *The Insect World of J. Henri Fabre*, p. 283.

31 See William L. Pressly, 'The Praying Mantis in Surrealist Art', *Art Bulletin* 55/4 (December 1973): pp. 600–615.

32 Roger Caillois, 'La Mante religieuse', *Minotaure* 5 (May 1934): pp. 23–26, and 'Mimétisme et psychasthénie légendaire', *Minotaure* 7 (June 1935): pp. 5–10.

33 Roger Caillois, 'The Praying Mantis: From Biology to Psychoanalysis', in *The Edge of Surrealism: A Roger Caillois Reader*, ed. Claudine Frank (Durham and London: Duke University Press, 2003), p. 77.

34 Ibid., p. 78.

35 Ibid., p. 79.

36 Insects and other arthropods similarly displaying an extraordinarily high degree of lyrical force and an ambiguous interpenetration of nature and representation include, according to Caillois, the death's head hawk-moth with its skull insignia on the thorax; the lantern fly with its protuberance shaped like a crocodile's head; and a particular trapdoor spider with uncanny markings resembling an Aztec mask. For Caillois, they testify to 'the existence of an underlying imaginary that is part of the real'. See Roger Caillois, 'The Natural Fantastic', in *The Edge of Surrealism*, ed. Frank, p. 357.

37 Jeannine Worms, *Entretiens avec Roger Caillois* (Paris: La Différence, 1991), p. 70 (my translation).

38 English-language publications discussing Duprat's *Trichoptera* (1980–2011) include *Hubert Duprat: Caddis, Crystal* and *Company* (Norwich: Norfolk Museums and Archaeology Service, 2011); Stephen Bann, *Ways Around Modernism* (New York and London: Routledge, 2007); Steve Baker, *The Postmodern Animal* (London: Reaktion Books, 2000); Jeffrey Kastner, 'Trichopterae', *Cabinet* 25: *Insects* (Spring 2007), pp. 102–3; and Hubert Duprat and Christian Besson, 'The Wonderful

Caddis Worm: Sculptural Work in Collaboration with Trichoptera', trans. Simon Pleasance, accessed February 20, 2013, http://www.leonardo.info/isast/articles/duprat/duprat.html.

39 Theory proposed by the Austrian physicist and philosopher Paul Feyerabend, who argued against prescriptive scientific method, in favour of multiple ways of knowing. See Raffles, *Insectopedia*, p. 37.

40 Poul Beckmann, *Living Jewels: The Natural Design of Beetles* (Munich, Berlin, London and New York: Prestel, 2001) and *Living Jewels 2: The Magical Design of Beetles* (Munich, Berlin, London and New York: Prestel, 2007).

41 Her work has certainly gained new relevance in the face of the Fukushima nuclear disaster caused by the earthquake and tsunami that hit Japan in March 2011.

42 See, for example, Cornelia Hesse-Honegger and Peter Wallimann, 'Malformation of True Bug (Heteroptera): A Phenotype Field Study on the Possible Influence of Artificial Low-Level Radioactivity', *Chemistry and Biodiversity* 5 (2008): pp. 499–539.

43 Engaging, scholarly literature on Hesse-Honegger is scarce; a welcome exception is Peter Suchin, 'Forces of the Small: Painting as Sensuous Critique', in Cornelia Hesse-Honegger, *The Future's Mirror* (Newcastle-upon-Tyne: Locust, 1997), n.p. See also Raffles, *Insectopedia*, pp. 15–40, and Siân Ede, 'Fission or Fusion: Art, Science and Society', in *Strange and Charmed: Science and the Contemporary Visual Arts*, ed. Siân Ede (London: Calouste Gulbenkian Foundation, 2000), pp. 164–83, and Siân Ede, *Art and Science* (London and New York: I.B. Tauris, 2005), pp. 176–77.

44 Haraway, 'A Cyborg Manifesto', p. 153.

4

Grotesque! Strategies of Figurative Genesis in the Sixteenth Century and in the Surrealism of the 1920s and 1930s

Maria Fabricius Hansen and Camilla Skovbjerg Paldam

Introduction

A striking similarity exists between the art of the sixteenth century and the Surrealist art of the twentieth century. Both periods abound with imagery of metamorphosis, of flux in the passages between figurations, leading to monstrous hybridity – and of flux in the passages of meaning, resulting in ambiguities and paradoxes in the pictorial content.

By focusing on that which may be termed 'the grotesque', we hope to identify some notions inherant to the conception and creation of images that were dominant in the two historical periods. This comparative analysis will entail an understanding of the sixteenth century as an early stage of modernity (Figure 4.1), and it will contribute to nuancing conventional Bürgeresque ideas about the avant-garde as radically breaking with the past.[1]

Yet the interest in figurative genesis informing the grotesque imagery is not identical in the two periods. Therefore, we will look not only at similarities but also try to illuminate the different approaches to image making, by, for instance, taking the understanding of nature and technology prevalent in the two periods into consideration.

By way of introduction, we will explore the notions of grotesque monstrosity in antiquity and the Middle Ages, when the ancient paradigm of a close association between art, nature and technology was still prevalent. We will then offer an account of the monstrous and of figurative genesis in the fifteenth and sixteenth centuries, when art and technology gradually began to be defined as contraries, and, finally, in Surrealism, when the modern condition of the mechanical reproduction of images had had its great impact on the concept of art (Figure 4.2). This will be followed by an analysis of the grotesque in the two periods, including its status in the creation of images. The hybridity of this essay (making it similar to the field it discusses) is intended

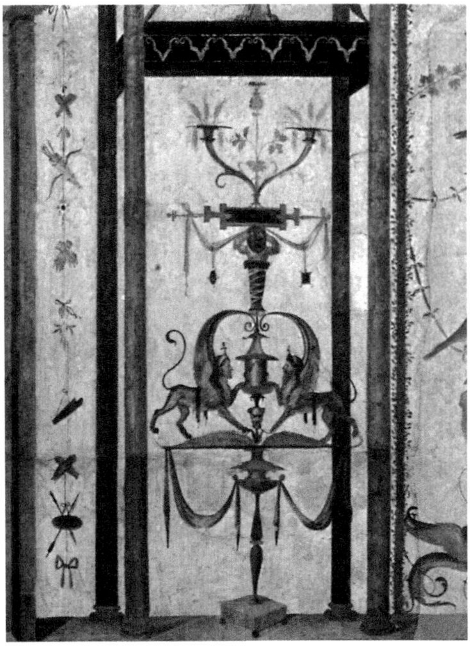

4.1 Giovanni da Udine/workshop of Raphael, detail of grotesques in the Loggetta of the Cardinal Bibbiena (1517). The Vatican

to generate new understandings of the grotesque and of the ways in which artistic innovations relate to the past, understandings of which a focus on an individual period of time would not make possible.

Monsters in Antiquity and the Middle Ages

For Aristotle (384–322 BC), monstrosity in terms of grotesque creatures did not go against nature, but just against what normally happened in nature. Sometimes nature did not reach its intended goal, and the result was called a monster. Also for Pliny the Elder (AD 23–79), monsters were part of the generosity of nature. In his *Natural History* (*Historia Naturalis*), he wrote about different monsters living at the borders of the world, such as Africa, India or Scandinavia: these peoples had no heads, or perhaps had dog heads, and some had eight toes on each foot, their feet pointing backwards. The descriptions are fascinating because they simultaneously sound like sensationalist stories while being presented as indispensable truths. As in all narration, the trustworthiness lies in the detail, which the following example illustrates:

> Ctesias writes that also among a certain race of India the women bear children only once in their lifetime, and the children begin to turn grey directly after birth; he also describes a tribe of men called Monocoli who have only one leg, and who move in jumps with surprising speed; the same are called the Umbrella-foot tribe, because in the hotter weather they lie on their backs on the ground and protect themselves with the shadow of their feet; and they are not far away from the Cavedwellers; and again westward from these there are some people without necks, having their eyes in their shoulder.[2]

Characteristically, Pliny's monsters were marginalized to the outskirts of the world. The monsters were no longer part of life, as in a cultic past, but images of the Other – the exotic, foreign and dangerous.

Not just in life but also in art, monsters were marginalized. An often-quoted warning against the grotesque in art was formulated by the Roman poet Horace, who, in *The Art of Poetry (Ars Poetica)*, AD 10, reflected on unity in artworks and on the relation between the different parts of an artwork. Horace rhetorically asked if one would be able to refrain from laughter if a painter united a horse's neck to a human head, or let the upper part of a beautiful woman terminate in an ugly fish below. According to Horace, images like these are all vain and fictitious. However, 'Painters and poets … have always

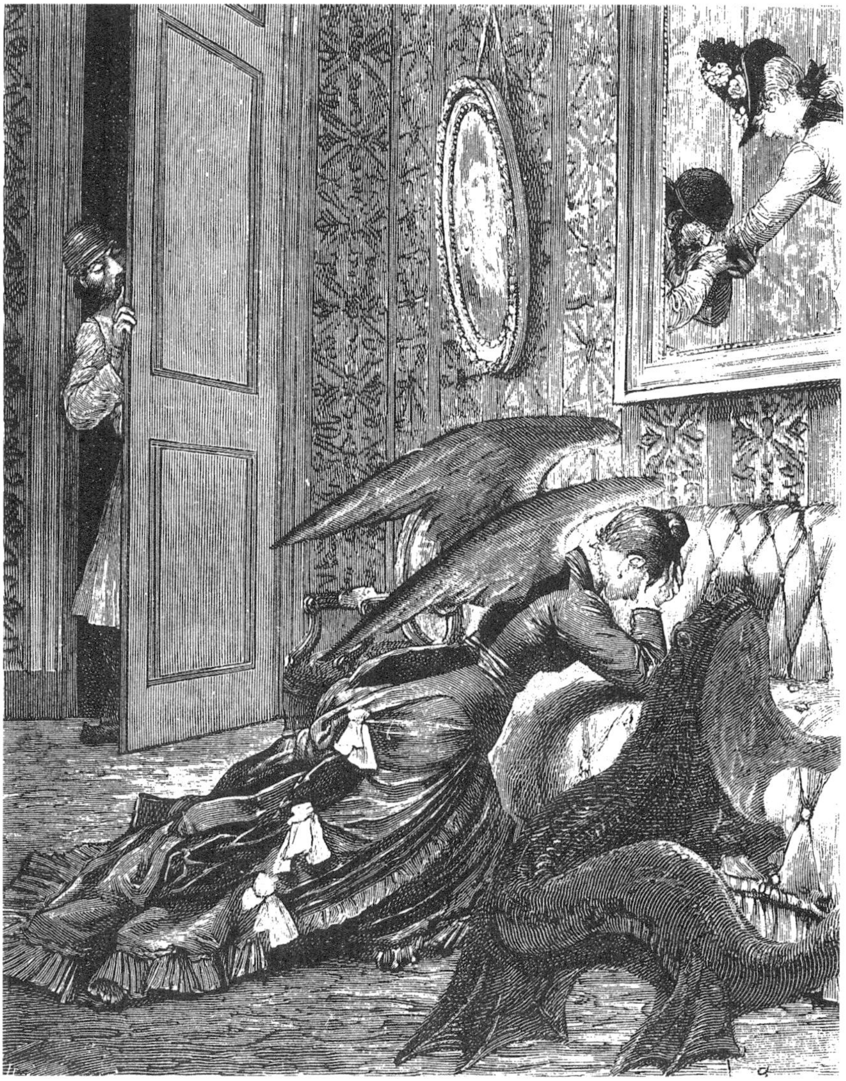

4.2 Max Ernst, illustration from *Une semaine de bonté* (1934), collage. © Max Ernst/ billedkunst.dk

had an equal right in hazarding anything'. Horace was conscious of this, and he demanded and allowed this privilege: 'but not so far that savage should mate with tame, or serpents couple with birds, lambs with tigers'.[3]

When Horace stated that such monstrous compilations as mermaids or centaurs – which later, in the terminology of the sixteenth century, were called 'grotesques' – would cause laughter, it was not positively meant. Even though painters and poets had the right to do as they pleased, they needed to avoid wild compositions because the work had to form a coherent unity. Horace, arguing that art influences society, compared the grotesque to feverish phantasms and found this madness of the poet dangerous.[4] This same chain of reasoning, but with the opposite outcome, was used by the Surrealists, when they tried to reintroduce madness and irrationality to society by means of hallucinatory, grotesque art.

Similar to Horace, Vitruvius in his treatise on architecture wrote derogatively on the wall paintings of his time (first century BC) because they no longer took 'their models from real things'. Instead 'monsters are now painted in frescoes rather than reliable images of definite things'. The problem with these frescoes was their representation of impossibilities: 'Now these things do not exist nor can they exist nor have they ever existed'.[5] Like Horace, Vitruvius was critical of imagery not founded in naturalism, thereby contributing to later discussions of the legitimacy of art transcending reality.

Based on the writings of Pliny the Elder, St Augustine (354–430) included the tradition of strange, monstrous people in his description of the world. Augustine, whose writings had an enormous influence on the Christian world, raised doubt about the existence of these creatures. He nevertheless concluded that if they existed and were men, defined as rational and mortal animals, then they were also part of God's marvellous creation, descending from Adam.[6] In Christian culture, monsters – for instance, hybrids of man and animal, which were not men – would otherwise be stamped as the work of Satan. The striking consequence was the transformation of Dionysian satyrs to goat-footed devils.[7] During the medieval period, the monsters increasingly appeared in the margins of illuminated books or in the sculptural decoration of cloisters, much to the chagrin of the Cistercian abbot St Bernard of Clairvaux (1090–1153), who, in the twelfth chapter of his *Apology* to Abbot William of Saint-Thierry, regretted the grotesque ornaments:

> in the cloister, under the eyes of the Brethren who read there, what profit is there in those ridiculous monsters, in that marvellous and deformed comeliness, that comely deformity? to what purpose are those unclean apes, those fierce lions, those monstrous centaurs, those half-men, those striped tigers, those fighting knights, those hunters winding their horns? Many bodies are there seen under one head, or again, many heads to a single body. Here is a four-footed beast with a serpent's tail; there, a fish with a beast's head. Here again the forepart of a horse trails half a goat behind it, or a horned beast bears the hinder quarters of a horse. In short, so many and so marvellous are the varieties of divers shapes on every hand, that we are more tempted to read in the marble than in our books, and to spend the whole day wondering at these things rather than in meditating the law of God. For God's sake, if men are not ashamed of these follies, why at least do they not shrink from the expense?[8]

Bernard warned against the fascination of monsters, against their horrifying beauty and charming diversity – or, to borrow a term used by the Surrealists for something similar, their 'convulsive beauty'. But contrary to the Surrealists, Bernard was opposed to being attracted to the monsters, seeing in them the risk of false deities that steal attention from the true God. Monsters were not devotional, and they served no purpose in the cloisters. Yet the quotation reveals indirectly that they were studied and admired.

Grotesque Grotesques: The Art of the Sixteenth Century

Horace's indication of the limits of artistic licence thus echoed through the centuries. To a certain extent, a centaur was never thereafter merely a centaur, that is, one of many mysterious creatures; just as a mermaid was not merely any arbitrary monster from the depths of the sea. These hybrid figures described by Horace as models of exaggerated artificiality came to be associated with reflections on the possibilities of art in relation to the imitation of nature and on the legitimacy of transcending these possibilities. The relevance of monstrosity in the sculptural or painted decorations of churches and cloisters was continuously discussed. Apparently monstrosity had a strong appeal, but simultaneously it triggered intense criticism, precisely from ascetical theologians and ecclesiastical reformers like Bernard.

However, in the sixteenth century, with artists like Michelangelo we encounter readings of Horace different from the antique and medieval ones. The ancient authors remained admired authorities, yet Michelangelo succeeded in reversing the argument to make Horace's observation appear an unequivocal defence of limitless artistic licence.[9] Michelangelo is claimed to have argued for the artistic right to create monstrosities, not understood derogatively as fictions, but instead as good inventions and thus in a sense living up to the demand that art should be *real*:

> But if, in order to observe what is proper to a time and place, he change the parts or limbs (as in grotesque work, which would otherwise be without grace and most false) and convert a griffin or a deer downward into a dolphin or upward into any shape he may choose, putting wings in the place of arms, and cutting away the arms if wings are better, this converted limb, of lion or horse or bird, will be most perfect according to its kind; and this may seem false but can really only be called well invented or monstrous. And sometimes it is more in accordance with reason to paint a monstrosity (for the variation and relaxation of the senses and in respect of mortal eyes, that sometimes desire to see that which they never see and think cannot exist) rather than the accustomed figure (admirable though it be) of men and animals.[10]

Evidently, Michelangelo – or the point of view represented as his – drew an analogy between reason and monstrosity. And the new view on art and the artist which corresponds with the licence to create monstrous figures and with the Michelangelesque reversal of Horace's statement was not only expressed theoretically. It was manifest in many artistic fields, especially in those involving grotesques. The grotesques, so fascinating to the people of the sixteenth century, seemed to include a vast potential for meaning.

In the strictest art historical sense, the term 'grotesque' designates the ornamental frescoes popular in the decoration of Italian palaces, villas and all other kinds of prestigious building (Figure 4.1).[11] Typical of the time, the grotesques were understood as imitations or revivals of the art of antiquity, in particular the ancient Roman frescoes found in the late fifteenth century in the Domus Aurea, the palace of the Emperor Nero.[12] But the claims by artists and art theoreticians are not necessarily an indication of the extent to which antiquity influenced the production of images at the time. Just as there were limits to how much Horace was used and respected, there were limits to how antique the grotesques of the sixteenth century were in their style and content.

In fact, monstrous and hybrid figures are relatively rare in the grotesques of antiquity, including the Domus Aurea. Hybrid creatures appear, but only sporadically, and they were pacified into ornamental borders through serial repetition. The sixteenth century, on the contrary, saw a striking inventiveness in the juxtapositions of peculiar and basically – seen from a naturalistic point of view – incompatible pictorial elements, sometimes incorporating strange technological machinery. The frescoes abound with hybrids and, consequently, the overall compositions break with all rationality, as, for instance, with the law of gravity emphasized by dramatic contrasts in scale. Characteristic of the decorations with grotesques is not only the single grotesque figure's combination of surprising parts but also their inclusion in more elaborate juxtapositions, in literally entangled compositions and in long series of figural developments, with indefinable passages from one motif or figure to the next.[13]

Figures of Invention

Remarkably, in sixteenth-century writings on art it was discussed whether the ancient grotesques found below ground had also been used originally in subterraneous rooms without daylight – the marginal, unlit or irrational spheres of the buildings. The grotesques were, indeed, understood as somewhat disturbing. Sometimes characterized as the dreams of the artists or even the chaos of the brain, the grotesques were, moreover, identified with the inventiveness of the artist.[14] The analogy to the Surrealist vocabulary and concepts of art and fantasy are, of course, evident here as dreams were used as a main reservoir for the production of art, in continuation of Freud's psychoanalytic investigations. The dramatic change which took place from antiquity to the sixteenth century in both the appearance of the grotesques and the interpretations of them in art theoretical writings appears to correspond with an equally dramatic shift in the understanding of the artist and of artistic practice, of how images come into being. Parallel to the insistence on artistic licence was the tendency to place artistic value within the field of the artist's creative imagination.

An appreciation of the grotesques as reflections of the artistic imagination does not contradict their roots in antique and medieval notions about the reality of monstrosity and about nature as a creator of images. That the creatures we understand as fantastic today were thought to really exist was a condition still valid through the sixteenth and seventeenth centuries and even later. Not least the mystic depths of the sea were thought to conceal innumerable strange and monstrous, often hybrid creatures. The common myth that unicorns actually existed was, for instance, generally accepted until as late as the 1630s when the Danish physician Ole Worm contributed to exploding it.

The hybrid figures composing the grotesques were thus closely related to ideas of the monstrous and marginal in general, just as the grotesques were a kind of marginal imagery: they literally inhabited the border areas, framing and delimiting the central, 'main' figural compositions. Unsurprisingly, the marginal creatures from the limits of the world described by Pliny the Elder were included among the grotesque imagery, as seen in the Tuscan Monte Oliveto cloister decorated with frescoes by Sodoma c. 1505–1508. Such continuations of subjects from antiquity and the Middle Ages abound in the pictorial universe of sixteenth-century grotesques.[15] The mermaid with a double split tail is another example among many, just as centaurs continued to fascinate as emblems of the artist or of his artistic licence.[16]

The Art of Nature

The attention invested in the grotesque and monstrous was interrelated with the notion of nature as producer of images, an understanding known from antiquity and up through the medieval period and the fifteenth century, when Leonardo da Vinci, among others, reflected upon it.[17] Leonardo wrote how contemplation of 'stains of walls, or the ashes of a fire, or clouds, or mud, or like things' could facilitate invention, arguing that if you only look long enough figures will appear. Notably, the examples he mentioned of such emerging figures were all of a rather grotesque kind: 'battles of animals and men, various compositions of landscapes and monstrous things, such as devils and similar creations, which may bring you honour, because the mind is stimulated to new *invenzioni* by obscure things [*cose confuse*]'.[18] For Leonardo these artistic inventions went hand in hand with inventions we today would place in the category of engineering work or technological instruments. At the same time, in the last decades of the century, Mantegna painted several pictures clearly turning nature's capacity for figurative genesis into a theme. In *The Adoration of the Magi*, for instance, a monstrous, scull-like mask appears in the rock above the otherwise idyllic scene celebrating the infant (1460s, Galleria degli Uffizi, Florence), and his *Pallas Expelling the Vices from the Garden of Virtue* (c. 1500, Musée du Louvre, Paris) is one of several examples in which the clouds materialize into a face.

The understanding of nature as a creator of images partly originated with attempts at explaining fossils, which were thought to have been produced by nature. The sixteenth-century fashion of painting on stone to exploit the natural patterns and colours of materials shows how the artists developed nature's activity.[19] With his painting, the artist concluded or perfected the figurations initiated by nature in the colours and structures of the stone.

The fascination of nature's potential for figuration is evident in the culture of the sixteenth century in general: the gardens of the period abounded with re-creations of the kind of often monstrous imagery traditionally ascribed to natural genesis, such as artificial grottoes, fountains and masks. The grotto being a space created by nature, an artificial grotto imitated nature's capacity for architectural activities.[20] Moreover, the water in the fountains contributed mobile, continually changing and unpredictable sculptural forms. Also here, the technological interest of the period visible in the construction of the fountains, in water organs and in playful machinery spraying the guests in the gardens, goes hand in hand with the artificial character of the 'nature' represented in the gardens as well as in painting.

The ideas on nature's creativity materialized on all scales, from monumental sculpture in the gardens to decorative art. Semi-precious, colourfully patterned stone was cut into sophisticated objects gathered in the curiosity cabinets. The material was elaborated into ornaments and figures as if these forms were present from the beginning in the stone and only needed to be completed by the artist, an idea expressed by Michelangelo in his poetry.[21]

It is remarkable that apparently even the most straightforward, objective representations of nature by no means excluded elaborate imaginative visual layers. There seems to have been an extremely well-developed predisposition to see metamorphoses in terms of observing ambiguities and potential figurative genesis in the surroundings. The combination in one work (painting, sculpture, gardens, decorative art) of, on the one hand, matter-of-fact, carefully observed and represented subjects from nature and, on the other, surprising, highly imaginative figurative formations within the natural forms, was apparently unproblematic. One example among many is the fruit garland framing the mythological figures of the Loggia of Psyche at the Villa Farnesina in Rome. The garland was painted by a member of Raphael's workshop, Giovanni da Udine, who specialized not only in precise representations of animals and plants but also in grotesques. In the garland at the Villa Farnesina, he combined fruits and vegetables to form sexy male and female genitals: a squash and an eggplant constitute the male, while an overly ripe fig stands for the female (Colour plate 7). The choice of this subject, which in its time was noticed, described and appreciated, was evidently not arbitrary.[22] By turning creation or genesis into an image through his metaphor of vegetables, Giovanni da Udine added a meta-reflective layer to the apparent naturalism.

Again, such imaginary or monstrous figurations appearing within seemingly straightforward recordings of our natural surroundings were no less 'real' than what we today regard as the real world, that is, the world as it appears according to the modern concept of natural science. The most advanced records of nature went hand in hand with the awareness of the self-generating imagery of nature. The fascination with automata – all kinds of mechanically moved figures or objects – points to the same pleasure of ambiguous phenomena.[23]

Systematic, scientific observation in the modern sense still lay far in the future, just as the split between mysticism and belief on the one hand, and factual, quantifiable knowledge on the other, typical of modernity, was foreign to a sixteenth-century world view. Similarly, the technological equipment facilitating the observation and mapping of the world and its phenomena was only gradually invented and developed. The technology available reflected the world view of its time; as it slowly became more sophisticated in a modern sense, with scientific inventions for observing and measuring facilitating the systematic mapping of physical reality, it worked back onto the representation of the world. Of great consequence was, of course, the possibility of mechanical reproduction of images, brought about by woodcuts and other methods of printing from the fifteenth century onward. The serial repetition and systematic logic of representation inherent in prints became decisive constituents of the modern concept of images.

The Movement of Artistic Creativity

The fifteenth century was still characterized by mistrust of observations too alienated from the imagination and from conventional ideas of the order of things. Just as scientists had a tendency to represent the world as they knew it from their reading of the ancient authorities (even though they may have found them contradicted by mere empirical observation), the artists were careful not to represent too much personal invention or artistic imagination in their art. In his highly influential treatise on painting, *De pictura*, from the mid-1430s, Leon Battista Alberti explicitly warned painters not to include too much movement in their painting, equating movement with the artistic fervour, the poetic fire from which the artistic creation grows.[24] The movement in the work, both literally and in a figurative sense, was directly linked with the movement in the mind of the creating artist, what we in a later terminology would call artistic genius. Alberti criticized figures turned so violently as to show the face and the toes of a figure pointing in opposite directions. According to Alberti, this was anatomically impossible, which meant unnatural. Since the imitation of nature was self-evident as a theoretical claim, the artist would consequently have to refrain from representing figures in this way.[25]

From Alberti to the time of Michelangelo a shift occurred in the attitude to artistic singularity and ferocity, so that the artist was no longer expected to control his violent mind.[26] Thus, contrary to Alberti's prescriptions, the visibility of artistic ferocity in the artwork had become very desirable by the time of Michelangelo. In the sixteenth century, the most bizarre or monstrous grotesques became associated with the most remarkable artistic genius, implying that these artists could surpass even the figurations made by nature. The figures depicted in vigorous movement and their inverted bodies that Alberti had criticized now became a leitmotif in art. Just as the period featured decorative frescoes with imagery more grotesque than anything ever produced, the turning, inversion or even perversion of figures and movements were cultivated more than ever before. The motion continuously coursing through the body in terms of a contrapposto – or a *figura serpentinata* – was not just an innocent way of positioning a figure to be chosen among many others.[27] It was a direct reflection of the artist's imagination, and the artists worked with contrapposto both literally and figuratively by means of twisted figures or a perverted form or content. Michelangelo, with his strongly spiralled corporeal representations, was just one representative among many of this tendency.

This rather dramatic development took place simultaneously with the development of not only the printed book but also technologically reproduced images. The systematic monotony of the printed book was thus counterbalanced by the unpredictability of personal artistic visions.

In the sixteenth century, the transcendence of mere naturalism was turned into the foremost ideal of artistic creativity, while at the same time the artists endeavoured to follow the principles underlying nature's creation, its predilection for metamorphoses. One example among many is Bronzino's strange painting, *An Allegory with Venus and Cupid* (National Gallery, London, c. 1545), a composition dense with enigmatic figures, such as the lovely girl who, on closer inspection, appears to have a serpent body, much like the hybrid figures criticized by Horace. Today's viewer is probably struck by the peculiar turning of corporeal, naked bodies into almost marble-like or metallic figures, the materials of nature turned into the materials of art. As in some Surrealist paintings, the meticulous description of all details with equal emphasis paradoxically undermines the reality effect in favour of a hyperreal or surreal effect. Similarly, the intimate kiss and touch of the ingeniously tangled figures tend to surprise a modern viewer, the main characters being Venus and her son, Amor. Such compositions, typical of the sixteenth century, may seem deliberately complex or even perverse in their strained artistry or artificiality, their anti-naturalism or archetypal investigations appearing related to the aim of Surrealism to reach beyond mere observation and examine the levels of reality within the human mind and thereby transcend the conscious mind.

The Ovidian metamorphoses as well, popular in the period, may illustrate the peculiar fascination with the limits of naturalism. Several of the mythological narratives deal with Jupiter's transformation into some animal based on the somewhat strange argument that it would help him seduce the women he desires. In one instance he takes the form of a swan. This was represented with great sensuality in the sixteenth century in the very act of sexual intercourse.[28] In Leonardo's representation (known from a copy after him) of *Leda and the Swan*, whose idyllic atmosphere makes it additionally grotesque, the couple is depicted with their dear offspring cracking their eggs.[29]

A variant of the theme of inversion in form and content is found in the paintings of Arcimboldo, whose anthropomorphic accumulations of vegetables, fruit or other objects illustrate the sensitivity towards the ambiguities of nature at the time. The *Still Life/The Gardener* is a variant of his more commonly known portraits (Colour plate 8). In this picture the straightforward, naturalistic, precisely observed bowl with vegetables was literally turned upside down to be transformed into a human portrait.

The sixteenth-century fascination with the grotesque, the monstrous and the hybrid as reflections of figurative genesis appears to express a self-conscious distance that is contrary to the attitude underlying the harsh critique of such pictorial endeavours raised by Horace, Bernard and even Alberti. To the people of the sixteenth century involved in artistic creation or patronage, nature's monstrous image-making was apparently no longer disturbing. The development was underway towards a modern understanding of the world, where nature was to a certain extent deprived of its soul, and technology was turned into neutral machinery. Gradually the notion of nature as a creative force that could actively move and form the surroundings was replaced by the idea of a passive, inactive nature. Movement and change were obviously not denied, but the mechanisms underlying the transformations of nature were governed by exterior natural laws. Apparently, the growing urban culture and scientific and technological development resulted in a distance from nature strong enough to provide humans with a rather objectively observing attitude towards it. Nature was, generally, demythologized, and the movement simultaneously began from the monstrosity of our exterior surroundings towards the interior, the one taking place in the mind of the artist. Parallel to this, the modern concept of the artist and the artwork developed. The appreciation of artistic imagination and outstanding inventions, of which the grotesques are only one, early example, points towards the modern era of the nineteenth and twentieth centuries.

Surrealism and Nature

The demythologizing of nature was evidently not clear-cut and definitive. Remnants of the former understanding of nature can, for instance, be found in Romantic pantheism and later in Surrealism, the latter of which we shall focus on for the purpose of this essay. The idea of the potentially evocative power of nature dominant from antiquity and the Middle Ages through the sixteenth century is consistent with Max Ernst's description of frottage. In his essay 'Au-delà de la peinture' ('Beyond Painting'), 1936, he related frottage to the observation made by Leonardo on finding bizarre inventions in spotted walls and clouds. In the grains of a wooden panel Ernst likewise saw all sorts of images, and he writes that the lesson of Leonardo first led him to the frottage technique and then to collage.[30]

The frottage technique 'objectively' captures the figuration of nature; rubbing with a soft pencil on a piece of paper, the texture of the underlying material is directly transferred in an almost mechanical process. The technology was however far from neutral machinery. Ernst related the texture of the material to his inner life, and the frottage bears witness to his more or less conscious choices, indicated also by his choice of title (Figure 4.3).

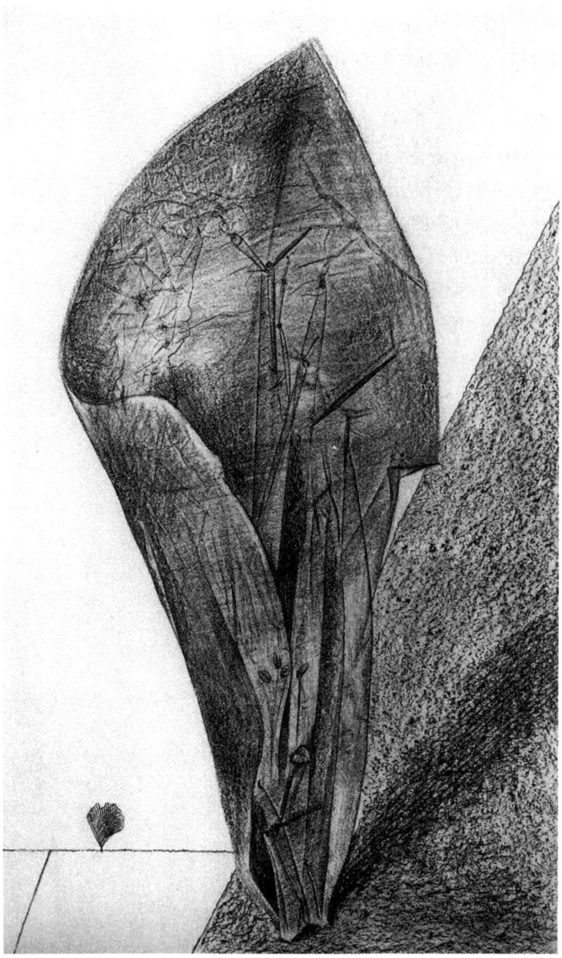

4.3 Max Ernst, *Elle garde son secret* (1925), from *Histoire Naturelle*, frottage. © Max Ernst/ billedkunst.dk

Frottage is thus simultaneously an indexical image of reality and a reflection of the artistic imagination; in Surrealism the separated domains of art and technology started to intermingle again.

In line with the art of the sixteenth century – and with explicit reference to it – Surrealist frottage was thus a way to let the inherent forms of nature become manifest. Yet this interest in nature and the organic is often excluded from descriptions and definitions of the avant-garde.

Whereas the sixteenth century represents a period when the ancient and medieval notions of nature and art were developing into the modern ones, in twentieth-century Surrealism the investigations into nature's potential for image-making were markedly different. And where the sixteenth century was still characterized by ambivalence towards the idea of nature producing images, modernity brought along a kind of loss of innocence together with ubiquitous meta-reflection on the figurative genesis on which pictures were based. In the early twentieth century, the uses of nature and of automatic or coincidental strategies of image-making were chosen very consciously. By applying these strategies, the artists hoped to regain an original approach to producing art, and to access their inner, mental world as well.

The interest in nature and the organic was present in not just the direct use of the inherent forms of nature in frottage but also the strong interest in metamorphoses and hybridity. Surrealism is loaded with monsters and chimeras (Figure 4.2). We find them in paintings, collages, literature, automata, mannequins and sculptures, by such artists as Louis Aragon, André Breton, Hans Bellmer, Max Ernst and Salvador Dalí. In answer to the question 'What is collage?', Max Ernst wrote that collage is 'the miracle of the total transfiguration of beings and objects with or without modification of their physical or anatomical aspect'.[31] Ernst saw collage as essentially attached to metamorphoses and transfigured creatures, with its significant use of technologically prefabricated imagery. And if we are to mention *one* privileged predecessor to Surrealist collage, it must be the compound organism, in other words the monster or the grotesque.

A characteristic variant of the Surrealists' monstrous literary or visual practice was the so-called exquisite corpses (*cadavre exquis*).[32] 'Exquisite corpse' designated a simple and popular game among Surrealists: they folded a piece of paper, wrote a word or drew a part of a figure without looking at the words and drawings made by the others. Finally the paper was unfolded and they could read or see the sentence or drawing made collectively. In line with the Surrealist fascination with automata, the game was a machine-like practice; an assembly line of collaborative imagination. The name of the game came from the first sentence made this way: 'Le cadavre – exquis – boira – le vin – nouveau'. In order to secure a minimum of syntactic coherence and readability, rules were made for the order of the words, and, similarly, all drawings were divided into an upper, middle and lower part of a body.[33] With the exquisite corpse, the Surrealists had 'an unfailing means for sending the critical spirit on holidays and setting the metaphorical activity of the spirit completely free', according to André Breton, who also claimed that 'all this is valid both in the graphic field and the verbal one'.[34] Hence, Breton did not see any significant difference between the textual and the visual version of the game, both resembling collage by being composed of different parts.

The game demonstrates a normalization of syntax: the product is not just a chaotic result of chance, but is made 'readable' through certain rules. Regardless of whether the *cadavre exquis* was constructed by heterogeneous collage elements or drawn in a harmonious style and colour scheme, they all represented recognizable bodies. A similar normalization of syntax is found in the collages: in textual collages, by maintaining the syntax no matter how exaggerated the verbal images are, or by making a text fairly readable despite breaks of syntax; in visual collages, by blurring the boundaries between the collage elements in different ways – for instance, by using materials that were homogeneous in style and scale, by cutting and pasting the materials meticulously or by letting one element cast a shadow on another. Just as the grotesques of the sixteenth century were characterized by a fascination with metamorphoses and hybridity, often involving continuously changing bodies (Figure 4.1), so too the Surrealists emphasized smooth bodily transitions rather than clear-cut, visible breaks (Figure 4.2).[35]

The Fusion of Humans, Nature and Mathematics

An example of smoothly united elements is a collage made in 1938 for the complete edition of Comte de Lautréamont's work (*Œuvres complètes*) (Figure 4.4). Here Ernst commented on the theme of metamorphosis that is recurrent in the writings of Lautréamont, especially *Les Chants de Maldoror* (1869), much admired by the Surrealists. The collage may either be seen as a reference to Daphne, who, according to Ovid, was transformed into a tree to escape Apollo's advances, or it may be seen as a reference to the beginning of Maldoror's forth canto, where it is said, 'A man or a stone or a tree is about to

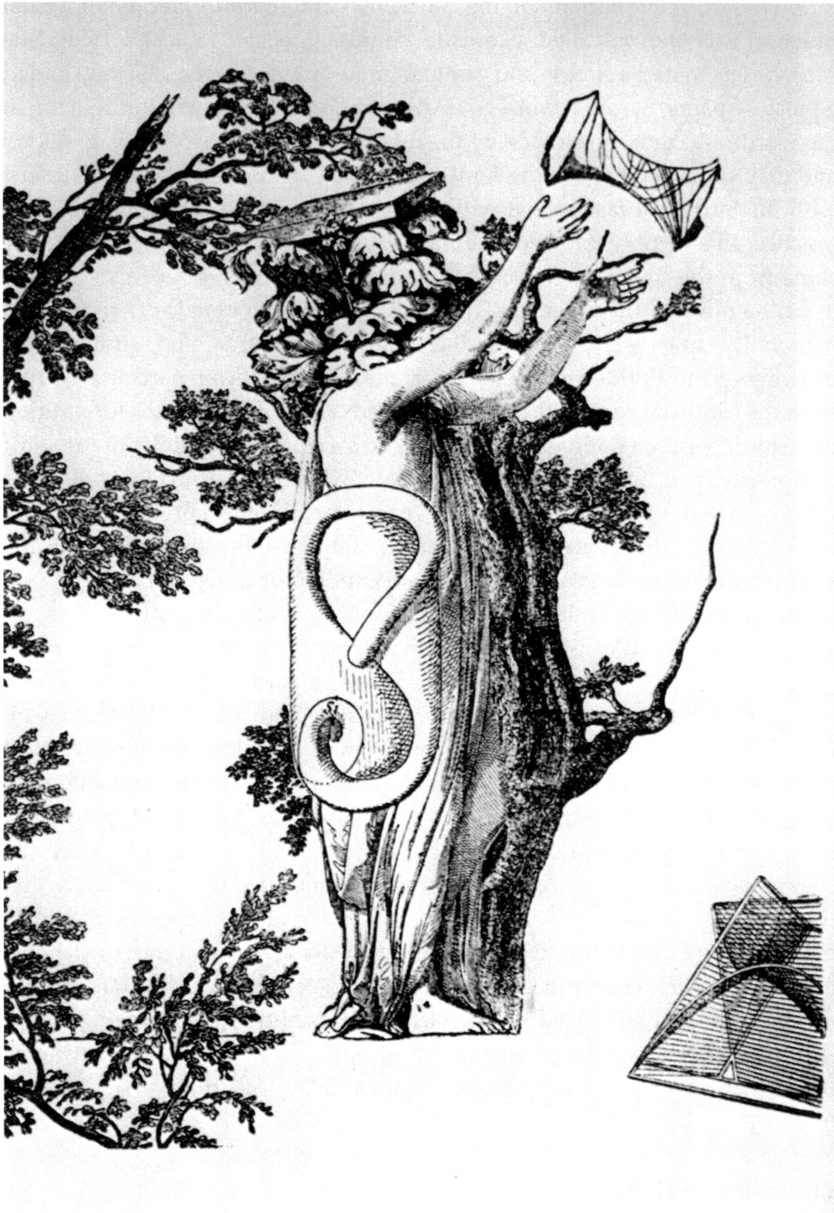

4.4 Max Ernst, illustration from Comte de Lautréamont, *Œuvres complètes* (1938), collage. © Max Ernst/ billedkunst.dk

begin the fourth canto'.[36] Humans and nature cannot be separated; they may both produce the lyrics.

This collage, however, contains elements other than just the fusion of tree and man. The androgynous creature is not just partly a tree but also partly a two-dimensional representation of a Klein bottle (a bottle with no boundary, just as the Möbius strip is a surface with no boundary), and illustrations of two mathematical wire models are evenly inserted in the collage. As Gavin

Parkinson shows in *Surrealism, Art and Modern Science* (2008), the Surrealists were generally very interested in this kind of mathematical model. Ernst discovered the models at the Institute of Poincaré in Paris in 1934, and soon many other Surrealists were inspired by them, not least Man Ray. The models were made in the nineteenth century to illustrate non-Euclidean geometry, and their purpose was, as Parkinson writes, 'to give in three dimensions the diversity of surface made available by geometry through the tracing out of straight lines in space'.[37]

The interest in mathematics and in different types of automatically functioning technology was in no way opposed to the interest in nature. As Parkinson writes, the Surrealists were inspired by morphology, in particular Édouard Monod-Herzen's *Principes de Morphologie générale* (1927), which contained 'photographs of mathematical models alongside rock and plant formations' and 'was passed around the Surrealist group in the 1930s, presenting a blueprint for the encounter between non-Euclidean geometry and natural morphology'.[38] The Surrealists used nature and mathematics according to their own mindset, just as they did with, for instance, psychoanalysis, ethnography or quantum mechanics. They valued the models for both their strange beauty and their break with Euclidean geometry, which previously had symbolized a fixed and logical order of the world. For instance, non-Euclidean geometry questioned Euclid's fifth postulate, the parallel postulate, which claims that lines remain at a constant distance from each other even if extended to infinity. By questioning such old 'truths', Surrealism regarded non-Euclidean geometry as an attack on the traditional limits of rational thinking.

One of the most dominant goals of Surrealism was to erase the borders between the interior and the exterior world; subject and object, dream and reality, rationality and irrationality. Both collage and frottage may be understood as a means to achieve this blurring: in both techniques the inner world of the artist (his unconscious) is found in the outer world. In frottage, the starting point was the imagery of nature; in collage, it was a union of reality fragments, which was claimed to reveal the deepest desires of the artist.[39] The Klein bottle, shown in the collage, is a materialization of this concept. It is a non-orientable surface with no boundary. Inside and outside are one and the same. In this manner the mathematical object substantiates Surrealist principles.

Organic Avant-garde Art

The fusion of different domains – humans, nature and science – is thus obvious in the collage. Many collages – Cubist, Dadaist as well as Surrealist – consist of elements from different domains. Specific to and characteristic of Surrealist collage is that the domains were not just opposed, but organically merged. Surrealist collage generally consists of surprising semantic compositions rather than of formal aesthetic ruptures. Rather than displaying each separate element, the collages emphasize how the elements merge and connect. By this means the collages approach what John Ruskin in *Modern Painters 3* (1856), called 'true grotesque'. John Ruskin differentiated between the organic coherent montage, which is the 'true grotesque', and the simple montage, the 'false grotesque', which consists of oppositions of uneven elements retaining their independent, fragmented character.[40] Following Jacob Wamberg, this last-mentioned 'simple montage' was the ideal of modernism.[41] Our claim is that Surrealism rejected this ideal in a number of ways. This does not necessarily imply that the organic avant-garde of Surrealism is regressive or reactionary. It just implies that a certain concept of nature, related to that of the sixteenth century, was the catalyst for the genesis of Surrealist imagery, a grotesque and twisted imagery that made use of metamorphoses and transformations rather than just simple, inorganic oppositions.

However, there are also some obvious differences between the monsters in the Surrealism of the twentieth century and the monsters of the sixteenth century. Even though Surrealists were inspired by nature, they were inspired by Freud as well, and in Surrealism, imagination and phantasms were strong mediators between nature and artworks. While many of the grotesques from the sixteenth century almost entail Freudian interpretation, as seen above in the erotized fruits of Udine, the Surrealist grotesques were made in conscious dialogue with Freudian psychoanalysis.

In Surrealism the creation of images was primarily located in the mind of the artist, as exemplified by the use of the mermaid as a recurrent motif. In Surrealism the mermaid was more than a mythological creature, who since Horace had incarnated free, artistic creation. She also became a deliberate Freudian symbol of a phallic, seductive woman. For instance, in Louis Aragon's novel *Le Paysan de Paris,* the Parisian roofed passageways are described as though they are underwater, and a mermaid is swimming in one of the shop windows.[42] This is clearly an inner image, which is connected, among other things, to the idea of women as especially close to the unconscious, and to an idea of the unconscious as an ocean into which one can dive with the help of various techniques, such as hallucination, automatism and collage.[43] The same sort of image is seen in André Breton's novel *Nadja* (1928), where the woman Nadja draws herself as a mermaid with horns that are similar to the horns in a monstrous tribal mask that she has never seen.[44] The woman was thus closely and instinctively attached to a 'natural' origin that, in the opinion of the male Surrealists, was shared by tribal people and children. Even though the Surrealist mermaids resemble those of the sixteenth century, their implications were thus quite different.

The grotesque of the sixteenth century was marginal, while in Surrealism it became a central motif. Moreover, with their collages, the Surrealists of the twentieth century questioned the concept of art's autonomy, a questioning that was not relevant in the sixteenth century. The monsters are products of different times and contexts, and the reception of them is likewise also dependent on time.

Despite these and many other differences, the parallels between the sixteenth century and Surrealism help to emphasize the interest in nature and the organic often dismissed from avant-garde studies, not least due to Peter Bürger, who in his canonized definition of the avant-garde saw it as fundamentally non-organic. This critical approach to Bürger is shared by the avant-garde theorist Hubert van den Berg. Based on Dadaist works by Hans Arp, van den Berg also pleads for a more organic notion of avant-garde art. Van den Berg convincingly shows how Arp let nature influence his visual and textual art.[45] That Arp is oriented towards nature does not mean that his art becomes a *re*presentation of anything given. The orientation towards nature should rather be seen as a genetic principle, as it is by Ernst. Arp himself writes, 'we do not want to copy nature. we do not want to reproduce. we want to produce. we want to produce like a plant that produces a fruit and not to reproduce'.[46] Van den Berg writes further about Arp's understanding of nature: 'nature is not understood as a fixed entity, as a given condition (as *natura naturata*), but, on the contrary, rather as a process or a set of creative devices (as *natura naturans*). Arp's nature is characterized by growth, change, perpetual motion … by metamorphosis and indeterminacy'.[47]

According to van den Berg, it is possible to mention quite a few avant-garde artists – among these Max Ernst – who were inspired by nature in a similar way, and who created works that through their biomorphic forms and organic character are closely related to Arp.

Thus, the Surrealists were obviously not just rejecting nature and the organic. But if this is so obvious, why is this hardly ever mentioned in avant-garde studies? Avant-garde as a concept is the epitome of critical reflection. But when the 'true grotesque' collage hides its breaks, it has a taste of illusionism or kitsch, and Bertolt Brecht, Clement Greenberg, Theodor Adorno and many other theorists have made it clear that this suppresses critical reflection. It might be such a fear of illusionism that keeps scholars from dealing with the organic aspects of avant-garde art. However, Surrealist collages are never just illusionistic totalities; there are always clashes of heterogeneous elements on the semantic level that refuse to settle and that make the collages stay open for interpretation. The Surrealist collages never offer one illusionistic truth. Therefore, Surrealist collage might form a kind of totality, but in our view this totality never becomes totalitarian.

Conclusion

In this essay we have argued that grotesques in terms of figures of monstrous hybridity and ambiguity were the object of considerable interest from antiquity and the Middle Ages through the sixteenth century and into the twentieth century with Surrealism. Often the sixteenth-century grotesques are seen as a revival of the antique grotesques, contributing to the enduring picture of the Renaissance as a period breaking with the medieval past. Our claim is that the ancient frescoes, eventually named grotesques circa 1500, were not revived at the time since they had survived in uninterrupted transformations from antiquity through the Middle Ages. In the Middle Ages the themes of change and metamorphosis were further developed, making the monstrous hybrids of the marginal imagery of the Romanesque, Gothic and Renaissance periods more grotesque than they had ever been in Roman antiquity. With the printed book and the mechanically reproduced image as crucial inventions, regularity and monotony became basic conditions within all domains of art from the High Renaissance and onwards. But at the same time, the imaginative qualities of art received new attention. The grotesque is a key figure within this paradoxical development.

Likewise, our historical perspective diverges from Peter Bürger's notion of the collage as a purely modern phenomenon rooted in a paradigm shift from the beginning of the twentieth century. By linking Surrealist collage to Mannerist grotesques, we have attempted to show how a certain concept of nature, with roots in the premodern period and perhaps culminating in the sixteenth century, became a catalyst for the genesis of Surrealist imagery.

Metamorphoses and hybrid creatures are well-known motifs in the Surrealist collage. What characterizes most such creatures – in Surrealism as well as in the sixteenth century – is that their heterogeneous parts form coherent, although monstrous and disturbing, organisms. Likewise, Surrealist frottage, which Max Ernst closely linked to the concept of collage, was a way to let nature's hidden and strange forms manifest themselves. Here, too, a similarity to the past is apparent and, as we have seen, Max Ernst explicitly referred to Leonardo in his writings on the subject. This does not mean that Surrealism was just a continuation of the late fifteenth- and sixteenth-century concept of nature and art. When Ernst saw hidden images in the grain of a wooden panel, these images were strongly influenced by his reading of Freud. The intention of Surrealism was not to fulfil what God had started in nature, but to free the Surrealist's own mind. Imagination and phantasms were thus to a greater extent deliberate mediators between nature and artworks than they had been in the sixteenth century. The Surrealists of the twentieth century also questioned the concept of art's autonomy, a questioning that was not relevant 400 years earlier. When attempting to describe the actual innovations occurring through history the obvious emphasis in the present context would be the changes in the concepts of art and the artist, including the understanding of personality and individuality, where both the sixteenth century and Surrealism seem to reflect new world views.

Moreover, the parallels between the sixteenth century and Surrealism help to emphasize an interest in nature and the organic. The essay has thus argued that at least this branch of the historical avant-gardes is much more organic than it is often perceived to be.

Especially in the last decades, many attempts have been made to develop alternative understandings of history that depart from the notion of radical breaches with the past which, surprisingly, still largely dominates our common, art historical representation of periodization.[48] While recognizing that historical developments and changes actually took place, a critical attitude to the fundamentally very modern topos of revolutionary innovation still seems necessary. In this way, it is our hope that the present analysis of the relation between the art of the sixteenth century and Surrealism, regarding both the concepts of nature they were based on and their dependence on the past, offers new insights into both periods, as well as into art historical processes and connections in general.

Notes

1 Peter Bürger, *Theorie der Avantgarde* (Frankfurt am Main: Suhrkamp Verlag, 1996 [1974]).

2 Pliny the Elder, *Natural History*, trans. H. Rackham (vols 1–5 and 9), W.H.S. Jones (vols 6–8), and D.E. Eichholz (vol. 10), Loeb Classical Library (Cambridge, MA: Harvard University Press; London: William Heinemann 1947–80), vol. 2, p. 521.

3 Horace, *Ars Poetica*, in *Satires, Epistles and Ars Poetica*, trans. H. Rushton Fairclough (Cambridge, MA: Harvard University Press, 1970).

4 Horace, *Ars Poetica*, p. 489.

5 Vitruvius, *Ten Books on Architecture*, trans. Ingrid D. Rowland, commentary and illustrations by Thomas Noble Howe (New York: Cambridge University Press, 1999), book 7, chapter 5, 3–4, p. 91.

6 St Augustine, *City of God* [*De Civitate Dei*, AD 426], in *A Select Library of the Nicene and Post-Nicene Fathers of the Christian Church*, vol. 2, ed. Philip Schaff (New York: Christian Literature Publishing Co., 1890): http://www.ccel.org/ccel/schaff/npnf102.toc.html), book XVI, chapter 8.

7 Jacob Wamberg, 'Det monstrøse', in *Monster* (Albertslund Rådhus, 2002), p. 22.

8 Bernard de Clairvaux, quoted in Caecilia Davis-Weyer, *Early Medieval Art, 300–1150: Sources and Documents* (Toronto: University of Toronto Press, 1986 [1971]), p. 170.

9 David Summers, *Michelangelo and the Language of Art* (Princeton: Princeton University Press, 1981), p. 135; for an earlier example (c. 1400) of this way of thinking, cf. Cennino Cennini, *Il libro dell'arte*, 2 vols, chapter I. English translation: *The Craftsman's Handbook*, trans. Daniel V. Thompson, Jr. (New York: Dover Publications, 1960 [1933]), p. 2: 'the poet … is free to compose and bind together, or not, as he pleases, according to his inclination. In the same way, the painter is given freedom to compose a figure, standing, seated, half-man, half-horse, as he pleases, according to his imagination'.

10 This statement attributed to Michelangelo is from the third of three dialogues on painting by Francisco de Hollanda, a Portuguese painter who was in Rome in 1538, Summers, *Michelangelo and the Language of Art*, pp. 135–36.

11 Alessandra Zamperini, *Le Grottesche. Il sogno della pitura nella decorazione parietale* (San Giovanni Lupatoto, Verona: Arsenale Editrice, 2007); André Chastel, *La Grottesque* (Paris: Le Promeneur, 1988); Philippe Morel, *Les Grotesques. Les*

figures de l'imaginaire dans la peinture italienne de la fin de la Renaissance (Paris: Flammarion, 1997).

12 Nicole Dacos, *La Découverte de la Domus Aurea et la formation des grotesques a la renaissance* (London: Warburg Institute; Leiden: E.J Brill, 1969).

13 Maria Fabricius Hansen, 'Maniera and the Grotesque', in *Manier und Manierismus* (Untersuch ... Band 106), ed. Wolfgang (Tübingen: Max Niemeyer Verlag, 2000), p. 252.

14 Anton Francesco Doni, third part of *Disegno* (Venice, 1549), Venedig 1549, ff. 18–24v, in *Scritti d'arte del cinquecento*, ed. Barocchi, vol. I (Milan and Naples: Riccardo Ricciarcli, 1961), p. 585; Giovan Paolo Lomazzo, 'Composizione de le grottesche' (from *Trattato dell'arte della pittura, scoltura et architettura*, Milan 1584), in *Scritti d'arte del cinquecento*, ed. Paola Barocchi, vol. III (Milan and Naples: Riccardo Ricciarcli, 1977), p. 2695; Giorgio Vasari, 'Introduzione alle tre arti del disegno: Della pittura', in *Le Vite de' più eccellenti pittori, scultori e architettori*, ed. Rosanna Bettarini and Paola Barocchi, vol. I, Testo (Florence: Sansoni, 1966–87), p. 85; David Summers, 'Michelangelo on Architecture', *Art Bulletin* 54 (1972): pp. 146–57.

15 Morel, *Les Grotesques*, pp. 15–20.

16 Cf. Donatello's north pulpit at San Lorenzo, Florence, mid-1460s, where two centaurs support an emblem with the signature of the artist; or Filarete's bronze doors, 1445, at St Peter's, Rome, with a self-portrait held by centaur-like, leafy creatures.

17 Leon Battista Alberti, *On Painting and On Sculpture: The Latin Texts of* De Pictura *and* De Statua, ed. and trans. Cecil Grayson (London: Phaidon, 1972), pp. 64–65, 120–23; H.W. Janson, 'The Image Made by Chance in Renaissance Thought', in *De artibus opuscula XL: Essays in Honor of Erwin Panofsky*, ed. M. Meiss (New York: New York University Press, 1961), pp. 254–66.

18 Robert Zwijnenberg, *The Writings and Drawings of Leonardo da Vinci: Order and Chaos in Early Modern Thought*, trans. (from the Dutch original) Caroline A. van Eck (Cambridge and New York: Cambridge University Press, 1999), pp. 60–61.

19 See, for instance, Hans van Aachen, *The Liberation of the Wind*, 1603–1604, Ambras/Wien: Kunsthistorisches Museum.

20 On the relation between nature and artistic genesis, see Jacob Wamberg, *Landscape as World Picture: Tracing Cultural Evolution in Images*, 2 vols (Aarhus, Denmark: Aarhus University Press, 2009), vol. 2, pp. 247–363.

21 Quoted and translated by Summers, *Michelangelo and the Language of Art*, pp. 206–7.

22 Giorgio Vasari, *Lives of the Painters, Sculptors and Architects*, trans. Gaston du C. de Vere, introduction and notes by David Ekserdjian (New York: Everyman's Library, 1996), vol. 2, p. 493: 'he made, to represent Priapus, a pumpkin entwined in bind-weed, which has for testicles two egg-plants and near the flower of the pumpkin he depicted a cluster of large purple figs, within one of which, over-ripe and bursting open, the point of the pumpkin with the flower is entering; which conceit is rendered with such grace, that no one could imagine anything better. But why say more? To sum the matter up, I venture to declare that in that kind of painting Giovanni surpassed all those who have best imitated Nature in such works '.

23 Horst Bredekamp, *Antikensehnsucht und Maschinenglaube. Die Geschichte der Kunstkammer und die Zukunft der Kunstgeschichte* (Berlin: Verlag Klaus Wagenbach, 1993), pp. 11–17.

24 Alberti, *On Painting and On Sculpture,* pp. 84–85.

25 Ibid.

26 This development has been carefully analyzed by David Summers, 'Contrapposto: Style and Meaning in Renaissance Art', *Art Bulletin* 59 (1977): pp. 336–61.

27 Ibid.

28 For example, the sixteenth-century copy after a lost painting by Michelangelo, National Gallery, London, or the representation by Giulio Romano in the frescoes of Palazzo Te, Mantova.

29 *Leda and the Swan,* copy after Leonardo by Cesare da Sesto, early sixteenth century, Wilton House, Salisbury.

30 Max Ernst, 'Au-delà de la peinture' (1936), in Max Ernst, *Écritures* (Paris: Gallimard, 1970), pp. 241–42, and Max Ernst, *Beyond Painting: And Other Writings by the Artist and His Friends* (New York: Wittenborn, Schultz, 1948), pp. 4–7.

31 Ernst, *Beyond Painting,* p. 12.

32 *Cadavre exquis* is also termed 'monster' by André Breton in 'Le cadavre exquis, son exaltation', in *Le Cadavre exquis (1925–1936),* ed. Arturo Schwarz (Saint Etienne: Musée d'art et d'industrie, 1975), p. 8.

33 Breton, 'Le cadavre exquis, son exaltation', and Tristan Tzara, 'Recette du "cadavre exquis" écrit' and 'Recette du "cadavre exquis" dessiné', in *Le cadavre exquis (1925–1936),* ed. Arturo Schwarz, pp. 23 and 32.

34 Breton, 'Le cadavre exquis, son exaltation', p. 12.

35 Despite these very characteristics of the Surrealist collage several theorists have insisted on defining collage by unmediated insertions, see, for instance, Hanno Möbius: *Montage und Collage: Litteratur, bildende Künste, Film, Fotografie, Musik, Theater bis 1933* (München: Wilhelm Fink Verlag, 2000), pp. 80, 82, 85, and Bürger: *Theorie der Avantgarde.*

36 Comte de Lautréamont, *Œuvres complètes, Les Chants de Maldoror. Lettres. Poésies I et II* (Paris: Poésie/Gallimard, 2001), p. 153. Translation by the authors.

37 Gavin Parkinson, *Surrealism, Art and Modern Science* (New Haven and London: Yale University Press, 2008), p. 71.

38 Parkinson, *Surrealism, Art and Modern Science,* p. 87.

39 Ernst, *Beyond Painting,* pp. 8, 14.

40 John Ruskin, *Modern Painters 3* (Whitefish: Kessinger Publishing, 2005 [1856]), pp. 95–103.

41 Wamberg, 'Det monstrøse', pp. 31–32.

42 Louis Aragon, *Le paysan de Paris* (Paris: Folio, Éditions Gallimard, 1998 [1926]), pp. 30–32.

43 Camilla Skovbjerg Paldam, 'Woman as Cohesive Agent: Surrealist Perception and Montage as Reconstructive Strategy in Louis Aragon's *Le paysan de Paris*', in *From Homer to Hypertext: Studies in Narrative, Literature and Media,* ed. Hans

Balling and Anders Klinkby Madsen (Odense: Syddansk Universitetsforlag, 2002), p. 99.

44 André Breton, *Nadja* (Paris: Folio, Gallimard, 1998 [1928]), 126, 132.

45 Hubert van den Berg, 'Poetry as a Reification of Nature: On the Role of Nature in the Poetry of Hans Arp', in *Nature. Literature and Its Otherness/La littérature et son autre*, ed. Svend Erik Larsen (Odense, Denmark: Odense University Press, 1997), 128.

46 Jean Arp, *On My Way. Poetry and Essays 1912–1947*, ed. Robert Motherwell (New York: Wittenborn, Schultz, 1948), p. 70. English translation by Ralph Manheim. Arp only uses lowercase letters. This practice is kept in the quote.

47 Berg, 'Poetry as a Reification of Nature', p. 133.

48 The idea of drastic paradigmatical changes in world views through history has, of course, found one of its most eloquent advocators in Thomas S. Kuhn, *The Structure of Scientific Revolutions* (Chicago: University of Chicago Press, 1962 [2nd ed., 1970]), and in Michel Foucault's writings in general.

PART II

Hermeneutics: Art and Technology Representing Nature

5

Applied Science in the Renaissance Art Academy

Bjørn Okholm Skaarup

In classical Greek antiquity the opposing concepts of *techné* and *episteme* – roughly consistent with the later Latin *ars* and *scientia* – represented a dichotomy between manual activity and a priori knowledge. The former category included painting, sculpting, cooking or any other 'mechanical' art involving the human hand, whereas the 'liberal' arts represented theoretical and unchangeable knowledge crowned by geometry and pure mathematics. Plato and Aristotle rigidly consolidated this division, while a few Roman authors, such as Vitruvius and Pliny, were more accommodating in their emphasis on geometry as essential for both the architect and the painter. During the Renaissance, the direct artistic study of nature and its underlying physical and mathematical structure would unite the two categories in new and unprecedented ways, though rarely on an institutional level. The late sixteenth-century art academy stands out as an exceptional educational setting which brought together practitioners of art and science under the same roof.

This study focuses on the particular role of applied science in the earliest formal art academies, founded in Florence (1563), Perugia (1573), Bologna (1582) and Rome (1593). From the outset these institutions pursued a strategy of social and intellectual promotion of the visual arts, and their teaching of anatomy and geometry (formerly associated with university education in medicine and the liberal arts) was integral to this objective. The artistic academies gradually incorporated not only geometry but also arithmetic and astronomy (omitting only music) from the liberal arts quadrivium. This integration was in continuity with Leon Battista Alberti's ideal educational programme for the pictorial arts, which considered established university knowledge essential to the formation of intellectually enhanced and socially viable artists. Alberti's *De pictura* (Florence, 1435/36) was programmatic for later artistic academies with its emphasis on anatomy studies and its unprecedented theory of the mathematical foundation of the visual arts.

This fusion of craft and theory was not without effort, however, and tension between aesthetic and scientific concerns would surface in the early

Renaissance workshop as well as the late Renaissance academy. Several fifteenth-century artists who mastered both perspective and anatomy occasionally downplayed this knowledge in their work, depending on the character of their commissions and the taste of their patrons. Even in the mid-sixteenth century, Giorgio Vasari's *Vite* (Florence, 1550 and 1568) was ambivalent in its evaluation of systematic artistic scrutiny of nature. While Vasari heartily approved of the intellectual foundation of artists such as Brunelleschi, Alberti and Michelangelo, he objected to any eccentric scientific obsession that led the artist away from his primary goals. He openly dismissed Paolo Uccello's obsessive inquiries into the laws of central perspective, Parmigianino's futile studies of alchemy and Battista Franco's unwarranted investigations into 'the minutiae of muscles'. Michelangelo developed a similar scepticism towards rules and canons, as seen in his radical break with the Vitruvian rules of ideal human proportions. Despite Michelangelo's lack of interest in direct imitation of nature, and the Mannerist style of his mature works, he was canonized throughout the 1500s as the unprecedented artistic master of human anatomy. His own art theory warned against placing too strong an emphasis on the rules of proportion and geometry, and instead called for the artist to use the 'compasses in his eyes'.[1]

Vincenzo Borghini, the first *luogotenente* and overseer of the newly created Florentine academy, made the most explicit plea against what he considered 'excess knowledge' among his artists. Following a heated *paragone* dispute in 1564 on the importance of painting versus sculpture, Borghini warned the academicians to 'fare, e non ragionare', and not to engage in lofty theoretical disputes outside their own practical realm.[2]

An ambitious curriculum of studious activities was nonetheless incorporated into the 1563 statutes of the new institution, with Vasari as one of six artistic *riformatori* who outlined its organization and educational programme. Borghini was director, and Michelangelo and Duke Cosimo I de' Medici acted as the two official *capi*. The incorporating statutes outlined both a mathematical teaching course and the compulsory attendance of all academicians during an annual dissection carried out at the hospital of Santa Maria Nuova. In 1567 the academy moved into the Cistercian monastery of Cestello, where it remained until 1629. An account by the Florentine painter Vincencio Carducho described its interior facilities as they appeared around 1600, with a room for scientific collections 'of globes, spheres, and mathematical instruments'. The lectures presented here were 'read not only by painters' and included studies of mathematics, anatomy, life drawing, architecture, sculpture and engineering.[3]

Anatomy

The 1563 statutes of the Florentine academy represent the earliest known formal requirement for the anatomical education of artists. To facilitate this procedure, three consuls representing painters, sculptors and architects were elected among the academicians, who (among other duties) were charged with procuring bodies for the annual dissection.[4] Between 1567 and 1570 two of these consuls, Alessandro Allori and Vincenzo Danti, were granted permission to dissect recently executed criminals – ranging from local bandits to corrupt state officials – whose names and deeds are unfortunately better documented than those of the academy's first teachers of anatomy.[5] Both artists wrote treatises on artistic anatomy, of which Allori's *Delle regole del Disegno* (1565) has survived only as an unpublished manuscript. Allori's treatise praised the first academy physician, Alessandro Menchi da Montevarchi (also included in Allori's Montauti Chapel fresco, *Christ before the Doctors*), as the author's eminent teacher of anatomy: 'The little I know of anatomy, I owe to him'.[6] Of Vincenzo Danti's intended 15-volume *Trattato delle perfette proporzioni* only the first book was printed in 1567. This publication was allegedly based on no less than 83 human dissections performed by Danti himself: 'not including those anatomies I have seen carried out by others in different places'.[7] Danti and Allori obtained some of their anatomical knowledge from Andreas Vesalius's *De humani corporis fabrica* (Basel, 1543 and 1555), but were obviously more indebted to (and often quoted verbatim passages from) the Spanish anatomist Juan Valverde, whose work was published in a popular Italian edition (Rome, 1559).[8] A direct involvement in anatomical disputes can be seen in the optical studies carried out by Vincenzo Danti and his brother Egnazio Danti, in which they supported Valverde's correction of Vesalius's erroneous placement of the crystalline lens in the centre of the human eye.[9]

Allori's studies were limited to the description of superficial musculature, osteology and proportion. The artist was assisted by Ludovico Cardi (Cigoli), who later made a famous wax écorché praised by contemporaries and subsequently cast in several bronze copies. Another academician, Pietro Francavilla, produced anatomical 'flap-books' in the same three-stage tradition as the images in Allori's *Regole del disegno,* in which a naked figure, an écorché, and a skeleton were shown in similar poses. Of his three treatises on anatomy, geometry and cosmography, the first was entitled *Il microcosmo* and explained 'la fabbrica dell'uomo'.[10] Unfortunately, these books have been lost, but three of Francavilla's sculpted écorchés have survived in later bronze casts and one was sketched simultaneously by the two academicians Andrea Commodi and Andrea Boscoli. The academy inventory of the late 1500s did not refer to such figures, but a chapter in the statutes called for the creation of clay and wax figurines, and from 1570 the academy offered biweekly drawing classes of sculpted and draped models. Commodi and Boscoli's sketches of the same sculpted figure show how academicians from different workshops could gather and sketch from each others' maquettes during formal drawing sessions in the academy. The écorché models coincided with the first academies and enabled the study of human surface anatomy outside the annual winter dissection stipulated in the academy statutes; in Federico Zuccaro's Roman residence and artistic 'scuola' (Palazzetto Zuccari) several écorchés appear in

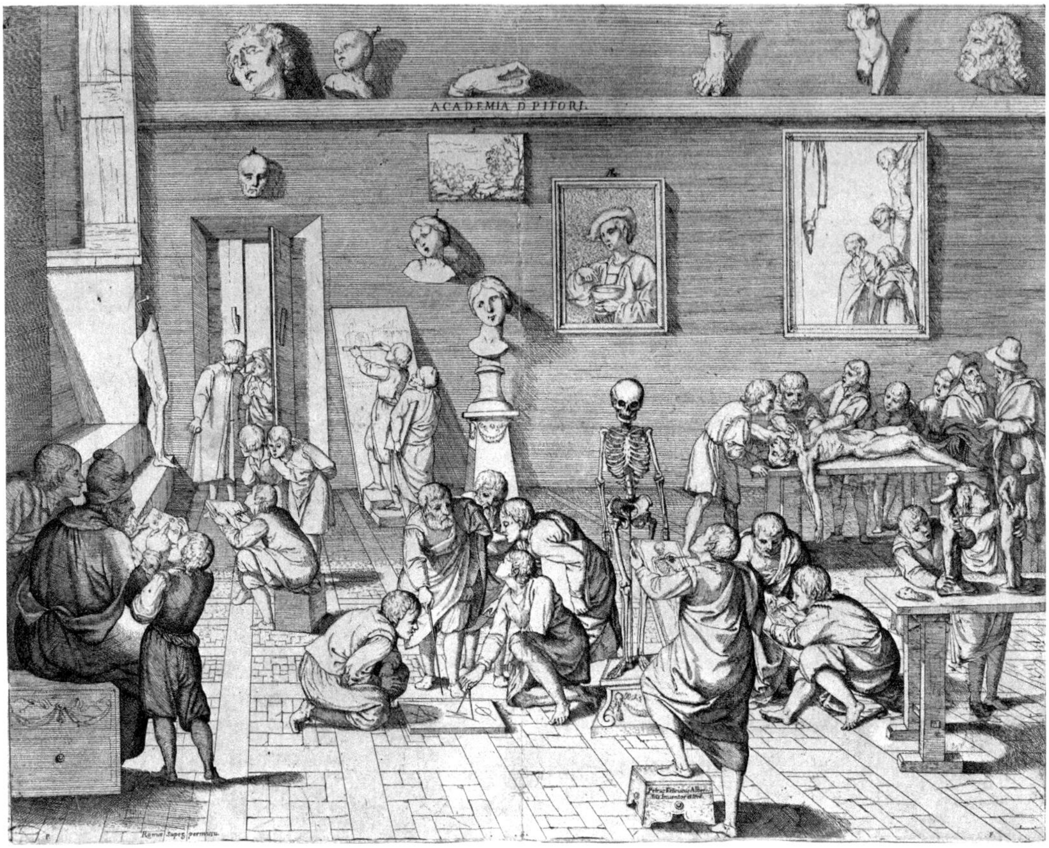

5.1 Pietro Francesco Alberti, *Academy of Painters* (c. 1600), etching. Metropolitan Museum of Art, New York. © 2014 Metropolitan Museum of Art/ Art Resource/ Scala, Florence. Elisha Whittelsey Collection, Elisha Whittelsey Fund, 1949 (49.95.12)

the fresco decorations, and also in the 1627 inventory of the Roman Accademia di San Luca. Pietro Francesco Alberti's illustration of activities in the Roman academy presented 'life-casts' of body parts as additional didactic tools for the study of anatomy and showed young artists gathered around a dissected corpse and sketching from a plaster cast of a human leg (Figure 5.1). The contemporary publication *Origine e progreso dell'Accademia de' Pittori, Scultori ed Architetti di Roma* (published in 1604 by the artist's uncle, the academy secretary Romano Alberti) related how the academy's first *Presidente*, Federico Zuccaro, obtained a dead body in 1594 and personally oversaw a two-week dissection within the academy. A subsequent mould and plaster cast were made of the dissected body, and exhibited on a 'piedestallo della notomia'.[11]

In Bologna, the Carracci Accademia degli Incamminati was engaged in similar activities, as described in Carlo Cesare Malvasia's account of Agostino Carracci and the surgeon Dr Lanzoni's joint dissections and production of terracotta casts from flayed body parts.[12] Like its Florentine and Roman equivalents, the Carracci academy offered drawing classes with an articulated human skeleton as the subject.[13] An articulated human skeleton can also be seen in Alberti's *Academia for Painters* (early 1600s) and Giovanni Stradano's

(Johannes Stradanus) *Allegory of the Florentine Drawing Acadamy* (1569), and his *Allegory of The Visual Arts* (1578) (Figure 5.2). For the artistic rendering of complicated anatomical features, a larger-than-life plaster ear was made by Agostino Carracci, who also produced a series of didactic step-by-step representations of the human eye, nose and limbs. Some of these illustrations were later included in Annibale Carracci and Luca Ciamberlano's *Scuola perfetta per imparare a disegnare tutto il corpo humano* (Rome, c. 1600), and in Odoardo Fiolatti's pattern book *Il vero modo ed ordine per dissegnar tutte le parti et membra del corpo humano* (Venice, 1608).

Vincenzo Danti's treatise on anatomy and proportion advocated systematic dissections of quadrupeds, birds and fish – 'a practice little different from that of the physicians' for any artist who wanted 'to imitate nature' and to represent its manifold creatures and structures.[14] Half a century later, the concluding chapter, 'Ai giovani accademici e pittori', to Pietro Accolti's *L'inganno degli occhi* warned against such excessive studies and claimed that only life drawing, not anatomy taught from dissection, would enable a proper artistic rendering of the human body: '(The young academician) needs first of all to carry out his particular studies through life drawing of the human body … Not those parts studied by the physicians and anatomists, but only the external parts that pertain to the study of painting'.[15] Life drawing, henceforth known as 'academies', would eventually become the bedrock of academic art education, with studies based on direct life observation rather than on the systematic dissection of the human body.

Mathematics

The 1563 Florentine statutes stated that 'Euclide, Vitruvio et altre matematiche' should be taught to the young academicians in a way useful to painters, sculptors and architects alike. The initial classes were held infrequently, however, and requests were made for more regular teaching.[16] The name of the first paid professor of mathematics, the Bolognese mathematician Pierantonio Cataldi, appears in archival records in 1569.[17] Cataldi taught at the Florentine art academy until 1570 and later moved to the recently established Accademia del Disegno in Perugia, where he published two of his mathematical lectures. These *Due lettioni* represent an interesting case of theoretical teaching for an artistic audience, emphasizing the mathematical foundation of the arts, and of architecture in particular, 'as Vitruvius has demonstrated'. The lectures presented both arithmetic and geometrical theorems as 'necessary and useful for the studies of perspective, architecture, and fortifications, and any of the other sciences dealt with in this honourable academy'.[18]

Among Cataldi's theorems, the calculations of surface areas of geometrical bodies were dedicated to chosen *amatori* (prominent academy members from the nobility). Theoretical teaching was obviously intended not only for the improvement of artistic skills but also appealed to an audience of

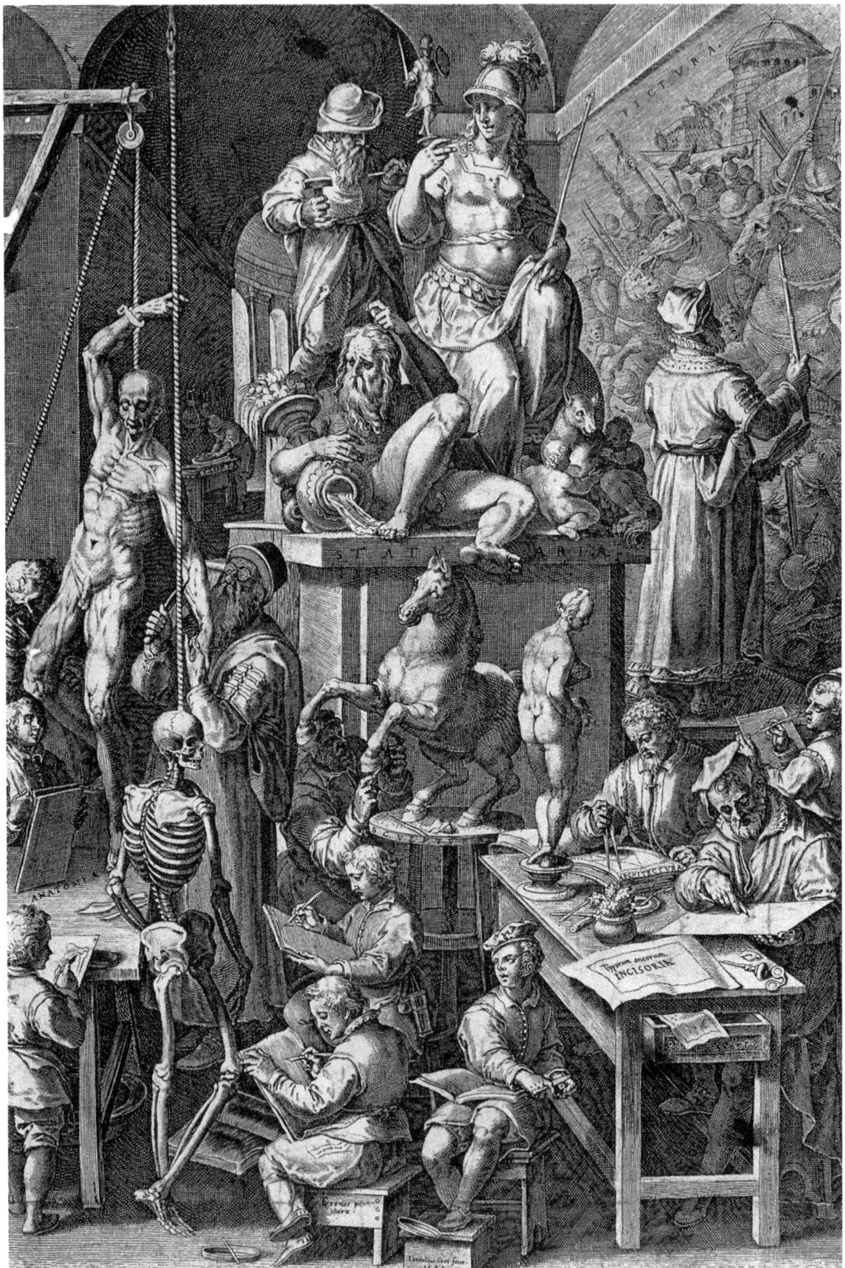

5.2 Johannes Stradanus and Cornelis Cort, *Allegory of the Visual Arts* (1578), engraving on paper. Department of Prints and Drawings, Staatliche Museen Berlin. Photo credit: Volker-H. Schneider/Art Resource, New York

noble, non-artistic members, who enrolled in the academies of Florence and Perugia during the 1570s, providing prestige and potential patrons for the new institutions. Cataldi addressed this group repeatedly and expressed his own desire to matriculate as an official academy member.[19] Other famous Renaissance mathematicians, such as Egnazio Danti, Ostilio Ricci and Galileo

Galilei, all considered membership of the Florentine art academy desirable both in their early and more advanced careers. A year before his 1589 appointment at the University of Pisa, Galileo corresponded with Cardinal and art patron Francesco Maria del Monte, seeking guidance on whether or not to apply for the Florentine academy lectureship of mathematics, which, as he emphasized, had been founded by Duke Cosimo I and was 'much favoured by the nobility'.[20]

Indicative of their scholarly aspirations, the lectures and treatises written by, or for, the first academicians of the 1560s and 1570s often referred to their content as *ragionamenti* (theories), thus defying Vincenzo Borghini's declaration that an academy of art was an inappropriate setting for theorizing. These artistic *trattati*, however, were often explicitly reduced in academic scope, and differed from contemporary Latin university treatises in language and style – and concentrated on scientific matters of direct relevance for practicing artists. Among the most ambitious examples of this genre are the early works by the cartographer, mathematician and instrument builder Egnazio Danti, who was an early member of both the Florentine and later Perugian art academies. In 1571 he was appointed court mathematician to the Medici princes and was further obliged to give one public lecture on mathematics each month – seemingly given at the Accademia del Disegno. Danti's publications repeatedly advocated mathematical teaching for artistic purposes and his annotated 1573 treatise on Euclid was dedicated to the artistic academy of his native Perugia. Its dedication mourned the lack of attention and esteem traditionally given to perspective studies in university contexts but envisioned their revival in the new Perugian art academy founded the same year.[21] Danti's treatise was unusually bold in anticipating that an expertise in perspective and geometry would make the new institution a well-known pilgrimage site of higher learning, on a par with the best Italian faculties of law and military studies.[22] An account written the same year by its founder, the architect Rafaello Sozi, affirmed the Perugian academy's ambitious objective to promote 'knowledge of painting, sculpture, architecture, and fortifications, and to present lectures on both architecture and mathematics'.[23]

Danti's treatise on Euclidean perspective and optics included several passages written for a purely artistic audience, with new observations and illustrations added to the original text. His account of the first theorem of Euclidean optics contained a detailed illustrated manual for the construction of a camera obscura and an enthusiastic description of the reversed reflected image within this black box. His later expanded version of Vignola's perspective treatise, *Le due regole della prospettiva pratica* (Rome, 1583), was conspicuous as the first printed and fully illustrated Italian treatise on perspective and included further instruments for illusionistic perspective drawing. Its illustrator, Tommaso Laureti, contributed his own devices for perspective projection, designs for illusionistic ceilings, stage scenery and anamorphic images – and later became the first teacher of mathematics at the Roman Accademia di San Luca and its second *Presidente* in 1594.[24]

Danti compiled written material from prominent artists and architects for the treatise and some of its illustrations from the book and drawing collections of the Florentine Accademia del Disegno.[25] Following his departure from Florence, Egnazio Danti was appointed professor of mathematics at the University of Bologna in 1576. In this position he was instrumental in the unprecedented consolidation of mathematics as a university discipline independent of astronomy and astrology, with which it had traditionally been associated. This remarkable ascent from a young art academy to a renowned university was repeated by another former academy mathematician, Pierantonio Cataldi, who succeeded Danti in 1584 and taught at the Bolognese University until his death in 1626. As a notable legacy of Cataldi's prior teaching at the Florentine and Perugian art academies, his university lectures and publications continued to be given in vernacular Italian and included references to his past employments. Cataldi's 1620 *Trattato geometrico* (on Albrecht Dürer's method for constructing a regular pentagon) was dedicated to the Florentine academy and opened with an appraisal of his first teaching experience there half a century earlier, 'when, as a youngster, from 1569 to 1570 I taught Euclid at the most renowned Accademia del Disegno'.[26]

A permanent Florentine academy lectureship in mathematics was finally established in 1589, and was bestowed upon the court mathematician Ostilio Ricci, who held the position until his death in 1603. Today Ricci is best known as the first tutor of Galileo, to whom he introduced the theories of Euclid and Archimedes and the writings of Leon Battista Alberti in the mid-1580s. During Ricci's tenure, the academy compiled a scientific collection of compasses, maps and globes, which had grown significantly by the time of Carducci's previously mentioned account from circa 1600. Ricci was furthermore mentioned in Filippo Baldinucci's artistic biographies as a teacher of geometry and perspective for aspiring artists in Bernardo Buontalenti's 'private academy', apparently during a period without similar activities at the Accademia del Disegno. A letter written by Federico Zuccaro to the academy's *luogotenente* in the late 1570s outlined a new educational programme for the institution, in order to 'get the studies of our academy back on their feet'.[27] Zuccaro called for the reintroduction of mathematical teaching, indicating some years of vacancy in this field, and warned that without mathematics any works by the young academicians would be deficient in execution, composition and grace.

Ricci's revival of mathematical instruction was praised by the Genoese painter (and Florentine academician since 1586) Giovanni Battista Paggi. In a detailed correspondence with his brother in 1591, Paggi eulogized the Florentine academy and 'all the lectures made there, on perspective, geometry, and other mathematical sciences from which painting derives'.[28] Paggi's praise for an institution without equal in his native Genoa showed a similar enthusiasm for artistic anatomy: 'I cannot stress enough how useful the theory of painting from anatomy is to this art. It is a study so profound that not even an entire life-span of study is enough to know everything'.[29] He furthermore proposed an (evidently successful) teaching programme of natural philosophy, physiognomy and astronomy to enable the academicians to understand and depict the influence of the four elements on the differing humours and appearances of the human body.[30] Lecture notes on this subject from around 1570 dealt with the most successful artistic rendition of all the human gestures and motions, as seen in works by Leonardo da Vinci and Andrea del Sarto.[31]

For the first time since Egnazio Danti, another renowned cartographer, instrument-builder and astronomer, Antonio Santucci de Pomerance, was among the Florentine academicians. Santucci is best known for his giant Ptolemaic sphere, which was made for Grand Duke Ferdinando I de' Medici, and for an illustrated treatise on astronomical and surveying instruments in the ducal collections. A 1595 reference to two spheres used for his academy lectures shows that both cartography and astronomy were being taught at the academy at that time.[32] In the mid-seventeenth century, studies of astronomy were expanded in a series of telescopic experiments in which the young academicians observed and drew the lunar surface to test Galileo's earlier discoveries and representations. A teaching programme of cartography and land measurement (*agrimensura*) was increased during the same period, and the studies of mathematics were expanded significantly, coinciding with a growing number of architects and engineers among the academicians. This renewed emphasis on mathematical instruction was emphasized in the detailed *Ordine da osseruarsi nella lettura delle Mathematiche* included in the 1639 statutes. Indicative of its success, the only and long-vacant Florentine university chair of mathematics was restored the same year and transferred permanently to the Accademia del Disegno. Teaching was intensified with two weekly lectures, which were supplemented by further instruction 'according to the needs of the students'.[33] This mathematical professorship, consolidated under Ostilio Ricci, was held throughout the late seventeenth century by Galileo's last apprentice and later biographer, Vincenzo Viviani. The applicable and vernacular mathematical education spearheaded by Cataldi and Danti was thus continued by both Galileo's initial tutor and his last disciple in the unlikely context of an artistic academy.

The vernacular scientific culture of the emerging art academies produced new didactic tools and treatises, and even influenced contemporary university education. Yet the incorporation of applied sciences in the academies did not always proceed easily, and various members questioned their relevance and usefulness. In 1582 one of the ageing founders of the Florentine academy, the sculptor Bartolomeo Ammanati, wrote a *Lettera agli Accademici del Disegno* in which he bemoaned his own and his colleagues' excessive studies and depictions of the naked human body. Similar warnings were heard from other academicians. While Zuccaro's reform programme from the late 1570s had encouraged mathematical teaching in the Florentine academy, his 1607 *Idea dei pittori* insisted that 'the art of painting does not derive its principles from the mathematical sciences and has no need of recourse to them to learn the rules and means for its practice'.[34] Despite such reservations, theoretical academy studies continued, enabling artistic elites to imitate a scholarly culture from which artists had traditionally been excluded. The leading European art institution of the subsequent century, the Parisian Académie royale de peinture et de sculpture (1648) credited the studious curriculum of the Florentine and Roman Renaissance academies in its own statutes – and surpassed any of these precursors with its extensive scientific courses and collections embracing mixed mathematics and comparative anatomy. Over the course of the next two centuries, new artistic academies across national borders embraced similar studies. Only the developing modernism and anti-academism would gradually put an end to this enduring educational legacy of the Renaissance art academy – and thus contribute to the institutional and intellectual division of art and science.

Abbreviations

AASL Archivio dell' Accademia di San Luca
ASF Archivio di Stato di Firenze
BNCF Biblioteca Nazionale Centrale di Firenze

Notes

1 Cited in Martin Kemp, *The Science of Art* (New Haven: Yale University Press, 1990), p. 74.

2 A. Lorenzoni, *Il carteggio artistico inedito di Don Vincenzo Borghini* (Florence: Succ. B. Seeber, 1912), p. 83.

3 Vincencio Carducho, *Dialogo de la pintura* (Madrid: Fr. Martínez, 1633), f. 11v.

4 Capitoli et ordini dell'Accademia et Compagnia dell'arte del Disegno. BNCF, cod. Magliabecchiano II, I, 399, fol. 1v. A more detailed, undated draft of the statutes written a few years later no longer referred to the hospital; seemingly by then the dissections had been relocated to the academy itself: ibid., Cap. XII.

5 BNFC, FN II.I.138, f. 98r, 98v, 99r.

6 Alessandro Allori, *Le regole del Disegno,* BNCF, f. 68r.

7 Vincenzo Danti, *Trattato delle perfette proporzioni* (Florence, 1567), introduction, n.p.

8 This *Nuttumia di Valverde* was also found among the titles in the 1608 book inventory of the Roman Accademia di San Luca. Archivio di Stato di Roma, TNC, uff. 11, 1608, pt. I, vol. 76, f. 818.

9 Vincenzo Danti, op. cit., p. 3.

10 Baldinucci, *Notizie de' professori del disegno da Cimabue …* (Florence: S. Franchi, 1681), vol. 3, p. 65.

11 AASL. Entrata e uscita del camerlingo, 1593–1625, vol. 42, f. 84v.

12 Malvasia, *Felsina pittrice, vite de' pittori Bolognesi* (Bologna, 1678), p. 485.

13 Ibid., p. 565.

14 Ibid.

15 Pietro Accolti, *Lo inganno de gl'occhi, prospettiva pratica* (Florence: P. Cecconcelli, 1625), p. 145.

16 Vincenzo Danti, op. cit., p. 23.

17 BNCF, cod. Magl. II, I, 399, c. XXXII, f. 21v.

18 ASF Accademia del Disegno, no. 24, f. 25v.

19 Cataldi, *Due lettioni di Pietr' Antonio Cataldi fatte nell' Academia del disegno di Perugia* (Bologna: Giouanni Rossi, 1577), n.p.

20 Cited in Karen-edis Barzman, *The Florentine Academy and the Early Modern State* (Cambridge: Cambridge University Press, 2000), p. 331.

21 Egnazio Danti, *La prospettiva di Euclide* (Florence: Stamperia de' Giunti, 1573): 'Alli molto magnifici nostri osser. Gli Accademici del Disegno di Perugia', n.p.

22 Ibid., n.p.

23 Giovanni Cecchini, *L´Accademia delle belle arti di Perugia* (Florence: Le Monnier, 1954), pp. 14–15.

24 *Le due regole della prospettiva pratica di M. Iacomo Barozzi da Vignola con I comentarij del R. P. M. Egnatio Danti* (Bologna: Francesco Zannetti, 1583), p. 39.

25 Ibid., n.p.

26 Pietro Antonio Cataldi, *Trattato geometrico* (Bologna: Sebastiano Bonomi, 1620), n.p.

27 Cited in Barzman, op. cit., p. 245.

28 Bottari and Ticozzi: *Raccolta di Lettere sulla Pittura, Scultura, e Architettura* (Rome: Eredi Barbiellini, 1754), vol. VI, p. 64.

29 Ibid., p. 85.

30 Ibid., p. 87.

31 *Lettione nell´Accademia del Disegno*, BNCF, Magl. XVII, 7.

32 ASF Accademia del Disegno, no. 27, fol. 29r.

33 ASF Accademia del Disegno, no. 5, fol. 29r.

34 Federico Zuccaro, *L'idea dei pittori* (Turin: A. Disserolio, 1607), p. 101.

6

Artisanal Epistemologies and the Artless Art of Post-Tridentine Painting

Claire Farago

The Council of Trent was suspicious of artistic licence, understood as an explicit display of artifice for its own sake, contrived composition and artistic ingenuity. As writers allied with the Catholic Reformation emphasized in the sixteenth century, the more closely the artist is allied with God, the 'truer' his artistic representation of sacred subject matter. For the same reason, Protestant theologians such as Ulrich Zwingli identified the *Abgott* in the patron's soul as anything that displaces God, be it money, glory or another deity: as one becomes reliant on those 'strange gods', the source of idolatry finds its external and indeed very monstrous expression *in* and *as* works of art.[1]

What Catholic and Protestant reformers shared is the understanding of *artifice as a form of truth.* In the mid-sixteenth century, the Catholic Church tried to redirect the increasing freedom recently accorded the artist, recognizing it as quite literally dangerous because it diverted the attention of worshippers away from their devotions and towards admiration for the artist and his magnificent artifice. In issuing its famous 'Decree on Images' (1563), the ecclesiastical establishment attempted to impose an absolute *decorum* – an improved and proper fit between the subject of representation, the style in which it was presented and the audience for whom the representation was intended – which was intended to focus on the artifice as an optimal instrument for enlivening the beholder by appealing to his imagination through his senses. There are many indications that the Catholic Reformation provided conditions for scientific literature on art to flourish – at least of a certain kind. The accelerated publication of treatises written by and for artists, both illustrated and unillustrated, printed editions and circulating manuscripts, is a notable development during this period.[2] A significant case in point is the Italian version of Leon Battista Alberti's *De pictura* originally written in 1435, which first appeared in print in 1547 and 1568, in two separate Italian translations.[3]

The complex entanglement of art, nature and technology during this formative period before art was defined in modern secular terms suggests that many factors were involved. Three of these are the subject of the following essay: the classification of knowledge, the significance attributed to artifice and the kinds of knowledge that circulated through the workshops of artisans. The connections pursued in this brief essay circle around the one fundamental methodological issue of how interests that individuals do not literally share may still draw their actions together in a sociocultural network that produces concrete effects. The approach used here to locate art, nature and technology in a 'field of cultural production' proceeds at three different scales.[4] The argument begins with a discussion of the historical classification of knowledge in medieval Latin sources and then dives briefly into the historical record to highlight specific issues about the significance of artifice debated in the fifteenth century. The first, fundamental point to be made is that early humanist arguments for eloquence were offered with Scholastic criteria for achieving certain knowledge in mind.

The same situation of shared attitudes towards artifice also characterizes arguments taken up in the third part of the essay through a close reading that relates an artist's text to workshop procedures. I will be arguing that artists verbalized their craft traditions in terms that echoed the significance attributed to artifice in these other arenas. In fact, humanist, Scholastic and workshop discussions of artifice cannot be divorced from issues of methodology, and all centre on the relationship between art and nature. Only with such an understanding can we approach questions of broad historical significance such as the core issue of how seemingly opposed Protestant Reformation and Catholic Reformation positions sustain the same ideas about the value of naturalism, a point to which I will return at the end of this essay.

Classifications of Knowledge

In the last three decades, historians of science joined by cultural anthropologists, art historians and other specialists of material culture have opened up the question of how scientific knowledge was produced during a key period of the emergence of modern science in later sixteenth- and seventeenth-century Europe.[5] The diverse practices denoted by the newly introduced term 'techno-science' cover a widely expanded range of vernacular subjects including some which possessed recognized textual traditions and others that were transmitted primarily orally but nonetheless involved the mastery of specialized knowledge and skills. The key role of the productive sciences has long been recognized, but the latest generation of thinking questions the very idea of the scientific revolution in favour of a slower, more widespread phenomenon of technological change that led to modern attitudes privileging experimentation and empirical investigation, the basis for an inductive rather than a deductive approach to the study of nature.

At the centre of the new historical initiatives are concrete relationships between making objects and knowing nature. The present chapter is intended as a contribution to these investigations. During the early modern period (c. 1450–1750), the significance attributed to natural knowledge varied widely. Indeed, the term 'science' was employed in many different ways, with some modern historians of science preferring to describe science during this transitional era as simply indicating how humans interacted with their natural environment in order to understand it.[6] Even though long-standing Scholastic classifications of knowledge were contested and bent to new purposes, it is crucial to understand the medieval tradition to appreciate what was being challenged, how and why. The medieval Latin term *scientia* meant something quite different from either the wide-ranging seventeenth-century attestations or our modern cognate, science. As John Weisheipl explains, in medieval texts *scientia* was 'used to designate a discerning, penetrating, intellectual grasp of a situation or of a given subject. Technically it was employed … in every field of intellectual endeavor in which true causal explanations could be discovered'.[7]

In keeping with this understanding of *scientia*, Thomas Aquinas defined the discursive power as a kind of logic or act of reason 'resembling nature', specifically the resolution of effects into causes. He took his definition of *discursis* from Aristotle's *Posterior Analytics* (95a25), construing effects in the physical world to conform with the structure of a mathematical demonstration if the premise is taken as that which in nature is more known to us (and unlike a syllogism, the terms do not necessarily exist simultaneously): we begin with the immediate, and the cause of one thing will be the coming to be of another.[8] Following Aristotle's account of human ways of knowing, Aquinas established a hierarchical system for classifying human knowledge beginning with mathematics, the only form of *scientia* capable of achieving certainty. All the 'sciences' were interrelated in a hierarchy ordered by the certainty of their results, rather than autonomous, each with its own principles and broader generalizations, as the physical and biological sciences are considered today.[9] The higher, purely mental mathematical sciences contained the principles that the subordinate, more 'physical' sciences demonstrated through experience. Thus geometry provided the principles for the subaltern science of optics concerned with the action of light. In Aquinas's scheme, the rational sciences were divided into three branches arranged in descending order of certitude according to the kinds of logic they employ. Poetry was considered a 'science' because it is concerned with causal explanations. Poetry uses inventive logic: poetry's method of proof is the *exemplum*, which concerns two particulars, and therefore its judgements cannot rise above the level of opinion. Thus poetry is classified as *scientia*, but it occupies a low position in the scale of the rational sciences classified according to their degree of demonstrability.[10]

The Value of Artifice

Aquinas and his Scholastic successors objected that poetry, associated by Aristotle at the beginning of the *Metaphysics* with myths that produce wonder, cannot rationally answer questions about nature: 'poets lie' in the sense that the content of poetry does not offer logical explanations.[11] In the mid-fourteenth century, humanists such as Boccaccio (*Geneologica deorum gentilium,* 1350–62) and Petrarch (*Invective contra medicum,* 1352–53) challenged the Scholastic system of ranking the liberal arts by arguing that the figured language of poetry is a form of 'revealed knowledge' granted to inspired poets and therefore poetry deserves to be repositioned next to theology, at the pinnacle of the whole system.[12]

Claims for the divine inspiration of figured poetic language were repeated throughout the fifteenth century, but arguments for the status of the *studia humanitatis* were also adjusted to Scholastic criteria that the higher human science is the one that achieves greater certainty. In associating rhetoric or eloquence with moral philosophy (partly to raise its status), in the early fifteenth century Florentine humanists Lorenzo Valla and Coluccio Salutati proposed attaching positive value to ordinary, 'natural' language aimed at a universal audience, similar to the terms in which Alberti described narrative painting as the imitation of nature.[13] In 1474, Cristoforo Landino described a scientific philological method for extracting the truth from the figured language of poetry in his *Disputationes Camaldulenses*.[14] In 1490, Angelo Poliziano, the greatest philologist of his time who still served as a model for editorial practices in the mid-sixteenth century, wrote an introduction to his lectures on Aristotle's *Nicomachean Ethics,* a work concerned with practical questions of how people should live, in which he associated poetry with theology, defined as a semi-divine *scienza* at the top of the hierarchy of knowledge.[15] In the same year, and perhaps in direct response to Poliziano's *Panepistemon,* the Dominican preacher Girolamo Savonarola delivered sermons on the abuses of poetry in which he followed Thomas Aquinas by associating poetry with the exemplum, a form of induction that induces virtue by delighting the individual soul through a concrete image.[16] Like his mid-fifteenth-century predecessor the Cardinal Giovanni Dominici, Savonarola asserted that only Scripture has true allegorical meaning and only natural things not created by art are perfect.

Late fifteenth-century humanists also devised new classifications of knowledge. Poliziano devised a new, universal system of classification indebted to Lorenzo Valla that associated the menial arts based on *graphike* with mathematics, which he derived from an ancient neo-Pythagorean treatise on arithmetic by Nicomachus of Gervasa.[17] Responding to Scholastic criticism of the arts of imitation, Giorgio Valla proposed one of the most interesting revisions to Scholastic classifications in his encyclopedia *De expetendis et fugiendis rebus,* published in 1501. He developed a new classification system for the theoretical sciences ordered to practical use that would replace

Hugh of St Victor's twelfth-century scheme of the seven mechanical arts parallel to the seven liberal arts. Valla grouped the practical liberal arts by their communicative function under the unifying mathematical principle of *armonia*. He described the persuasive mode of argumentation associated with rhetoric using mathematical terminology and propositional forms of argumentation borrowed from Scholastic logic that could be applied in the common language of orators and poets.

What all these late Quattrocento humanist schemes shared was an interest in making rhetorical modes of inventive argumentation capable of achieving greater certainty or truth value, hence higher status as a form of knowledge, like mathematics. In fact, the discussions of artifice cannot be divorced from these issues, and both humanist arguments and Scholastic classifications of knowledge centre on the relationship between art, nature and certainty. Important to bear in mind in the following section of my argument, which treats the knowledge that artists used in their workshop practices, is the manner in which artists verbalized their craft traditions in the same terms that the classification of knowledge and the significance attributed to artifice were debated in these other arenas. In Italian universities, such as Padua, methods used in the natural sciences had been in a state of flux and disagreement since the early fourteenth century because they dealt with probability and conjecture.[18] The theoretical sciences ordered to practical use were originally defined as such by Aristotle, who described medicine in these terms (*Physics* 200a15–25; *Nicomachean Ethics* 1112b15–25, 1032b5–15). The second-century Alexandrian physician Galen and his medieval commentators such as Avicenna developed discussions of method in the natural sciences that came to be known as the 'double procedure' of analyzing effects into their causes and causes into their effects. Many documents of the modern, fifteenth-century controversy over method were first printed in the 1480s, when they became available to artists like Leonardo da Vinci, who probably owned at least one of the most important publications while he was part of a group of artists, musicians, poets, natural philosophers and other luminaries active at the court of Milan.[19]

Leonardo derived some of his ideas on painting as a *scientia media*, a term first used by Thomas Aquinas to refer to a science grounded in both theory and experience,[20] directly from John Pecham's *Perspectiva communis*, published in Italian translation in 1480 by Fazio Cardanus in Milan; from the writings of Galen, known to Leonardo in vernacular epitomes by circa 1488; and from his contemporaries such as the Pavian mathematician and physician Giovanni Marliani, whose *Tractatus de proportionibus*, circa 1482, developed connections between natural causes and mathematical formulae.[21] The artist had access to the recently deceased Marliani's library through the latter's two sons teaching at the University of Milan, and referred to borrowing books from this library on a wide variety of subjects including algebra, anatomy, optics and the science of weights.[22] Leonardo's earliest notes on Pecham date from circa 1488–90. The most mature version of his definition of painting as a 'mixed science' that involves both the knowledge of geometric principles and direct experience forms the opening section of his treatise on painting compiled posthumously from autograph writings by his student Francesco Melzi (c. 1491–c. 1570). Features such as the double movement of argument between cause and effect, the use of geometric analysis, the treatment of differentiating characteristics of the whole and its parts, the statement of the results of demonstration by definition, all identify Leonardo's concept of painting with contemporary Aristotelian methods for ordering the operations of an investigative science based on both theory and experience.

To summarize, Leonardo's richly conceived discussions of painting as a scientific art that results from the artist's understanding of the causes of nature through observation and ingenious mental discourses translating nature's ornaments into painted ones, are informed by Scholastic classifications of knowledge, deeply engaged with humanist debates over the value of poetic artifice, and strongly reminiscent of contemporary issues regarding methodology in the natural sciences.[23]

In Milan, Leonardo's views were closely related to those of his friend Luca Pacioli, who wrote about music, perspective and architecture as grounded in geometric proportion, specifically in the laws of harmonic proportionality that can be worked out precisely with a knowledge of geometry and drawing. Leonardo and Pacioli were at the centre of a circle of artists who wrote about painting on the foundation of both mathematical principles and experience.[24] We turn to this formative period of Leonardo's artistic career now.

Artisanal Epistemology in Action

Leonardo's first documented apprentices also date from the beginning of the 1490s, when his research on optics, anatomy and many other subjects including theoretical considerations of painting intensified. It is likely that he began the second, London version of the *Virgin of the Rocks* at this time, although this dating is controversial because the documents are so difficult to reconcile: in any case, by 1490 Leonardo and his associates had been paid nearly in full for the first version of the painting originally commissioned in 1483.[25] Regarding the attribution of the second version, despite the unevenness of its finish, it has been proposed by Luke Syson and others that the quality of the paint handling and the coherence of the composition had been underestimated prior to its recent cleaning. Furthermore, the surprising underdrawing related to Florentine studies revealed by reflectography two layers beneath the executed composition shows that Leonardo himself was directly involved in its production over a protracted period.[26]

The relationship between Leonardo's writings about painting and his actual paintings has long been a fraught question. His statements are generally abstract and generic – no matter how carefully observed the phenomena described are, it is difficult to pin them down to specific projects. Leonardo's writings are also indebted to textual sources far outside the sphere of actual painting practices, in ways that the preceding discussion has described. Even his practical advice to painters omits the details of how to fabricate actual paintings. In fact, the *lack* of technical advice – procedures which would vary from one workshop to another – must be partly responsible for the widespread reception and utilization of his treatise on painting which circulated in manuscript for 80 years before the first of many printed editions appeared in 1651.[27]

Bearing these caveats about relating Leonardo's words and images in mind, great advances in non-invasive imaging technology and recent scientific analyses of the material evidence suggest new ways to interpret the manuscript evidence which, along with drawings and paintings by Leonardo and his associates, might enable concrete connections to be made between his recorded observations and experiments, on the one hand, and on the other hand, actual works of art such as portraits and commissioned paintings produced in Milan by Leonardo and the artists who worked for, with or directly after him in the 1480s and 1490s. Quite aside from matters of dating and attribution, what might be reconstructed are the types of knowledge that percolated in the workshop. Beyond manual skills and technical procedures, artisanal epistemology can embrace many kinds of knowledge transmitted orally as well as through texts like those documented in Leonardo's autograph manuscripts which might have been used in the fabrication process. Leonardo's treatise on painting itself became such a text for practicing artists. The chief surviving autograph source of Leonardo's practical workshop advice to students is 'Ms. A', which can be firmly assigned to the years 1490–92.[28] Ms. A records *in nuce* most of the subjects that Leonardo explored throughout the three decades of his literary activity. It is also worth noting at this juncture that these early notes on painting are closely related to Alberti's *della Pittura*, as has long been recognized.[29]

In-between prescriptions for figurative decorum modelled after Alberti's concerns, Leonardo recorded observations on light and shadow along with practical advice to the painter that suggest he had specific problems of representation in mind. Such are the passages devoted to ways of lighting the *corpo ombroso*, a term Leonardo derived from formal treatises on optics meaning 'bodies capable of receiving shadow'. For example, on folio 91 recto, accompanied by a diagram, Leonardo developed an extensive set of considerations regarding the different ways in which bodies receive light and shadow if they are seen through a window in a dark dwelling and arranged at different positions equidistant from a single external light source.

In this and numerous other passages, Leonardo explored the multiple reverberations of reflected lights and shadows on *corpi ombrosi* further and recorded how bodies fill the surrounding air with their 'infinite similitudes' which carry with them 'the quality of body, colour and shape, from their source' (f. 93r). Gradations of shadow on bodies are examined on the adjacent pages, 93 verso, 94 recto and 94 verso, where two short passages on painting

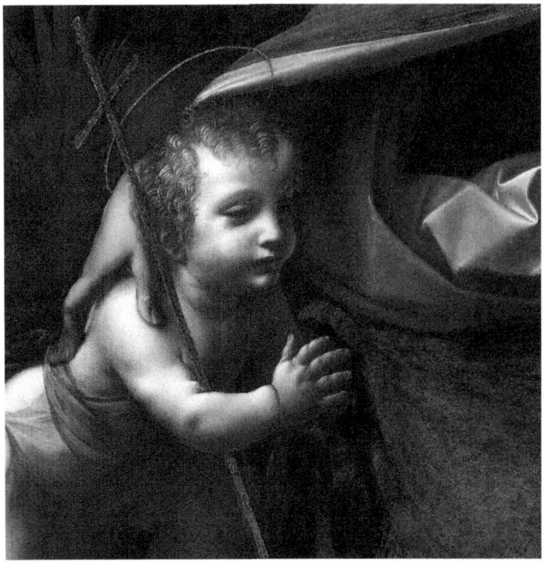

6.1 Leonardo da Vinci, detail of infant John the Baptist, *Virgin of the Rocks* (1490–1508), oil on wood. National Gallery, London

develop the implications of these lighting conditions. The first says to consider 'with the greatest diligence' the boundaries of bodies, the way they twist (*serpeggiare*), and this twisting is discerned (*giudicato*) if 'the faces take on a circular curvature or concave angles'. The second passage says that shadows which are discerned with difficulty do not have boundaries that can be recognized, so they are understood with 'confused judgment', and this too should be transferred into the work, because if the painter makes the edges finite or bounded, the work will be 'wooden'.

What is so striking about the situations that Leonardo investigated is their straightforward resemblance to situations depicted in the London *Virgin of the Rocks* (Colour plate 9), where the figures of the Baptist, the Virgin, the Christ Child and the Angel are separated from one another and seen as if through an Albertian window emerging from shadow, with the figures merging with shadow in the extremities of the picture space, which is depicted as lit from a single central source in the viewer's space. The varied and complex pictorial resolutions to individual instances of the figures' curved and concave surfaces as they emerge from shadow are virtuoso demonstrations of the artist's understanding of the laws governing phenomenal appearances that he investigated in Ms. A. For example, where the Baptist turns in three-quarter view towards the Virgin (Figure 6.1), the far edge of his left cheek is rendered as a highlighted contour, which is justified by the action of light as it reflects off the satin-sheened, gold-coloured lining of the Virgin's robe. A soft, not sharply defined ridge of highlighting extends vertically as the gold lining curves around her body and gradually becomes lost in the shadow. The play of reverberating light is differentiated from the sharper modelling of light and shadow on the drapery folds of the same gold lining that is facing forward to receive light from a source in the viewer's space.

A second area with a very complex play of light/dark juxtapositions occurs where the Christ Child's right arm, raised in a gesture of blessing, recedes behind Christ's neck and head (Figure 6.2). The curving planes overlapping in recession create a pattern of light/dark oppositions, but in this case the source of the lit fleshy part of the Child's upper arm is not so effectively justified on optical principles. What accounts for the highlight in this section of the painting? The pictorial solutions to problems of representation follow patterns established in the first version of the painting, but in the second version the head is not tilted so the chin obscures the upper arm. At the same time a more difficult problem of representation is avoided – namely, that of accounting on

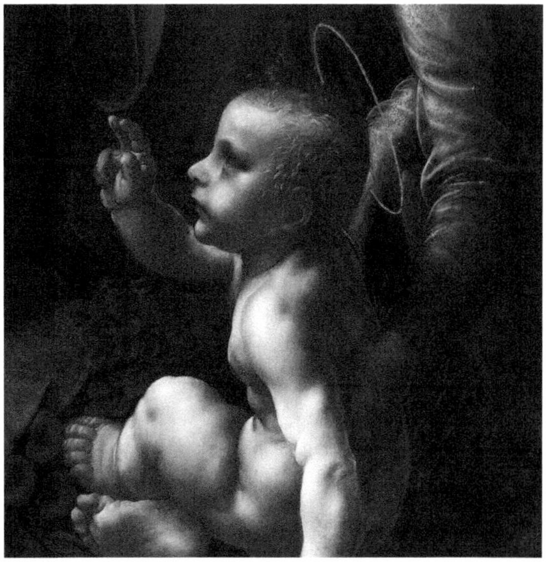

6.2 Leonardo da Vinci, detail of Christ Child, *Virgin of the Rocks* (1490–1508), oil on wood. National Gallery, London

optical grounds for both the tilting surface receding along a diagonal *and* the direction of light falling perpendicularly from the front, outside the window of the picture. This is the kind of situation Leonardo described in Ms. A, in the cited passage on 94 verso that begins: consider 'with the greatest diligence' the boundaries of bodies, the way their twisting is discerned if 'the faces take on a circular curvature or concave angles'.

The reason for spending so much time describing these details is to suggest how a collaborative effort entailing the problems of representation that Leonardo articulated in his discussion of the *corpo ombroso* in Ms. A might have proceeded, without parsing out individual figures to individual painters. The most challenging problems of representation, as Leonardo described in graphic and verbal detail in Ms. A, involve the successive reverberations of light and the gradual merging of curved surface into shadow so that 'shadows which are discerned with difficulty do not have boundaries that can be recognised'. This visual and textual evidence is supported by the new information about the underdrawing and superimposed paint layers. Conservators Larry Keith and Jill Dunkerton describe how Leonardo achieved infinitely subtle gradations of modelled flesh by extraordinarily simple means. It has been supposed that the creation of subtle *sfumato* transitions in the final blending of light and shade was accomplished with many superimposed, fine translucent layers. Yet the London *Virgin of the Rocks* does not bear this out: it consists of just two basic layers, opaque leaded paint, comprised largely of white pigment with small amounts of colour, laid over the monochromatic understructure of the composition. The dark tones were created in the first tonal underpainting.[30] The same structure is clearly visible in the unfinished *Musician* panel (Colour plate 10), where the edge of shadow cast by the nose is created within the undermodelling, a technique that ultimately derives from Verrocchio's tempera painting procedure in the late 1460s of shading with washes in the underdrawing.[31]

We can also examine the kinds of drawings and paintings that Leonardo's students and collaborators produced. There are examples by Boltraffio and others which show that Leonardo's techniques for establishing *rilievo* through drawings of individual elements such as faces and draperies were mastered and used by other artists who employed techniques favoured by Leonardo, such as metalpoint on tinted paper with white heightening. Their methods for assembling compositions from elemental cartoons also followed from Leonardo's practice in the 1490s, as is the case with both versions of the

Virgin of the Rocks. Among the most important recent discoveries is the reuse of drawings that Leonardo shared with his workshop, which then appear in independent compositions: notably, the modelled head (pricked to use as a cartoon) of the young St John the Baptist in the Paris *Virgin of the Rocks* was reused by Boltraffio for the Christ Child in a Virgin and Child panel, now in the Poldi Pezzoli Museum in Milan.[32] Francesco Napoletano used a drawing similar to Leonardo's surviving Florentine pen-and-ink sketches of Madonna and Child themes of the 1480s, confirming the presence of an archive of Leonardo's Florentine drawings in his Milanese shop.[33] However, none of the surviving evidence suggests that the artists who might have been involved in the *Virgin of the Rocks* commission ever rose to the level of invention required for integrating the figures into new, emotionally charged and unified narratives.

Fast Forward to the Catholic Reformation

Three interrelated issues regarding artisanal epistemology emerge from the case study of Leonardo's *Virgin of the Rocks* in relationship to Ms. A: (1) how Leonardo achieved the strong *rilievo* of his figures, both in terms of painting technology and in terms of the conceptual framework that supported it; (2) how and what he taught his students, a question related to how he might have organized a team of workers; and (3) how the writings compiled on these subjects entered the historical record and became available to later generations of practicing artists who had access to Leonardo's writings in the abbreviated version of his treatise on painting. No matter how these questions are addressed, there can be no doubt that the craft tradition included knowledge of theory, or *scienza*, to use the more common historical term.

In fact, what is most significant about these connections on a broader historical scale is the manner in which Leonardo's theoretically grounded and technologically experimental painting practice, geared to imitate natural appearances with great visual force, conformed to the ontological requirement stressed by ecclesiastical writers that sacred Christian images be reliable and true, as artless or 'perfect' as possible, to recall Savonarola's understanding of poetry's moral efficacy. For the age-old question is, given human fallibility, what guarantees the truth of an artistic representation made by human hands? The short answer – given from St Thomas Aquinas to Ulrich Zwingli to Archbishop Gabriele Paleotti to Federico Zuccaro, first director of the Roman Accademia di San Luca and instrumental in the development of the Florentine Accademia that preceded it – is that the mind of the artist is in contact with the Divine when he receives his exemplum, or works out his *concetto*.[34] The same logic guarantees the truth of a miraculous painting, a contact relic like the Shroud of Turin or any other *acheiropoietos*. A great deal of tension in early modern writings about art revolves around the double-sided issue of the artist's share, his freedom to interpret Scripture versus the limits of his human understanding. Significantly, this is also the substantive issue driving humanist and Scholastic disagreements since the thirteenth century over the value of poetic figured language, disagreements that were translated directly to fifteenth- and sixteenth-century discussions of visual artifice by writers – and also by many artists who left no verbal traces in the historical record who nonetheless understood the ontological requirements of Christian sacred images as their workshop practices attest.

To return to the core mystery, how can the Protestant Reformation and Catholic Reformation alike sustain the same general trend towards naturalism? It is crucial to paint the historical picture regarding such profound questions with a broad brush – one capacious enough to include Lombard and Emilian as well as Florentine, Venetian and Roman naturalism before and after Correggio, the Carracci, Caravaggio and others built on the established foundation of scientific naturalism that Leonardo's combined visual and verbal legacy documents in midstream. It is also crucial to include within the same compass the parallel artistic phenomenon in northern Europe – such as German, English, Dutch and Flemish naturalism before and after the heated iconoclastic events of the Protestant Reformation. The *differences*, which are many, pertain above all to matters of individual and collective style, or to use the period term for style, matters of *elocutio* that elicit discussion, debate and judgement about the value of this or that demonstration of pictorial artifice.

When ecclesiastical writers such as Johannus Molanus, Gabriele Paleotti and Carlo Borromeo first began to flesh out the Catholic-Reformation position on the kinds of artistic images suitable for use in religious devotion, they supported a return to naturalistic observation. It is not entirely coincidental that the first official art academy, the Florentine Accademia del Disegno, was established in 1563, the same year that the Council of Trent issued its Decree on Images at its final session. Nor is it simply a fluke that the Roman Accademia di San Luca, established by artists in 1577, was authorized by Pope Sixtus V in 1588 and finally realized in 1593 under the protection of Pope Clement VIII, two pontiffs active in establishing Catholic-Reformation policies governing intellectual activity.[35] And it should not be surprising that these times also supported a revival of interest in Leonardo's scientific approach to painting.[36]

Many other case studies could be cited to support the claim that Counter-Reformation writers sustained a move in the visual arts towards a 'scientific' approach to representation that defined the value of pictorial artifice in terms of its fidelity to natural appearances. Paradoxically, the pressure on practicing artists to acquire knowledge of geometry, optics, anatomy, human movement, impetus mechanics, natural philosophy and other subjects requiring theoretical understanding imperative to the Catholic Reformation's redirection of the visual arts, was in fact initially formed within craft traditions. We do not usually think of Western styles of 'naturalism' as ideologically freighted by the ontological concerns of Christianity, but it was precisely on this basis that academic instruction developed the teaching methods that came to be in common use throughout Europe and beyond since the seventeenth century.

The legibility of scientific naturalism and the teachability of artifice rendered on these terms made it an ideal vehicle to express many different forms of institutional authority.

Acknowledgements

My thanks to Jacob Wamberg, Helen Hills, Angela Cerasuolo and Donald Preziosi for reading earlier drafts of this essay. I am also grateful to Luke Syson, Jill Dunkerton, Larry Keith and Elisabeth Walmsley for their critique and encouragement when I presented the arguments about Leonardo's workshop practices at a National Gallery research seminar (London, December 12, 2011) and a conference on *Leonardo's Legacy* organized by Francesca Fiorani at the University of Virginia (Charlottesville, April 12–14, 2012). Their perceptive comments have improved the final results, although I remain solely responsible for the shortcomings.

Notes

1 Carlos M. N. Eire, *War against the Idols: The Reformation of Worship from Erasmus to Calvin* (Cambridge: Cambridge University Press, 1986), p. 84, citing *Huldreich Zwinglis Sämtliche Werke*, vol. 4, p. 133. The clearest, most detailed exposition appears in Zwingli's *Commentary on True and False Religion*, 1525 (Eire, *War against the Idols*, p. 85). Calvin also located the root of idolatry in man himself – in his efforts to domesticate God, man transferred the honour due to God to material objects. Calvin's most popular published work against idolatry, *The Inventory of Relics* (1543), compares idolators to unreasoning beasts and considers idolatry an ever present danger because men are by nature metaphysically deranged. See further discussion in Eire, *War against the Idols*, pp. 211–20.

2 On other publication projects of the time dealing with the arts by members of the Florentine Accademia, see Zygmunt Wazbinski, *Accademia medicea del disegno a Firenze nel Cinquecento: idea e istituzione*, 2 vols, (Florence: L.S. Olschki, 1987), vol. 1, pp. 229–38.

3 Alberti's *De pictura*, published in Latin in 1540 (Basel: Westheimer), newly translated into Italian by Lodovico Domenichi and published in Venice in 1547 (Venice: Gabriel Giolito de Ferrari); and translated and published again by Cosimo Bartoli in 1568 in Alberti, *Opuscoli morali* (Venice: Franceschi, 1568).

4 The term (but not the idea of scale) derives from Pierre Bourdieu, *The Field of Cultural Production: Essays on Art and Literature* (New York: Colombia University Press, 1993).

5 See Pamela Smith, 'Science on the Move: Recent Trends in the History of Early Modern Science', *Renaissance Quarterly* 62 (2009): pp. 345–75.

6 Early modern definition proposed by Smith, 'Science on the Move', p. 346.

7 John Weisheipl, 'Classification of the Sciences in Medieval Thought', *Medieval Studies* 27 (1965): pp. 54–90, citing p. 55.

8 George P. Klubertanz, *The Discursive Power: Sources and Doctrine of the Vis Cogitativa according to St Thomas Aquinas* (St Louis, MO: Modern Schoolman, 1952), p. 206ff. See also discussion in David Summers, *The Judgment of Sense: Renaissance Naturalism and the Rise of Aesthetics* (Cambridge: Cambridge University Press, 1987), pp. 211–27.

9 On what constitutes a science in the modern sense, contrasting biology with physics, chemistry and other physical sciences, see the historiographical study by Ernst Mayr, 'The Autonomy of Biology: The Position of Biology among the Sciences', *Quarterly Review of Biology* 71/1 (March 1996): pp. 97–106.

10 Thomas Aquinas defined *exemplum, a particular representation*, as a form of induction based on Aristotle, *Prior Analytics*, 68b38–69a19, writing that poetry induces virtue by delighting the soul through a concrete image (Thomas Aquinas, *The Division and Methods of the Sciences*, 4th ed., ed. Michel Doyne, trans. rev. W.H. Kane [Montréal, 1956]).

11 Jerrold E. Seigel, *Rhetoric and Philosophy in Renaissance Humanism: The Union of Eloquence and Wisdom from Petrarch to Valla* (Princeton: Princeton University Press, 1968), p. 63ff.; Thomas Aquinas, *Commentary on the Metaphysics of Aristotle*, trans. John P. Rowan, 2 vols (Chicago: Henry Regnery, 1961), I.L.3: C32–68.

12 On Savonarola and Thomas Aquinas, see Concetta Carestia Greenfield, *Humanist and Scholastic Poetics, 1250–1500* (Lewisburg: Bucknell University Press; Toronto: Associated University Presses, 1981), pp. 249–51.

13 Lorenzo Valla, *Dialectica libri tres*, c. 1440, proem and chapter 2. Alberti's *della Pittura* has been cited as participating in the same trend of devising more powerful methods of rhetorical argumentation, on which see Hanna H. Gray, 'Renaissance Humanism: The Pursuit of Eloquence', *Journal of the History of Ideas* 24 (1967): pp. 497–514, citing p. 501. Many writers followed Valla: see David Marsh, *Quattrocento Dialogue: Classical Tradition and Humanist Innovation* (Cambridge, MA: Harvard University Press, 1980).

14 See Charles Trinkhaus, *In Our Image and Likeness: Humanity and Divinity in Italian Humanist Thought*, 2 vols (Chicago: University of Chicago Press, 1970), vol. 2, pp. 683–721, with an overview of later fifteenth-century humanist issues and the *theologica poetica*.

15 See Cesare Vasoli, *La dialettica e la retorica dell'Umanesimo: 'Invenzione' e 'Metodo' nella cultura del XV e XVI secolo* (Milan: Feltrinelli, 1968).

16 Greenfield, *Humanist and Scholastic Poetics*, p. 251. See n. 7 above.

17 Poliziano classified painting as an art belonging to the mathematics of architecture, a mechanical art related to the physical sciences following from arithmetic (Angelo Poliziano, *Panepistemon*, in *Omnia Opera* [Venice, 1498], ed. I. Maier, 2 vols [Turin: Bottega d'Érasmo, 1971], p. 470). On Nicomachus, see Martin D'Ooge, introduction, in Nicomachus, *Introduction to Arithmetic*, trans. M. D'Ooge (New York: Macmillan, 1926), pp. 1 and 18, nn. 1 and 2.

18 John Hermann Randall, 'The Development of Scientific Method in the School of Padua', in *Renaissance Essays from the Journal of the History of Ideas*, ed. P.O. Kristeller and P.D. Weiner (New York: Harper and Row, 1968), pp. 217–51.

19 Albertus of Saxony, *Tractatus de proportionibus*, published with Walter Burleigh, *De intentione et remissione formarum* (Venice, 1496), is probably the publication listed on Leonardo's booklist in Madrid Codex II, c. 1503–1505, published in Ladislao Reti, 'The Library of Leonardo', in Leonardo da Vinci, *The Madrid*

Codices, 5 vols (New York: McGraw-Hill, 1974), vol. 3, pp. 91–108, citing p. 93, n. 10. Jill Pederson, 'Henrico Boscano's *Isola beata*: New Evidence for Academia Leonardi Vinci in Renaissance Milan', *Renaissance Studies* 22/4 (2008): pp. 450–75, publishes new evidence which supports long-standing speculation about an informal academy comprised of personalities associated with the Sforza court.

20 Joan Gagné, 'Du *Quadrivium* aux *scientiae mediae*', in *Arts Libéraux et Philosophie au Moyen Age* (Montréal: Bellarmin; Paris: Vrin, 1969), pp. 975–86.

21 Martin Kemp, 'Leonardo and the Visual Pyramid', *Journal of the Warburg and Courtauld Institutes* 40 (1977): pp. 129–49, suggests Cardano. The textual source for Leonardo's knowledge of double procedure, a translation of Pecham's definition of optics, is CA 203 r–a, c. 1490. See further Claire J. Farago, *Leonardo da Vinci's 'Paragone': A Critical Interpretation with a New Edition of the Text in the Codex Urbinas* (Leiden: E.J. Brill, 1992), p. 83ff.

22 CA 222a, c. 1492; see Martin Kemp, '"Il Concetto dell'anima" in Leonardo's Early Skull Studies', *Journal of the Warburg and Courtauld Institutes* 34 (1971): pp. 115–34.

23 See further, Farago, *Leonardo da Vinci's 'Paragone'*.

24 Farago, *Leonardo da Vinci's 'Paragone'*, pp. 86–87, on similar arguments advanced by Jacopo da' Barbari, Francesco di Giorgio Martini and others acquainted with Leonardo and Pacioli in Milan.

25 See Luke Syson with Larry Keith, *Leonardo da Vinci: Painter at the Court of Milan*, exh. cat. (London: National Gallery, 2011), for the most recent review of the documents and dating issues; further, Charles Hope in his review of the exhibition, *New York Review of Books*, February 9, 2012, amends one important date in the documents to 1489.

26 See n. 24; and further, *National Gallery Technical Bulletin: Leonardo da Vinci: Pupil, Painter, and Master* 32 (London, 2011).

27 See *Re-Reading Leonardo: The Treatise on Painting across Europe 1550 to 1900*, ed. Claire Farago (Farnham and Burlington: Ashgate, 2009). My thanks to Luke Syson, Larry Keith, Jill Dunkerton, Elizabeth Walmsley and Angela Cerasuolo for their comments on the possible relationship between Leonardo's written considerations and workshop practices discussed below. I remain solely responsible for the views expressed here.

28 Ms. A is dated on the basis of a memorandum dated July 10, 1492, on the verso of the final sheet, and on the basis of its close connections with other notes such as those in Ms. C, dated April 23, 1490, on folio 15 verso.

29 See Farago, *Leonardo da Vinci's 'Paragone'*, pp. 340–46, and passim.

30 Larry Keith, 'Leonardo's Painting Technique', in *Leonardo da Vinci: Painter at the Court of Milan*.

31 Jill Dunkerton, 'Leonardo in Verrocchio's Workshop: Re-examining the Technical Evidence', *National Gallery Technical Bulletin* 32 (London, 2011), suggests that Leonardo's preference for shading with washes rather than hatching derives from Verrocchio's egg tempera painting procedure in the late 1460s, which indicates his awareness of the innovative oil techniques of the Pollaiuolo brothers.

32 Syson, in *Leonardo da Vinci: Painter at the Court of Milan*, cat. nn. 41 and 63.

33 Syson, in *Leonardo da Vinci: Painter at the Court of Milan*, cat. n. 48.

34 See further, Donald Preziosi and Claire Farago, *Art Is Not What You Think It Is* (Oxford: Wiley-Blackwell, 2012).

35 See Paul Grendler, *The Roman Inquisition and the Venetian Press, 1540–1605* (Princeton: Princeton University Press, 1977), pp. 260 –61.

36 There is no space to develop this line of argumentation here, but I will pursue it in a future publication, *The Critical Edition of Leonardo da Vinci's Trattato della Pittura, 1651*, forthcoming from E.J. Brill Press, c. 2016. Abridged copies of Leonardo da Vinci's treatise on painting were circulating by 1580: see *Re-Reading Leonardo*, ed. Claire Farago.

7

Printing Plants: The Technology of Nature Printing in Eighteenth-Century Spain

Alisa Luxenberg

The Question

During the age of enlightenment and exploration, new demands were made on art's representations of nature, often in the service of science, and technology increasingly played a role in these concerns. To explore the shifting relationships between art, nature and technology, this essay examines a little-known printmaking process in Spain intended for botanical illustration. First demonstrated in 1777 by the Spanish sculptor Celedonio Nicolás de Arce y Cacho (1739–95), his process was said to 'print from nature all kinds of botanical leaves without the use of plates'.[1] It was a remarkable claim, since, at that time, most printing processes required plates, and in turn, the manual skills to design them. Impressed by the apparently automatic nature of Arce's process and unaware that nature prints had been made in Europe for centuries, the Spanish crown first awarded him an honorarium for his invention, then additional support for its perfection.[2] Despite providing a technological solution to making botanical imagery, Arce's process was never implemented in the way that he had envisioned. This essay will argue that it ran up against aesthetic conventions, disciplinary boundaries and professional rivalries that vexed the status of his prints as nature/science, art or both.

The Objects

Most of Arce's prints (Figure 7.1) are flat, monochromatic images, large as life, ungracefully arranged on the page, and simply labelled. In their crisp articulation of the plants' contours, intensely graphic patterns of the veins and unadorned presentation, these prints diverged sharply from conventional botanical illustrations. As one contemporaneous Spanish printmaker

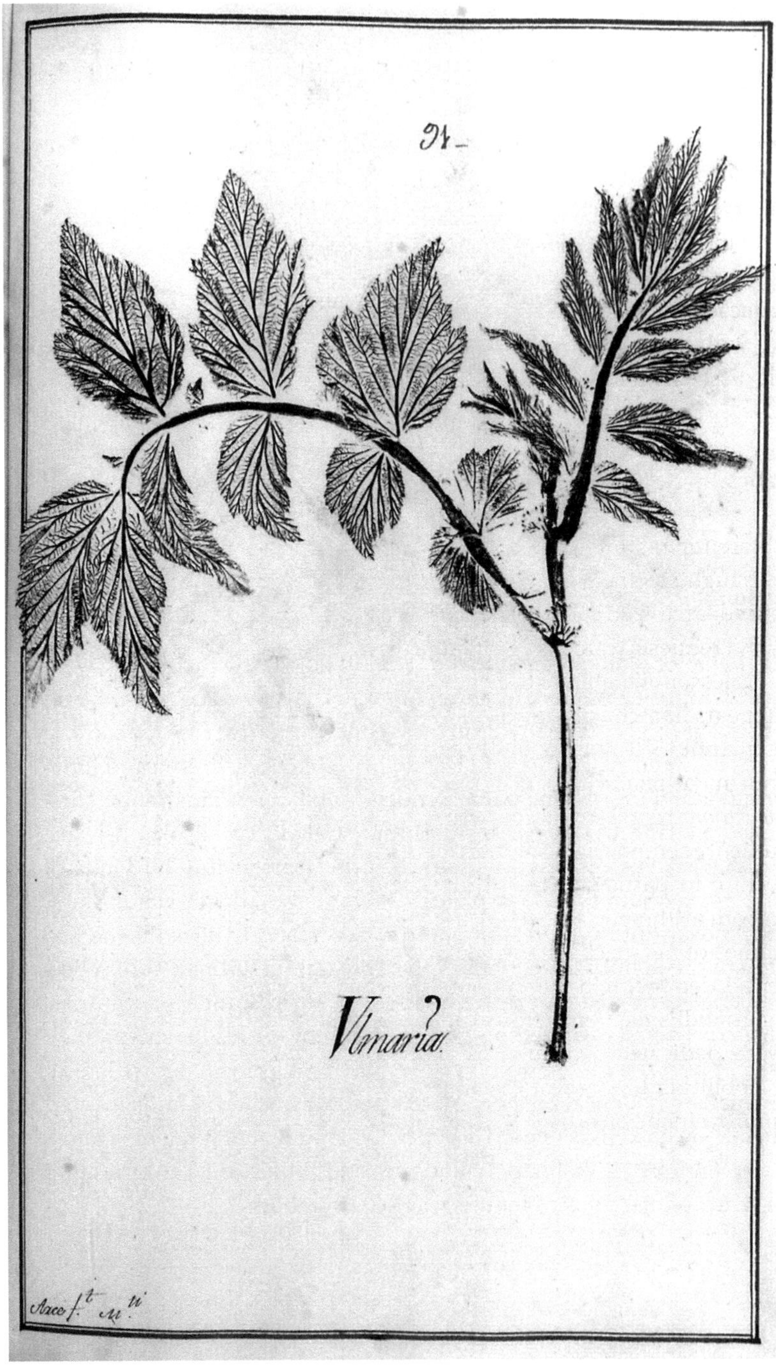

7.1 Celedonio de Arce, *Ulmaría* (c. 1777), nature print, from Arce, *Plantas Nat.*, pl. 91. Palacio Real, Biblioteca Real, Madrid. © Patrimonio Nacional/Photo credit: Patrimonio Nacional

described nature prints, '[they] distinguish themselves from drawings that imitate something, [by] appear[ing] to integrate accuracy into [their] forms'.[3]

Visual Conventions for Botanical Illustration

By Arce's time, conventions for botanical prints had been codified: sharp focus and detail, illusionistic tonal modelling, artful composition, pleasing colour, blank background and identifying inscription. Precisely how the flowering plant emerges from its stem would be elegantly and naturalistically depicted, with its leaves gracefully folded to expose both sides to the viewer. These specimens always appear perfectly intact and consistently idealized, unlike Arce's prints that reproduce broken tips of leaves, wormholes and other damage. Another formal device in botanical illustration was the anatomy, or isolated parts of the flower or fruit, positioned to the side or below the fuller depiction of the plant.[4] The anatomy was nearly impossible to achieve in nature printing, which coated the plant's surfaces with an oily substance and pressed it flat, thereby denying the third dimension and obliterating the more fragile features. Whereas Arce's images are true to life-size, botanical illustrations were designed to fit comfortably on the page, regardless of their actual dimensions. In sum, traditional botanical illustrations provided a generalized, legible, aesthetic image, while Arce's prints insisted on the particularities and imperfections of nature.

Most botanists needed drawing skills to represent plants in the field or to have them illustrated for publication. Those botanists who did not possess such skills or those who planned a deluxe publication often turned to artists to make their images. Notwithstanding the established conventions in botanical illustration, tensions could arise between what botanists wished these images to convey, and what artists hired by them actually represented. Nature printing offered a way to bridge this content-format gap.

Arce's pictures were exceptional among nature prints, too. From the earliest examples of the sixteenth century, nature prints were typically a leaf printed in a manuscript or in the margin of another illustration; they were not stand-alone pictures or made in large numbers for a coordinated project like Arce's.[5] Many nature prints were made by coating a sheet of paper with oil, then holding it in the smoke of a candle or oil lamp to blacken all over. A leaf was laid on the blackened sheet, a clean sheet was laid on top, and the leaf was burnished through the top sheet by rubbing a smooth stone or tool. Removing the blackened leaf, placing it between two clean sheets and repeating the burnishing resulted in a nature print. Others were fashioned from coating a smooth metal plate with printer's ink, and placing the leaf on that for the initial burnishing.

But some eighteenth-century botanists, like Johann Hieronymus Kniphof in his book *Original Botanical Specimens*,[6] did apply nature printing to a series of illustrations. Like Arce, Kniphof aimed to reproduce plant forms for botanical study, without drawing, and he similarly concentrated on leaves. The German inscribed each print with identifying names and captions, and added colour, both conventions for legibility and naturalism in botanical illustration. However, the appearance of the same plant varies from copy to copy, as some specimens became worn or damaged from multiple printings and were replaced with others; so, too, the application of colour, done manually, differs among examples. Yet, such discrepancies did not seem to undermine his prints' status as accurate images.[7] Nevertheless, the small number of complete sets of his book attest to the difficulties of this method.

The Technology of Nature Printing

During his appeal for government support and recognition, Arce agreed to demonstrate his print process to Casimiro Gómez Ortega, a chair at the Royal Botanical Garden in Madrid, who recorded the recipe and printing method.[8] He described the preparation of an oil-based medium that was spread on the surface of the plant. The plant was then laid on moistened paper, covered with another sheet of paper and pressed; the plant was removed, placed on a third sheet of paper and pressed again. The making of the special oil involved various organic and inorganic ingredients – most of which were not included in other nature printing instructions – including some from the artist's studio (linseed oil, oak galls, pigment) and kitchen (garlic, onions and bread crumbs!). All were heated together, producing sooty air and a foul aroma.[9] Neither the preparation of the oil nor the pulling of the prints called for any drawing or engraving skills.

In a slightly earlier report on other nature prints, Gómez Ortega identified the advantages of nature printing: its speed and precision in recording the form and vein structure of leaves, its elimination of drawing skills, the availability of the necessary materials and its convenience to field work.[10] The Spanish nature printer Andrés Muela attested that he could make dozens of prints in half a day, while drawing them would take three to four months.[11] To Gómez Ortega, who apparently lacked drawing ability, it was a breakthrough, 'For those who do not know how to draw, this process is even more recommendable, as I consider it to be a lower and imperfect level of the Art of drawing'.

What else could this have signified if not technological progress? The term 'technology' did not exist in Arce's time, but our dictionaries define it as the application of scientific knowledge to practical purposes, usually through industrial tools and processes, to achieve progress. In nature printing, these Spaniards recognized two consequences that are always equated with technology and progress: savings of time and labour.

Nature printing held great promise for botany, according to Gómez Ortega, and Arce pitched his invention in 1777 to the Spanish Minister of the Indies José de Gálvez as an efficient and effective method for recording plant specimens.[12] At the time, botany provided critical material and information to important disciplines like pharmacology, medicine, agriculture and horticulture, and to all the trade that was generated by these fields.[13] In that era of imperial expansion, voyages of discovery, development of royal botanical gardens and increased publishing, botanical illustration flourished as never before, and any promising new invention related to it would certainly have attracted official and financial interest. It was no coincidence that just a few months before Arce wrote to Gálvez, the Spanish crown announced a scientific expedition to its viceroyalty of Peru for the primary purpose of gathering botanical specimens and knowledge. Gálvez was to oversee the mission, and Gómez Ortega was named its scientific advisor. Like the native people there, the Peruvian flora was to be categorized, controlled and exploited by and for the Spanish kingdom.

After Gómez Ortega recommended Arce's invention to Gálvez, the artist received a royal honorarium and further funds to continue making prints.[14] With the collaboration of the botanist Antonio Paláu i Verdera (1734–93) and painter-scribe Manuel Muñoz Matarranz, Arce produced *Libro de Plantas* (*Book of Plants*), six bound volumes, each containing approximately 65 to 100 nature prints,[15] dedicated to the prince of Asturias (the future Charles IV), who had a keen interest in natural history. Some prints, especially at the beginnings of the first five volumes (Figure 7.2), accommodate conventions of botanical illustration, displaying more graceful and centred arrangements, anatomies, multiple colours and calligraphic inscriptions, including their 'virtues', or the effects that the plant could have on human health and behaviour. As one progresses through the volumes, however, such additions disappear, and the plain, undecorated prints that are the direct result of Arce's process dominate the later pages of these volumes and the other four tomes. Perhaps Arce intended eventually to elaborate more of these prints, but by making and binding so many into books before they were reworked, he showed an appreciation for their startlingly simple, graphic patterns. Others did, too, as attested by numerous unsigned nature prints that survive in the Real Jardín Botánico archives, and by a 1784 periodical.[16]

The Rise of the Artisan-Technician

To understand what would have motivated a sculptor like Arce to invent a botanical printing process, it is useful to consider Pamela H. Smith's study of the rise of naturalism in the visual arts in early modern Europe. She contends that nature had become *the* site of knowledge, challenging or replacing texts as the potential source of data against which to test scientific theories, and artisans were thus motivated to represent nature in a convincing manner.[17] Smith calls these artisans 'technicians … [who] saw natural knowledge as an arena in which they could gain new authority and legitimacy'.[18] But she finds strong resistance to this new scientific philosophy and its representations even

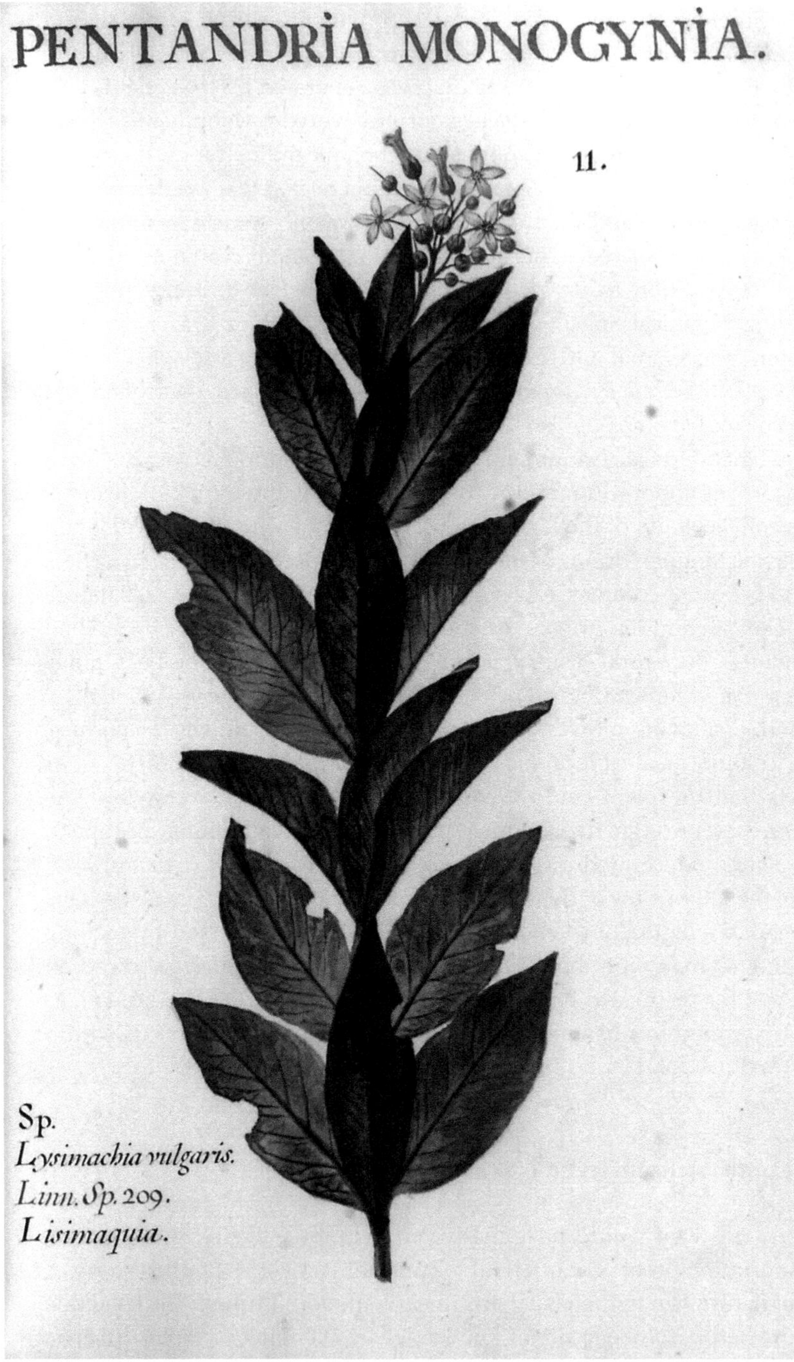

7.2 Celedonio de Arce, *Pentandria Monogynia* (*Lysimachia vulgaris*) (1781), nature print, from Arce et al., *Libro Quinto de Plantas*, pl. 11. Palacio Real, Biblioteca Real, Madrid. © Patrimonio Nacional/Photo credit: Patrimonio Nacional into the eighteenth century among Europeans who equated the human body with subjectivity. For those thinkers and patrons, the technician's labour, enacted through the body, could not provide reliable knowledge.

Shifting Standards of Scientific Objectivity

A third way that could reconcile those two divergent approaches to scientific understanding and illustration was to replace human labour with technology. The scholars Lorraine Daston and Peter Galison trace a paradigm shift in the concept and standards of scientific objectivity that occurred around the mid-nineteenth century. In earlier botanical and other scientific atlases, their typically idealized illustrations and standardized format were read as highly objective due to their naturalism, legibility and consistent viewpoint, even though all this evidenced much human intervention. What changed by the 1850s, according to Daston and Galison, was a prevailing notion of objective scientific illustration that aimed to avoid errors or distortions resulting from the essentially interpretative practice of drawing, through processes that reduced human intervention and offered more particularized representations of nature, warts and all.[19] As a solution, scientists and publishers increasingly turned to mechanical devices and processes like photography and photolithography.

The technology of photography figured significantly in meeting the new standards of objectivity. The discourse that developed from the 1830s onwards to characterize photography and explain its special qualities was predicated on its technology. Inventors and critics of photography regularly referred to the chemical reaction of its materials to the action of light as 'spontaneous reproduction' and 'nature drawing itself'. In addition, the photographic apparatus of the camera, fitted with a calibrated lens, was a mechanical device believed to obey predictable laws of optics and geometry.[20] Nature, of course, was essential to the new medium, both in its unceasing role as subject and in the action of light and chemical reaction that created the image. As nature and technology were complicit in making these pictures, the resulting prints were invested with the authenticity of nature and the unmediated truth of technological reproduction.[21]

But many early photographs were made without a camera and its lens. Photograms like those made by the English inventor of photography William Henry Fox Talbot look remarkably akin to Arce's unadorned prints in their artless, isolated presentation of botanical forms in flattened patterns and monochromy. Both representational methods insisted on the embodiment of nature, as the specimen itself was laid on the paper and helped enact the process, either through the transfer of oil or the blocking and filtering of light.[22] As someone keenly interested in botany,[23] Talbot would have known of nature prints, and it seems reasonable to posit Kniphof's and Arce's prints as earlier expressions of a shared desire for images of nature that were produced without manual drawing or engraving, through natural, spontaneous forces.[24]

Objectivity and Aesthetics within the Academic Communities of Spain

Tensions over standards of scientific objectivity emerged within the Spanish expedition to Peru. The stated goals of the mission were predominantly scientific: to know and study the natural resources of the land to advance the physical sciences, medicine, painting and other important arts; to expand commerce; to form a herbarium and to enrich the royal cabinet of natural history and the Royal Botanical Garden in Madrid.[25] As scientific advisor, Gómez Ortega helped select the botanists for the expedition (Hipólito Ruiz and José Pavón), and wrote instructions for the artists (José Brunete and Isidro Gálvez),[26] who were hired to make drawings of the flora under the botanists' supervision in Peru. Given to the draughtsmen prior to their departure, these instructions explicitly ordered them to study only fresh plants, in situ, to draw every part, and 'to copy nature exactly without presuming to correct it or decorate it as some draughtsmen are used to doing, adding colours and ornaments drawn from their imagination'.[27] These directions imply that artists were inclined to enhance their representations, and that scientific standards of objectivity had to be imposed on them. While these instructions reflect a desire for greater objectivity in botanical images, they offer little description of a concrete practice for attaining it.

In 1780, Gálvez asked Gómez Ortega to evaluate Arce's prints. The minister's choice of reviewer indicated that he considered Arce's prints to be scientific, for artistic matters were assigned to the Royal Academy of Fine Arts of San Fernando, whose charge it was to review important and public issues of aesthetics, such as church architectural plans. In contrast, Gómez Ortega recommended that the academy's vice-protector be put in charge of Arce's project in order to harmonize its team of artists and botanist to help improve the quality of the prints.[28] It is true that, since the 1760s, the academy had promoted the study of flowers and fruits among its students as beneficial to industry and natural history.[29] Nevertheless, Gómez Ortega's recommendation was not followed, and the reason for that may be found in the academy's response eight years later to another plan for botanical engravings, presented by a former student of Gómez Ortega, who supervised his project. The academy, led by the same vice-protector as in 1780, recused itself from judging the project, explaining that its opinions were limited to the three fine arts, and that only a trained botanist was capable of evaluating the prints' content and function.[30] Despite sharing close study from nature, the scientist and the artist-academicians considered botanical prints from distinct perspectives, the former seeking aesthetic advice, the latter reading only science there. Like photography, nature printing was believed to provide direct evidence of the world, but its reception could vary dramatically, depending on a viewer's understanding of the process or aesthetic tastes.

In his two reports of 1780, Gómez Ortega enumerated errors and weaknesses in Arce's prints, upholding the conventions of botanical illustration:

it is good if [the plant] is complete and ... most beautiful and luxuriant. One should not be satisfied only representing a leaf, as was done ... and take care not to omit the essential parts of the flower and fruit that were left out [... .] Above all, [the authors] must add at the bottom of each plate the anatomy of the flower, following the example seen in the modern collections of this type.[31]

Still, the botany professor saw great potential in Arce's invention, 'I have no doubt that if this work comes to acquire proportionately in all its parts the degrees of perfection of which it is conceivably capable, it will be able to rival and perhaps lead all the works of this kind that have been published by other Nations up to now'.[32] The report's ambivalence reflects the author's own uncertainty about judging Arce's prints, and perhaps, too, his sense of responsibility for having urged the royal purse to support this invention.

Despite claiming in 1777 that Arce's invention was the first of its kind, Gómez Ortega reversed his opinion in 1780. In fact, just a month before pronouncing on Arce's method in 1777, the botany chair made the same claim for the nature prints of another artist, Andrés Muela of Madrid.[33] Yet, Gómez Ortega chose to promote Arce rather than Muela.[34] In his comments on Muela's prints, Gómez Ortega sounded disappointed with their aesthetics, 'the trace or portrait that is printed ... looks like a uniform stain of the plant's flat parts, ... for it lacks the light and shade that are needed to represent the dimensions of objects ... [as well as the] context [of] their natural situation'. While Arce's prints had similar deficiencies, Gómez Ortega distinguished them as superior when he wrote, 'today in Paris, Turin, Holland, Germany, Rome, and London, works of this kind are being made, but neither in this genre, nor in such a singular and exquisite method'.[35]

In a reactive letter to Gálvez, Arce complained that Gómez Ortega's report had merely taken Arce's own critical assessment of his prints in order to denigrate his achievements. As evidence, the artist recited his comments so that they could be compared to those of Gómez Ortega, and pointed out that this primarily formal and aesthetic criticism was beyond the ken of the botany chair, who had no artistic formation or drawing skills.[36] Furthermore, Arce's letter implied that Gómez Ortega, by not focusing his report on the prints' botanical content, shirked his professional duty, or worse, was incapable of executing it. Indeed, Gómez Ortega never trained as a botanist, and scholars harbour doubts about his qualifications for his first career in pharmacy.[37]

It is evident that Gómez Ortega and Arce had a falling out, which reached an impasse in 1780. Arce had already felt betrayed by his so-called protector, who never lobbied for his process to be used on the Peru expedition, and then, adding insult to injury, gave younger artists access to his invention to make prints, without Arce's permission.[38] Arce attested that Gómez Ortega had withdrawn from the project, probably by 1779, when the artist sought Paláu's advice on his nature prints.[39] The widely acknowledged hostility between these two botany chairs has been attributed by one scholar, Puerto Sarmiento, to the creation in 1780 of a new position on the Spanish medical licensing board, which both men sought.[40] In fact, Gómez Ortega mentioned the new position as the cause of friction between him, Paláu and Arce,[41] and clearly, Arce became a pawn in the chairs' rivalry. When the position went to Gómez Ortega in September of that year, it consolidated his power over botany and pharmacy in Spain.

Arce's Motivations

Like the artisans of Pamela H. Smith's argument, Arce recognized that by leaving his principal medium of sculpture and engaging with nature and science, he might elevate his status. Working on small-scale sculpture and in non-noble materials, Arce had little chance for advancement at court, the academy or the market. That did not stop his ambitious petitions to be made a member (1772) and professor (1782) of the Royal Academy of San Fernando, the most prestigious body of artists in Spain.[42] Undeterred by those rejections, Arce tried to establish his intellectual credentials in other ways; he was one of the few Spanish artists (and fewer sculptors) to write a book on art.[43] Also to that end, he became a technician, inventing a printing process for government-sponsored botanical expeditions, linking his work to science and public utility. Not only did his prints focus on nature, they preserved botanical specimens as images for study, a much-desired alternative to transporting unfamiliar plants on long sea journeys and growing them in Spain's different soil and climatic conditions. Perhaps to perform the rise in standing he desired, Arce even named a plant after himself, an honour usually reserved for royalty, high-ranking officials and leading botanists.[44]

The Failure of Arce's Project

Despite continued funding, Arce's process was never employed on the expedition, which ended in 1788. Clearly, the rough living and travelling conditions in Peru would have complicated this relatively straightforward printing process, but more consequential to its neglect was the absence of a protector bending a minister's ear. By 1789, Arce seems to have abandoned nature printing; curiously, around the same time, Gómez Ortega lost his clout. In 1790, when the official publication on Peruvian and Chilean flora opened an office in Madrid, it neither located in the Royal Botanical Garden nor implemented Arce's printing method. More than 500 drawings were selected from more than 2,000 supplied by the expedition's draughtsmen, and 51 different engravers created prints from them. The concept of scientific accuracy that then reigned allowed for multiple translators and translations between the subjects, drawings and prints. Although this notion would change by the mid-nineteenth century, it was not soon enough to spare Arce's prints from obscurity.

The brief flourishing of nature printing in Spain suggests that, despite its ease of execution, striking forms and timeliness for botanical expeditions, the existing concepts and practices for scientific accuracy still satisfied. The disappearance of Arce's process resulted from a nexus of factors: its unusually graphic form, animosity between Gómez Ortega and Paláu and the material difficulties encountered in Peru. Arce's prints seemed to allow nature to speak directly, with the authority of autogenesis, auguring later notions of naturalism and objectivity in scientific illustration by presenting imperfections, damage and varying compositions, dependent on the plant's size and mass. Hovering between science and art, Arce's nature printing participated in a growing desire to illustrate the natural world without drawing and through increasingly mechanical means, foreshadowing photography.

Acknowledgements

Research and travel for this essay were made possible by grants from the University of Georgia Research Foundation and the Program for Cultural Cooperation between Spain's Ministry of Culture and U.S. Universities. The author also thanks the staff of the archive of the Real Jardín Botánico in Madrid, and Arline Meyer for helpful improvements to the original paper, which are retained here.

Notes

1 José de Gálvez, letter to Miguel de Muzquiz, September 18, 1777, San Ildefonso; Archivo de Simancas [AS], Hacienda, Legajo 61. The sole publication on Arce's prints is Santiago Castroviejo and Pilar Coello, 'Sobre la "estampación natural" y noticias de los "libros de plants ..." de Antonio Palau, Celedonio de Arce y Manuel Muñoz y Matarranz', *Anales del Jardín Botánico de Madrid* 54/1 (July 1996): pp. 595–607.

2 See the royal payments from 1780–88, Archivo General del Palacio Real, Madrid [AGPR], Carlos IV, Principe, Leg. 47/1, Expediente 3.

3 Andrés Muela [a.k.a. de la Muela], letter, May 9, 1777, Madrid; Archivo del Museo Nacional de Ciencias Naturales, Madrid [AMNCN], Expediciones, Exp. 12.

4 For examples, see Wilfrid Blunt and William T. Stearn, *The Art of Botanical Illustration*, rev. ed. (London: Antique Collectors Club, 1995).

5 See Roderick Cave, *Impressions of Nature: A History of Nature Printing*, rev. ed. (London: British Library and Mark Batty Publishers, 2010), and Isabelle de Conihout, *De Botanica in originali: Livres de botanique réalisés en impression naturelle du XVIe au XIXe siècle* (Paris: Bibliothèque nationale, 1993). Benjamin Franklin had the idea of printing leaves on paper money as an authentication device.

6 The definitive version is Johann Hieronymus Kniphof, *Botanica in originali seu Herbarium vivum ...*, 3 vols (Halle an der Saale: J.G. Trampe, 1757–64).

7 Kniphof's book was known outside of Germany. See Cave, *Impressions*, p. 44.

8 Casimiro Gómez Ortega, report to Gálvez, June 10, 1777, Madrid; AGPR Exp. Personal, Leg. 3879/3.

9 Experiments to recreate Arce's process with Dr John Bordley, professor emeritus, University of the South, produced these results.

10 Casimiro Gómez Ortega, report, May 20, 1777; AMNCN, Expds., Exp. 19: 'In travelling, when there is not always time or convenience to draw some plant with perfect accuracy, this kind of printing could suffice, and it is unthinkably more expeditious than drawing, and in one go, represents a number and the exact contours of different parts of plants … and it will not weigh down the draughtsmen in Peru to resort to this process to help them later finish and perfect their drawings in their lodgings'.

11 Muela, letter, May 9, 1777; AMNCN, Expds., Exp. 12.

12 Arce, letter to Gálvez, June 6, 1777, Aranjuez; AGPR, Leg. 3879/3: 'Having heard that Your majesty was sending Botanists to America to search for Plants and make prints of them, [I], hav[e] a profitable Secret for this, … so that they can achieve a greater perfection in their prints, as exemplified by the Book, made by [my] hand'.

13 From his initial career in pharmacy, Gómez Ortega steered the Royal Botanical Garden and the expedition to Peru towards medicinal botany. Antonio González Bueno and Raúl Rodríguez Nozal, *Plantas americanas para la España ilustrada: Génesis, desarollo y ocaso del proyecto español de expediciones botánicos* (Madrid: Complutense, 2000), pp. 38–39, 66–68. Arce's nature printing work seemed to respond to and was judged from that outlook.

14 It is possible that Arce modified his process from the one that was recorded in 1777.

15 The volumes are in the Royal Palace Library, Madrid. By 1782, the prince possessed all six volumes. Arce, letter to the academy, September 16, 1782, San Ildefonso; Archivo de la Real Academia de Bellas Artes de San Fernando, Madrid [ARABASF], Documento 172–2/5.

16 Some of these prints in the Real Jardín Botánico archive are labelled *negro de humo*, made in the traditional way by heating oil or printer's ink. 'Real Academia de las tres nobles artes', *Memorial literaria* (May 1784): p. 19, announced that Bartolomé Vázquez used Arce's process to make a botanical print. Arce, letter to Gálvez, July 29, 1780, San Ildefonso; Archivo de Indias, Sevilla [AI], Indiferente, Leg. 1544, mentioned that Gómez Ortega offered 24,000 reales to Vázquez to make nature prints.

17 Pamela H. Smith, *The Body of the Artisan: Art and Experience in the Scientific Revolution* (Chicago: University of Chicago Press, 2004), pp. 8–20.

18 Ibid., p. 9.

19 Lorraine Daston and Peter Galison, 'The Image of Objectivity', *Representations* 40 (Autumn 1992): pp. 81–128.

20 For the camera obscura as model of objective vision, see Jonathan Crary, *Techniques of the Observer: On Vision and Modernity in the Nineteenth Century* (Cambridge, MA: MIT Press, 1990), pp. 25–66.

21 Like nature printing, photography required the presence of a human operator to choose the materials, subject and the latter's position within a frame.

22 Carol Armstrong and Catherine de Zegher (eds), *Ocean Flowers: Impressions from Nature* (New York: Drawing Center; Princeton: Princeton University Press, 2004), reconsider early photography, especially photograms, as a practice of printmaking that developed in relation to natural history picture making. In her essay 'Cameraless: From Natural Illustrations and Nature Prints to Manual and Photogenic Drawings and Other Botanographs', pp. 160–61, Armstrong insists that we not overdetermine photography's history as a lens-based practice when so many of its first creators made photograms.

23 Larry J. Schaaf, *Out of the Shadows: Herschel, Talbot, and the Invention of Photography* (New Haven and London: Yale University Press, 1992), pp. 21 and 130–31, sketches out Talbot's interest in botany. Nature printing reemerged in mid-nineteenth-century England through Henry Bradbury and others.

24 The most compelling study of this desire is Geoffrey Batchen, *Burning with Desire: The Conception of Photography* (Cambridge, MA: MIT Press, 1997).

25 In April 1777, the Royal Botanical Garden moved into its present quarters in Madrid to form a science hub with the royal observatory, hospital and academy of sciences. In 1776, the Gabinete de Historia Natural opened in the same building with the art academy.

26 A memo of November 25, 1776, states that Gómez Ortega selected the botanists, and Ignacio de Hermosilla's letter to Gálvez of March 19, 1777, describes the selection of the draughtsmen; AMNCN, Expds., Exps. 7 and 10.

27 Note, November 25, 1776, Madrid; AMNCN, Expds., Exp. 7. These instructions were approved April 9, 1777.

28 Gómez Ortega, Memo to Gálvez, August 1, 1780; AI, Indif., Leg. 1544: 'All these talents rest, in my opinion, in the Marqués de la Florida Pimentel, who has furthered the progress of the Young [artists] of the Academy of Fine Arts of which he is Director, or Vice-Protector, through the means of prizes in the same genre of Flowers and fruits. I have heard his extremely judicious, useful and adaptable comments on [Arce's printing], which in his hands would appear without the errors it has today, and with the greatest possible accuracy'.

29 At the Junta Ordinaria of April 10, 1763, the marqués de la Florida Pimentel recognized the benefits to industry and natural history from the study of flowers; at that of December 4, 1774, the conde de Baños suggested that the academy establish a special room for such study. ARABASF, Docs. 3/82 and 3/83, respectively.

30 Junta Ordinaria, August 3, 1788; ARABASF, Doc. 3/85.

31 Gómez Ortega, Report, June 6, 1780; AI, Indif., Leg. 1544.

32 Ibid.

33 Gómez Ortega, Report, June 3, 1777; AMNCN Expds, Exp. 19. Also see Muela's letter of May 9, 1777; AMNCN, Expds., Exp. 12.

34 Two days after Gómez Ortega wrote his June 3 report on Muela, a note was added to it in another hand, 'Please let Muela know that this discovery was already made by another who … presented a finely printed Book of Plants, and demonstrated the secret for the use of the Botanists who are going to Peru. June 5'.

35 Arce, letter to Gálvez, July 29, 1780; AI, Indif., Leg. 1544.

36 Ibid.

37 Francisco Javier Puerto Sarmiento, *Ciencia de Cámara: Casimiro Gómez Ortega (1741–1818), el científico cortesano* (Madrid: Consejo Superior de Investigaciones Científicas, 1992), pp. 32–43, argues that Gómez Ortega's positions and power were due primarily to his social and political connections.

38 Arce, letter to Conde de Floridablanca, July 18, 1780, Madrid; AI, Indif., Leg.,1544: '[Gómez Ortega] elicited another [nature print] … by don Juan Bru, Mounter of Plants of the Royal Cabinet, publishing it openly. Since [I] demonstrated [my invention] under the protection of Your Excellency, [I] deserve the merit'.

39 Arce, letter to Gálvez, July 29, 1780; AI, Indif., Leg. 1544. The first volume of *Libro de plantas* is dated 1779.

40 Puerto Sarmiento, *Ciencia*, p. 139. The position was Chief Pharmaceutical Examiner.

41 Gómez Ortega, letter to Gálvez, June 6, 1780, Madrid; AI, Indif., Leg. 1549.

42 The title 'Academician' brought with it genteel social status; very few artists from Arce's native Burgos ever attained it. From 1769, Arce worked for the crown prince but never obtained a position as court sculptor. In 1788, when Carlos IV became king, Arce was finally accorded an honorary academic title, *Academico de mérito.* René Jesús Payo Hernanz, *El artista burgalés en la época ilustrada* (Burgos: Institución Fernán González, 2005), p. 14, documents the many artisans of Burgos who worked in artistic mediums other than the one in which they trained, as Arce did for nature printing.

43 Celedonio de Arce, *Conversaciones sobre la escultura* (Pamplona: Joseph Longa, 1786).

44 For an investigation of botanical nomenclature as Western imperialism, see Londa Schiebinger, 'Naming and Knowing: The Global Politics of Eighteenth-Century Botanical Nomenclatures', in *Making Knowledge in Early Modern Europe: Practices, Objects, and Texts, 1400–1800*, ed. Pamela H. Smith and Benjamin Schmidt (Chicago and London: University of Chicago Press, 2007), pp. 90–105.

8

The Microscope as a Musical Instrument: Art, Hermeneutics and Technoscience

Pernille Leth-Espensen

The artist Anne Niemetz and the nanoscientist Andrew Pelling pick up the sound of cells with the atomic force microscope (AFM) and transform it into music, the artist Paul Vanouse makes the result of gel electrophoresis resemble a world map and in the artwork *MEART – The Semi-Living Artist* by SymbioticA Research Group, rat neurons are cultivated in a Petri dish with electrodes in it, and the neural activity is sent over the Internet and transformed by a robotic arm into a drawing on paper.

These seemingly quite different works of art all share an interest in investigating scientific instruments and processes. What might be the motivation for this artistic exploration?

According to sociologist of science Bruno Latour, scientific knowledge is constructed by means of measuring devices, diagrams, text and so forth – what he terms 'inscription devices' and 'inscriptions'. The philosopher of technology Don Ihde also stresses the importance of technological instruments in the generation of scientific knowledge. He argues that since the main part of contemporary natural science is concerned with phenomena that are small, invisible, inside a body or out in space, almost all science today is technoscience. In continuation hereof Ihde argues that not only the humanities but also the natural sciences are hermeneutic, as scientific instruments always interpret the phenomena they represent. Phenomena are seldom just seen, they are 'made readable' through instruments. The art historian James Elkins has also discussed a range of scientific imaging technologies and shown how instruments in various ways interpret phenomena in order to visualize them.[1]

Moreover, the philosopher of technology Peter-Paul Verbeek emphasizes that technologies and scientific representations are not neutral; they have ethical implications as they affect our conception of our own body, of nature, of society, and ultimately they also affect our actions. Given the accelerating scientific and technological development, it is therefore not strange that artists find it interesting to investigate the implications of the use of technologies employed in science.

In the present essay, it will thus be discussed how artists in various ways interpret the role that instruments, inscriptions, technologies and metaphors play in the generation of scientific knowledge.

Bruno Latour – Inscriptions in the Laboratory

Bruno Latour's anthropological analysis of the scientific laboratory may help us to understand why artists are interested in working with scientific instruments, representations and inscriptions. Some of the main points from the book *Laboratory Life* written in collaboration with Steve Woolgar will therefore be outlined. According to Latour and Woolgar, knowledge developed in the laboratory is not just discovered 'facts'. What we consider scientific facts are constructed with what they call inscriptions and inscription devices.[2] The term 'inscription' is taken from the philosopher Jacques Derrida and covers various forms of text including traces, spots, numbers, spectra and so on.[3] Inscription devices are different forms of apparatus that produce these inscriptions.

A simple and illustrative example of this is an object that I encountered in my childhood. At home we had a so-called barograph, a device for measuring the pressure of the air. It consisted of a measuring device and a scroll of paper that was to be changed every week. The measuring device was connected to a pen that inscribed the air pressure upon the paper. The change of paper was done with ritualistic care and over the years my stepfather amassed large amounts of inscribed barograph paper. The inscribed barograph paper can be seen as an example of an inscription and the barograph as an inscription device. The inscription device turns something intangible, in this case the pressure of the air, into a concrete representation.

This focus upon inscriptions on the one hand, and apparatus on the other, was visible in the laboratory where Latour made his field study. It was divided into two areas: a section containing books, dictionaries and paper, and another section where people were working with different kinds of apparatus. Those in the first section were writing, reading or typing while those in the other section were cutting, sewing, mixing, shaking, screwing and marking. Latour was surprised by the amount of written material generated in the laboratory; to him the workers in the lab seemed to be almost manic writers.[4]

Afterwards, the inscriptions produced in the laboratory with the inscription devices are turned into diagrams and figures and taken as the starting point in another kind of writing: namely, the scientific article. There is furthermore what Latour characterizes as 'material dictionaries': refrigerators filled with beakers, test tubes and bottles, many of these provided with writing or long numbers.[5]

Don Ihde – Hermeneutics and Technoscience

Whereas Latour's point of departure is constructivist, Don Ihde's is the tradition of phenomenology and hermeneutics. Latour and Ihde thus have quite different epistemological points of departure. They share, however, an interest in the material aspects of science as they both analyze concrete artefacts and scientific instruments and technologies and the role these play in science. The actor-network approach, of which Latour is part, has been characterized as material semiotics,[6] while Ihde describes his approach as material hermeneutics.[7] Although their epistemological standpoints differ, they both focus upon technological mediation, as pointed out by the philosopher of technology Finn Olesen.[8]

We usually connect hermeneutics to the humanities, but Ihde argues that there are many hermeneutical aspects of the natural sciences as well.[9] Ihde stresses that since the major part of the natural sciences is concerned with phenomena that are small, invisible, out in space or inside a body and so on, almost all science today is technologically mediated.[10] In continuation of Ihde's argument Peter-Paul Verbeek writes, 'Scientific instruments constitute what scientists observe; they "interpret" reality before humans can observe it'.[11] The technological mediation means that interpretation plays an important role in science.

Don Ihde stresses that scientific representations have to transform what they represent in order to be useful. He writes, 'In older empiricist epistemology, representations could be considered to be correspondent if and only if they isomorphically replicated the thing being replicated. They were thought of as *copies* of the thing itself. Technologies transform all possible representations and are never purely correspondent. In this respect they are "hermeneutic"'.[12]

One of the examples given by Don Ihde is that of histological slides.[13] As most biological tissue is more or less transparent, the specimen has to be stained before we look at it under the microscope. But, not only do the specimens have to be stained; firstly, the decaying process has to be stopped by immersion in formalin, then they are infiltrated and embedded in wax, cut and then stained and mounted, before it is possible for us to observe the tissue under the microscope.[14] Another example is the sample preparation for the scanning electron microscope (SEM). As this microscope scans specimens with electrons, the specimens have to be covered in carbon or gold to prevent them from being electrically charged and to enable the electrons to bounce back and be collected by the microscope.[15] The gold covering can thus be considered a kind of technology of vision. It is a way of physically transforming the object in order to represent it or make it 'visible'. A third example is the atomic force microscope (AFM). With the AFM a probe is scanning the specimen and these 'tactile data' are then afterwards translated into a visual image.[16]

Technology and Ethics

Don Ihde and Peter-Paul Verbeek both stress that technologies and scientific representations have ethical consequences as they affect our view of nature, our own body and society. How scientists represent natural or social phenomena matters, as in any representation some aspects of reality are amplified while others are reduced. One of the examples Verbeek analyzes is obstetric ultrasound. As ultrasound is merely a representation, it might seem innocent as a technology. But as it shows us a part of the female body that we cannot see without the help of the technology, it may have ethical implications, as it provides us with a knowledge that might be the basis for subsequent action. As Verbeek argues, it visually isolates the fetus from the female body and gives the fetus a new status as a separate being. Furthermore, ultrasound frames the fetus in an environment of medical norms, as it is possible to screen for various birth defects. Ultrasound thus turns pregnancy into a process of choices and the fetus into a possible patient.[17] The praxis of ultrasound today embodies two conflicting views of the fetus, as the anthropologist Janelle S. Taylor points out. On the one hand, the child is seen as a sort of 'commodity' that may be deselected. On the other, ultrasound stresses fetal personhood and is seen as a bonding experience.[18] Latour also stresses that technological artefacts are not neutral; he considers artefacts to be nonhuman actors that mediate our actions.[19]

As scientific technologies thus have intentional as well as unintentional ethical consequences and philosophical implications, it is not strange that artists find it interesting to explore this area.

SymbioticA Research Group – Scientific Inscriptions

In the artwork *Fish&Chips* a part of a goldfish brain was connected with electrodes and wires to a robotic arm placed on a table. The robotic arm translated the electrical activity in real time from the brain into a drawing on paper, using pens of various colours. The work was created by SymbioticA Research Group at the University of Western Australia and exhibited at the Ars Electronica Festival in 2001. This artwork was later developed into the artwork *MEART – The Semi-Living Artist* in collaboration with the neuroscientist Steve Potter's Laboratory for Neuroengineering at The Georgia Institute of Technology in Atlanta in the United States.[20] In *MEART*, the fish brain was substituted with cultured embryonic rat neurons placed in a multi-electrode array. A multi-electrode array is a tissue culture dish with electrodes in it, and these electrodes are capable of picking up signals from the cells. Potter's laboratory cultivates the neurons in the multi-electrode array, and the neurons are then, via the Internet, connected to the location where the artwork is exhibited. At the exhibition space, the neural activity is embodied by means of a robotic arm making a drawing on paper with pens of various colours

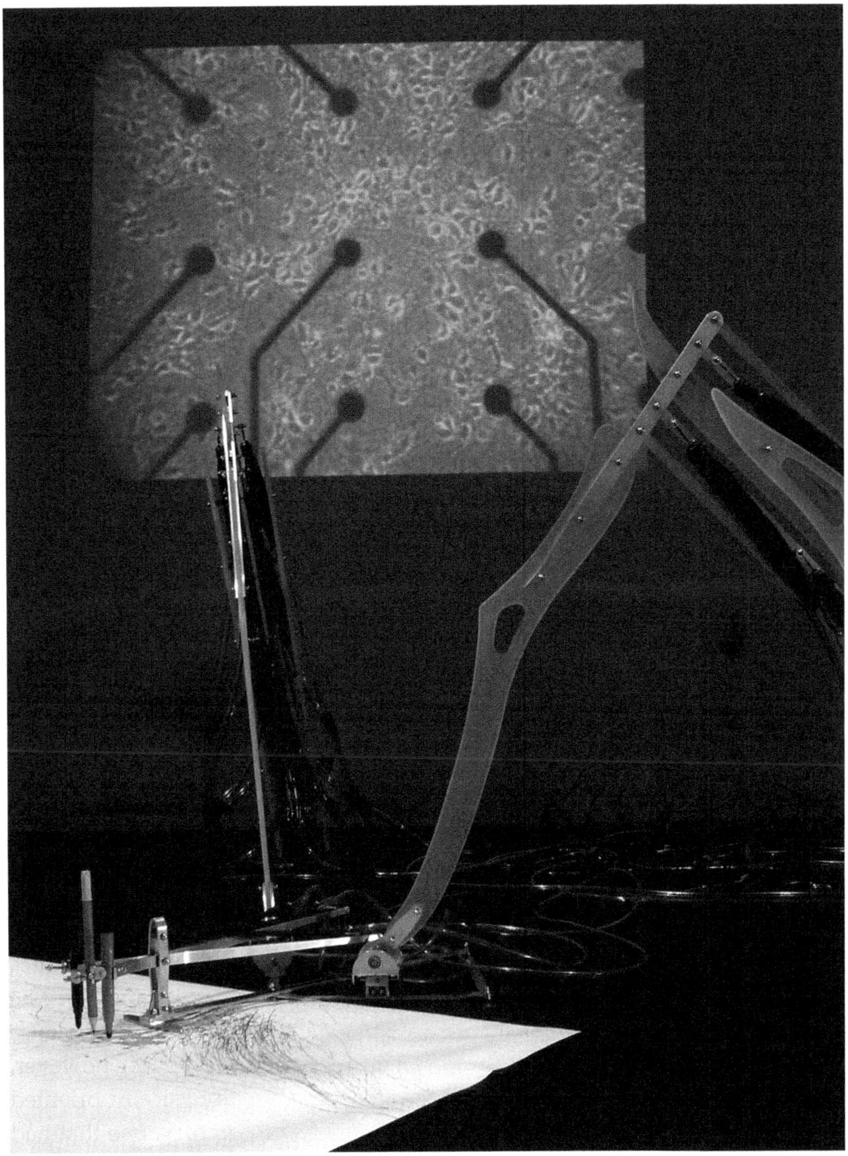

8.1 SymbioticA Research Group in collaboration with Dr Steve Potter's Laboratory for Neuro-engineering, The Georgia Institute of Technology, *MEART – The Semi-Living Artist*. Biofeel exhibition, Perth Institute of Contemporary Art, Perth 2002. Photo credit: SymbioticA Research Group

(Figure 8.1). This work has been installed a number of times in different parts of the world, for instance in Perth, Melbourne, Moscow and New York.[21]

Two of the artists behind the work, Guy Ben-Ary and Phil Gamblen, explained that they chose to embody the work by means of a robotic arm drawing on paper because the archetypical artistic product is a drawing or a painting. The subtitle of *MEART* is thus *The Semi-Living Artist*. As described above, Bruno Latour emphasizes the importance of inscriptions and inscription devices in the scientific lab. I will therefore add that *MEART* is not only interpreting the activity of an artist, it is also interpreting scientific processes and the importance of measuring devices and inscriptions in the laboratory. The combination of a multi-electrode array and a robotic arm can be interpreted as an inscription device, and the drawings made by the robotic arm can be interpreted as inscriptions.

MEART's translation of neural activity into a visual representation brings to mind the range of various neuroimaging technologies that have been developed in recent years, such as CT scans, PET scans, MEG, MRI and fMRI. The imaging technology fMRI, for instance, is not only used medically but is applied more broadly to look into which areas of the brain are active during specific mental processes,[22] and in the media we are often presented with these images of the brain with coloured areas representing activity in specific regions. We may ask how these imaging technologies shape our view of what consciousness is. In an article addressing brain visualizations, mathematician and computer scientist Britta Schinzel argues, 'Brain visualizations have established drastic changes in the way humans see themselves, their bodies and their "personhood" in general'.[23] With the multicoloured pens, *MEART* also produces coloured representations of neural activity, but in an unusual way. By representing neural activity in an unorthodox manner, *MEART* can be interpreted as a reflection upon how various neuroimaging technologies translate neural activity into a representation and how these representations of the brain make us understand the mind and consciousness.

The interpretation of scientific inscriptions and inscription devices is even more obvious in *Silent Barrage,* the development of *MEART*. *MEART* and *Silent Barrage* use the same multi-electrode array, but they are embodied in different ways and we consequently get a different impression of the neural activity. In *Silent Barrage,* instead of a drawing arm, the embodiment consists of a number of poles placed in a grid system. The poles are covered with paper and robotic devices make inscriptions on them. A robotic device moves up and down along the pole and when it is at a certain height, it will rotate while drawing.[24] Visually, *Silent Barrage* resembles various kinds of scientific apparatus even more. *MEART* has an expressionist flavour to it with the drawing arm and the multicoloured pens, whereas *Silent Barrage* expresses a more dry, 'scientific style' similar to the style of my stepfather's barograph. In fact, when I showed the picture of the barograph to the manager at SymbioticA, Jane Coakley, she immediately remarked, 'that resembles *Silent Barrage*'. I have been told that the neuroscientists find it important that the inscriptions and the generated data are very accurate – whereas this is less important for the artists. These differences in opinion show that the drawings/inscriptions are not only conceived as an artistic product but also as a scientific one. By representing brain activity in a novel way and by exhibiting it in an art context, the artwork both reflects on how different neuroimaging technologies represent the brain and on the use of inscriptions in science more broadly.

Paul Vanouse – Subtexts

The artist Paul Vanouse also explores technological processes in science. In the works that will be discussed below, he focuses upon instruments and metaphors in molecular biology. In the artwork *Latent Figure Protocol,* Paul Vanouse uses the technology of gel electrophoresis, one of the key technologies of molecular biology, and to understand the artwork, it is therefore necessary to delve into this technology for a moment.

Gel electrophoresis is a technology used to determine the size of DNA molecules, to determine the sequence of bases or to separate fragments of DNA.[25] Plasmids are also frequently used in molecular biology. A plasmid is a relatively small DNA molecule, often circular. In genetic engineering it is used as a so-called vector, which means that the plasmid can be genetically modified and afterwards inserted into another organism.[26] To cut out a part of a DNA molecule, restriction enzymes are used. A restriction enzyme recognizes a specific sequence of bases and cuts adjacent to it. If scientists want to cut out a specific fragment, in a plasmid for instance, they have to find the appropriate restriction enzyme to cut it with.[27] Two restriction enzymes are necessary, as the fragment has to be cut at both ends.

When the fragment has been cut from the plasmid, it is possible to separate the plasmid and the fragment by doing gel electrophoresis. The apparatus for doing gel electrophoresis consists of a container with an agar or acrylic gel in it. As the name indicates, the container is electrically charged. At the negatively charged end, a comb is used to make a series of small holes in the gel where the liquid containing the DNA can be inserted with a pipette. The DNA moves from the negatively charged end to the positively charged end, as there is a negatively charged phosphate group on the sugar backbone of DNA. In the gel are microscopic pores that allow the DNA to move slowly through the gel.[28] The speed with which the DNA fragment moves through the gel is inversely proportional to its size: the smaller the fragment the faster it moves. The DNA fragments can be stained with a fluorescent dye and observed under UV light.[29]

Latent Figure Protocol

In his work *Latent Figure Protocol,* Vanouse has used an industrially produced bacterial plasmid named PET11a. In this piece, Vanouse has inverted the normal procedure of electrophoresis. He began by deciding that he wanted the end result of the electrophoresis to resemble a copyright symbol. Subsequently, with the help of a computer programme, he calculated with which restriction enzymes to cut the plasmid inserted in the different lanes so that the fragments would be of specific sizes. These small fragments then functioned as 'pixels' that constituted an image of a copyright symbol after running the gel (Colour plate 11a).[30] Vanouse is thus doing what he calls 'molecular biology in reverse'.[31] Vanouse has made other versions of this work where the result resembles the ciphers 0 and 1 (Colour plate 11b), the abbreviation ID or a skull.

This artwork shows the inscription character of the electrophoresis gel, and how the electrophoresis apparatus is used as an inscription device. That Vanouse is familiar with the theory of Latour is indicated by the use of the notion 'inscription device' in one of his earlier works, *The Relative Velocity Inscription Device*.[32] As mentioned, Latour argues that what he considers inscriptions are not only texts; various sorts of blots, spots, diagrams, and so forth, can also be considered inscriptions. Hence, a regular electrophoresis gel with its band patterns can be considered an inscription. The artwork stresses the inscription character of the electrophoresis gel by constructing an image that we can easily decode as an inscription. Vanouse writes about the work: 'I am interested in creatively "hacking" them with the aim of forcing the arcane codes of scientific meaning "to speak" in a broader cultural language'.[33] When running the gel, the fluorescent yellow copyright symbol slowly emerges as though written with a Magic Marker. It appears as a Freudian slip of the technology, highlighting the commercial subtexts in biotechnology. In the other versions of this work, he instead investigates metaphors used in molecular biology.

The result of gel electrophoresis does not normally have iconic resemblance to the organism it is taken from; instead, it consists of a band pattern. From the result we can, for instance, infer the size of a particular DNA fragment or the sequence of bases in DNA. When a scientist ordinarily runs a gel in the lab, the result appears – in the vocabulary of the semiotician C.S. Peirce – as an index, as a trace.[34] But in *Latent Figure Protocol* the resulting images are also iconic signs. By constructing an image that does not tell us anything about the size of the DNA or the sequence of bases, the causal relation is deliberately blurred. By creating images with iconic traits that wilfully confuse the viewer, the artwork is pointing to potential fallacies following from various metaphors and analogies.

Like technologies, metaphors are also a key hermeneutic frame in science. The rhetorician of science Michelle Sidler writes, 'Metaphors, because they help conceptualize that which cannot be seen or touched, are playing a vital role in the information age of science. Nowhere is metaphor more evident than in molecular biology'.[35] Some scientific metaphors, such as 'genetic code', 'DNA profiling' and 'genetic mapping', are so commonly used that we tend to forget that they are actually metaphors. Vanouse's works make us reflect upon several of these metaphors. The electrophoresis image resembling the numbers 0 and 1 makes us reflect upon the metaphor of the genetic code, and the image resembling the letters ID prompts us to reflect upon DNA profiling methods as markers of identity. These metaphors are not neutral; they affect the way we perceive our own body or nature. The philosopher Max Black describes metaphors as filters that emphasize some aspects and suppress others aspects of the phenomena that they describe. According to Black, the metaphors thereby conceptually organize phenomena.[36]

Ocular Revision

Vanouse's work *Ocular Revision* is a development of *Latent Figure Protocol*. In this work he uses a custom-made electrophoresis container. The container is not only larger than usual, it also has a different shape. A regular container is rectangular while this one is circular, and instead of moving from one edge to the other, the DNA fragments move from the rim of the container to the middle. In *Ocular Revision,* Vanouse aims at making the result of the gel resemble a world map. He has run two different gels, one that is modelled on a map of the eastern part of the world, and another that is modelled on a map of the western part. Interestingly, he has explained that the electrophoresis process works differently in a round container; the pace with which the DNA fragments move in the circular container is not regular: some samples move faster than others.[37]

By using a custom-made circular container, this work points to the fact that the design of the scientific instrument, in this case a container, in itself structures the way scientific knowledge is generated. By aiming at making the result of gel electrophoresis resemble a world map, the work points to the metaphor of 'genetic mapping'. Vanouse makes us reflect upon the metaphor by making a literal and slightly absurd version of it. That Vanouse's work addresses the map metaphor is not accidental. The biologist and cultural theorist Donna Haraway points to the ubiquitousness of that particular metaphor. She writes, 'Cartography is perhaps the chief tool-metaphor of technoscience'.[38] Haraway discusses an example of a very literal use of the metaphor of genetic mapping in an advertisement.[39] In the advertisement we see a map of Africa and Europe covering a female body as if the map was made of a transparent fabric. Haraway argues that this advertisement makes a direct connection between the mapping of the world in the age of exploration and the Humane Genome Project today. In their book *DNA Mystique,* Dorothy Nelkin and M. Susan Lindee study the way DNA is depicted in popular culture. They write about the metaphor of mapping: 'Map imagery suggests that once a gene is located, its interpretation will be objective and independent of context'.[40] In the nature versus nurture debate, the map metaphor – as well as the code metaphor – might stress the inherent properties of the gene and downplay the context and environment. Other metaphors like DNA profile, DNA portrait and DNA fingerprinting stress the connection between a tiny part of a person's DNA and the identity of that person. Vanouse points out that a DNA fingerprint by the lay public often is assumed to be a unique human identifier, a sort of unique sentence written by Mother Nature.[41] A DNA fingerprint is not a single identifier or a sequencing of the full genome. It is an analysis of the number of repeats of short sequences in specific regions (in the U.S. CODIS system 13 different regions) of the genome. If there is a match between the DNA profile from a specific person and the DNA from a crime scene, for instance, the probability of a random match from an unrelated individual in the same population group is statistically calculated.[42]

Vanouse's artworks thus bring to our attention to the fact that scientific data are not immaculately conceived; scientific knowledge is generated by means of instruments and processes and understood by means of metaphors. Furthermore, Vanouse's artworks highlight that the instruments and metaphors used are not neutral mediation, but they contribute to shape the way we understand nature and the human body.

Anne Niemetz and Andrew Pelling – Blindness at the Nanoscale

The last subject which will be discussed are artists who deliberately use senses other than the visual sense. The art historian W.J.T. Mitchell argues, in relation to art, that it is a common misconception that purely visual media exist, that some media only involve the visual sense and not the other senses. Instead, he argues that 'All media are mixed media'. He writes, 'One corollary of the claim that "there are no visual media" is that all media are mixed media. That is, the very notion of a medium and of mediation already entails some mixture of sensory, perceptual and semiotic elements'.[43]

Similarly, Don Ihde writes that he considers all perceptual phenomena to be synaesthetic and multidimensioned. In connection with this, he argues that modern science has a visualist bias: 'The natural sciences are intensely visualist'.[44] He argues that the natural sciences have a habit of producing what counts as evidence by means of visual representations, and he believes that this cultural habit has accelerated in recent years as a result of the development of various imaging technologies. In many instances, the information of a scientific representation is not gathered visually, but it is presented visually. One example of this is ultrasound. As the name indicates, the data are generated through sound but presented on a screen. Another example is the atomic force microscope (AFM), where the data are generated by the touch of a probe, but translated into a visual representation.[45] I believe that a number of artists within this field address the senses in science by working explicitly with senses other than the visual sense and by translating information from one sense to another. An example of this is the work *The Dark Side of the Cell,* made in collaboration between the artist Anne Niemetz and the nanoscientist Andrew Pelling.

In the artwork, Pelling and Niemetz have recorded the sound of cells with the AFM and transformed these sounds into music/sound art. How is it possible to make music with a microscope? To most people it may sound like a contradiction in terms. It is possible, as the AFM is not an optical microscope, as previously described. The AFM makes an image of a specimen by touching it with a probe and subsequently translates these data into a three-dimensional image. However, when recording the sounds of cells, the AFM is not scanning the surface, as the microscope is not moving. The AFM instead acts as a motion detector and picks up the oscillations at the membrane of the cell.[46]

Niemetz and Pelling write, 'The tool with which the sounds of the cells are extracted – the atomic force microscope (AFM) – can be regarded as a new type of musical instrument'.[47] As the sound is within the human audible range, Niemetz has not changed the frequency, but only amplified the sound.[48] They have turned the recordings into a musical composition consisting of five parts: 'Observation', 'Taming', 'Subjugation', 'Calming' and 'Manipulation'. In most of the recordings they have used yeast cells, but in part of the composition they have used bone cancer cells as well. In the series, they have experimented with altering the sounds by affecting the cells in various ways, both mechanically and chemically, and the five parts of the composition are quite different.[49]

In the artwork, it is not only the sounds as such that are interesting but also the source of the sounds and the scientific instrument with which they are generated or perceived. Like the other artworks analyzed above, this work focuses on the technological mediation in science. By experiencing the cell through a sensory modality other than the visual,[50] the work makes us perceive the cell differently and makes us reflect upon the technological mediation in science.

There is something intriguing about the fact that, on the one hand the work is interesting because of the indexical aspect: it is not only the sound itself which is interesting but also the source of the sound, the cells. On the other hand, as it is not a sound that we can detect with the naked ear, we do not know how much the sound is affected by the technological mediation. The nanoscientist James Gimzewski, who in collaboration with Andrew Pelling discovered the sound of cells, says that he believes that it is the vibrations of the cells that we can hear, but that we at this point cannot exclude the possibility that it is an artefact of the instrument.[51]

There is a limit to the magnification delivered by an optical light microscope determined by the wavelength of visible light.[52] Therefore, other microscopes, such as electron microscopes and the atomic force microscope, are necessary when looking at phenomena at the nanoscale. In an interview with Pelling and Niemetz, Pelling expresses it this way: 'As scientists we are essentially blind at the nanoscale. And very much how like a blind person would read by feeling Braille by feeling bumps on the page, this instrument with a very tiny finger can feel the features of a surface'.[53] The title, *The Dark Side of the Cell,* may point to the fact that when investigating phenomena at the nanoscale, we are visually in the dark and have to rely on the other senses, such as touch or hearing.

Don Ihde argues that the function of many technological instruments is fundamentally to transform phenomena into a scale it is possible to perceive with the human body. A microscope enables us to perceive phenomena at a very small scale, and the telescope enables us to perceive phenomena that are very far away. Consequently, through the instruments we perceive large and minuscule phenomena at the same bodily scale. Ihde writes, 'The "image size" of galaxy or amoeba is the *same*'.[54] This relation between images obtained with a telescope and images obtained with a microscope is thematized in Niemetz's work, as there is an indirect reference to astronomy. The title, *The Dark Side of the Cell,* is a pun on the expression 'the dark side of the moon', and the image of a cell on the entrance page showing the work resembles an image of the moon with craters and mountains. By making this connection between the nanoscale and the astronomic scale, the work indirectly illustrates Don Ihde's point.

Like the other artworks discussed above, *The Dark Side of the Cell* thus emphasizes the technological mediation in science. That we experience the cells through another sensory modality makes us more aware of the technological mediation. Other examples of artists working with sounds in relation to science are Joel Ong, who makes us listen to music through a stethoscope; Cat Hope, who has recorded the sounds from a guinea pig in decay; and Boo Chapple, who has used cow bones as loudspeakers.

Concluding Remarks

The artworks discussed in this essay are in many ways different, but they all explore the part that technological mediation plays in the generation, interpretation and presentation of scientific knowledge. *MEART* and *Silent Barrage* interpret, amongst other things, the role inscriptions, diagrams and instruments play in science. The works furthermore make us reflect upon how various neuroimaging technologies shape our understanding of the mind. Paul Vanouse's works investigate the hermeneutics of technologies and metaphors in molecular biology. Anne Niemetz generates music in an unorthodox way and makes us experience aurally data that we would normally experience as visual representations.

Notes

1 James Elkins, 'Art History and Images That Are Not Art', *The Art Bulletin* 77/4 (December 1995): pp. 553–71. James Elkins (ed.), *Visual Practices Across the University* (Munich: Wilhelm Fink Verlag, 2007). James Elkins, *Six Stories from the End of Representation: Images in Painting, Photography, Astronomy, Microscopy, Particle Physics, and Quantum Mechanics, 1980–2000* (Stanford: Stanford University Press, 2008).

2 Bruno Latour and Steve Woolgar, *Laboratory Life: The Construction of Scientific Facts* (Princeton: Princeton University Press, 1986 [1979]), chapter 2.

3 Ibid., p. 45.

4 Ibid., pp. 45–48.

5 Ibid., pp. 48-52.

6 John Law, 'Actor Network Theory and Material Semiotics', in *The New Blackwell Companion to Social Theory*, ed. Bryan S. Turner (Malden, MA: Wiley Blackwell, 2009), pp. 141–58.

7 Don Ihde, *Postphenomenology and Technoscience* (Albany: State University of New York Press, 2009), pp. 63–64.

8 Personal conversation.

9 Ihde, *Postphenomenology and Technoscience*, p. 64.

10 Ibid., p. 45.

11 Peter-Paul Verbeek, *What Things Do: Philosophical Reflections on Technology, Agency and Design* (University Park: Pennsylvania State University Press, 2005), p. 141.

12 Ihde, *Expanding Hermeneutics: Visualism in Science* (Evanston, IL: Northwestern University Press, 1998), p. 166.

13 Ibid. p. 92.

14 Michael H. Ross., Gordon I. Kaye and Wojciech Pawlina, *Histology: A Text and Atlas*, 4th ed. (Baltimore: Lippincott Williams and Wilkins, 2003), pp. 2–5.

15 Ross, Kaye, Pawlina, *Histology: A Text and Atlas*, p. 17.

16 Peter Eaton and Paul West, *Atomic Force Microscopy* (Oxford: Oxford University Press, 2010), p. 1.

17 Peter-Paul Verbeek, 'Moralizing Technology: On the Morality of Technological Artifacts and Their Design', paper for workshop, Ethics Institute, Utrecht University, March 24–25, 2006, p. 10.

18 Janelle S. Taylor, 'Image of Contradiction: Obstetrical Ultrasound in American Culture', in *Reproducing Reproduction: Kinship, Power and Technological Innovation*, ed. Sarah Franklin and Helena Ragoné (Philadelphia: University of Pennsylvania Press, 1998), p. 24.

19 Bruno Latour, 'On Technical Mediation – Philosophy, Sociology, Genealogy', *Common Knowledge* 3/2 (Fall 1994): pp. 30–36.

20 SymbioticA Research Group: http://www.fishandchips.uwa.edu.au/exhibitions/ars.html, accessed May 5, 2015. Interestingly, this collaboration has not only been fruitful to the artists, it has also been interesting for the scientists working with the project. The scientists and the artists have published two scientific articles on the project, see: Douglas J. Bakkum, Philip M. Gamblen, Guy Ben-Ary, Zenas C. Chao and Steve Potter, 'MEART: The Semi-Living Artist', *Frontiers in Neurorobotics* 1 (November 2007): pp. 1–10, and Douglas J. Bakkum, Alexander C. Shkolnik, Guy Ben-Ary, Phil M. Gamblen, Thomas B. DeMarse and Steve M. Potter, 'Removing some "A" from AI: Embodied Cultured Networks', in *Proceedings of the Dagstuhl Conference on Embodied Artificial Intelligence*, ed. Luc Steels and Rolf Pfeiffer (Berlin: Springer, 2004), pp. 130–45. Douglas Bakkum, who collaborated with the artists on the project, remarks that *MEART* 'was one of the first closed-loop neurally controlled animats with a robotic body', Bakkum, Gamblen, Ben-Ary, Chao, Potter, 'MEART,' p. 1.

21 SymbioticA Research Group: http://www.fishandchips.uwa.edu.au/project.html, accessed May 5, 2015.

22 John Nolte, *The Human Brain: An Introduction to Its Functional Anatomy*, 6th ed. (Philadelphia: Mosby Elsevier, 2009), p. 137.

23 Britta Schinzel, 'Recognisability and Visual Evidence in Medical Imaging versus Scientific Objectivity', in *Dynamics and Performativity of Imagination: The Image between the Visible and the Invisible*, ed. Bernd Huppauf and Christoph Wulf (New York: Routledge, 2009), p. 340.

24 SymbioticA Research Group: http://www.silentbarrage.com/project.html, accessed May 5, 2015.

25 William H. Elliot and Daphne C. Elliot, *Biochemistry and Molecular Biology*, 4th ed. (Oxford: Oxford University Press, 2009), chapter 28.

26 Ibid., p. 465.

27 Ibid., p. 459.

28 Paul Vanouse, 'The Relative Velocity Inscription Device', in *Signs of Life: Bio Art and Beyond*, ed. Eduardo Kac (Cambridge, MA: MIT Press, 2007), p. 279.

29 Elliot and Elliot, *Biochemistry and Molecular Biology*, chapter 28.

30 Paul Vanouse, 'Discovering Nature, Apparently: Analogy, DNA Imaging, and the Latent Figure Protocol', in *Tactical Biopolitics: Art, Activism, and Technoscience*, ed. Beatriz da Costa and Kavita Philip (Cambridge, MA: MIT Press, 2008).

31 Paul Vanouse: http://www.paulvanouse.com/lfp3.html, accessed May 5, 2015.

32 Paul Vanouse, 'The Relative Velocity Inscription Device', p. 278.

33 Paul Vanouse: http://www.paulvanouse.com/lfp3.html, accessed May 5, 2015.

34 Charles Sanders Peirce, 'The Icon, Index, and Symbol', in *Collected Papers II: Elements of Logic*, ed. C. Hartshorne and P. Weiss (Cambridge, MA: Harvard University Press, 1965), pp. 156–73.

35 Michelle Sidler, 'The Rhetoric of Cells: Understanding Molecular Biology in the Twenty-First Century', *Rhetoric Review* 25/1 (2006): p. 61.

36 Max Black, *Models and Metaphors: Studies in Language and Philosophy* (Ithaca: Cornell University Press, 1962), p. 41.

37 Personal conversation with the artist.

38 Donna J. Haraway, *Modest_Witness@Second_Millenium.FemaleMan©_Meet_OncoMouse* (New York: Routledge, 1997), p. 163.

39 See illustration in Haraway, *Modest_Witness@Second_Millenium.FemaleMan©_Meet_OncoMouse*, p. 164.

40 Dorothy Nelkin and M. Susan Lindee, *The DNA Mystique: The Gene as a Cultural Icon* (New York: W.H. Freeman and Company, 1995), p. 8.

41 Vanouse, 'Discovering Nature, Apparently: Analogy, DNA Imaging, and the Latent Figure Protocol', p. 177.

42 John M. Butler, *Fundamentals of Forensic DNA Typing* (London: Academic Press, 2010), chapter 14.

43 W.J.T. Mitchell, 'There Are No Visual Media', *Journal of Visual Culture* 4/3 (August 2005): p. 260.

44 Ihde, *Bodies in Technology* (Minneapolis: University of Minnesota Press, 2002), p. 37.

45 Eaton and West, *Atomic Force Microscopy*, p. 1.

46 Anne Niemetz, 'Singing Cells, Art, Science, and the Noise in Between', MFA Thesis (UCLA Department of Design Media Arts, 2004), pp. 15–17. See also Anne Niemetz and Andrew Pelling: http://www.darksideofcell.info/about.html, accessed May 5, 2015.

47 Niemetz, 'Singing Cells, Art, Science, and the Noise in Between', p. 17.

48 Ibid.

49 Ibid, pp. 24–25.

50 Niemetz also mentions herself that she wishes to draw attention to the other senses and thereby challenge the visual dominance in art and science. Ibid, p. 17.

51 Clive Thompson, 'Sonocytology' Collision Detection – Science, Technology, Culture, May 12, 2004; http://www.collisiondetection.net/mt/archives/2004/05/sonocytology.php, accessed May 5, 2015.

52 Weilie Zhou, Robert P. Apkarian, Zhong Lin Wang and David Joy, 'Fundamentals of Scanning Electron Microscopy', in *Scanning Microscopy for Nanotechnology: Techniques and Applications*, ed. Weilie Zhou and Zhong Lin Wang, (New York: Springer, 2007), p. 1.

53 Andrew Pelling: http://www.darksideofcell.info/zkminterview10.mov, accessed May 5, 2015.

54 Ihde's italics. Don Ihde, *Technology and the Lifeworld: From Garden to Earth* (Bloomington: Indiana University Press, 1990), p. 79.

9

How to Experience and Relate to Climate Change: The Role of Digital Climate Art

Søren Bro Pold and Christian Ulrik Andersen

Greenwashing or a Toolkit for Survival?

Towards the end of the first decade of the new millennium, a wave of interest in the climate crisis culminated with the United Nations Climate Summit COP15 in Copenhagen in December 2009, and waned somewhat after the failure to reach a binding agreement. As part of this, exhibitions were taking place and in general there was a broad discussion about art's relation to climate change. Within this, digital art forms – including installations, performances and activism – played an important role, exploring the new and prevalent complex connections between technology and nature that climate change puts on the agenda. This chapter will discuss the role of digital climate art and the renewed artistic interest in the climate crisis. After introducing the discussion, the chapter explores digital climate art from an *epistemological* and a *political* perspective based on analyses of two art projects: Petko Dourmana's room installation *Post Global Warming Survival Kit* presented at Transmediale in 2009 (Colour plate 12), and *Planetary Pledge Pyramid,* a collaborative project presented during COP15 in Copenhagen in December 2009 (Figure 9.1).

In the discussions surrounding the focus on climate art – for example, in relation to the Copenhagen Rethink programme of exhibitions and events which accompanied COP15 – it was argued that the focus on climate change is damaging to art, reducing art to being subservient to a political agenda. Especially at COP15, where climate art was closely related to a prestigious international climate conference, it was tempting to disregard the art as just a politically correct way to brand Copenhagen, the sponsors and the summit.[1] This temptation indicates that climate change is a difficult subject for art. It is a highly complex issue that disrupts many of the traditional concepts and dichotomies according to which we understand, construct and order the world, such as nature versus culture, reality versus representation, ecology versus technology. As a result, it is difficult to set up the criteria for and

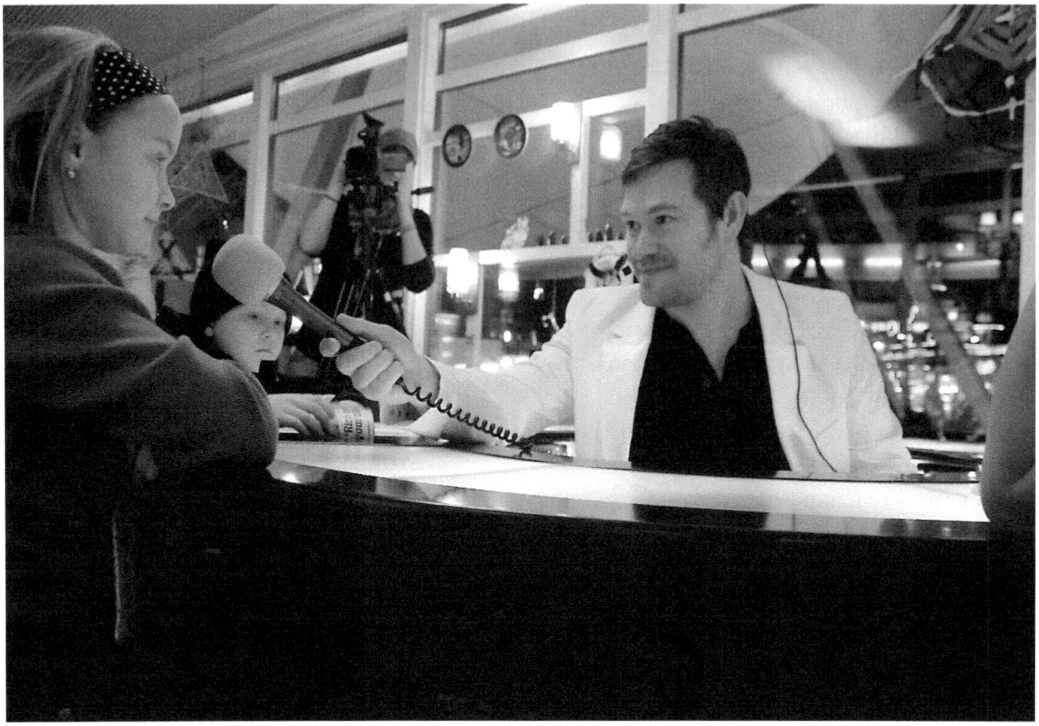

9.1 The People Speak, Talkaoke at City Hall Square (2009), Copenhagen, during COP15 with host Erik Pold. © The People Speak/ Photo credit: Hektor Kowalski/ The People Speak

discuss the qualities of climate art. Does climate art have to be green itself, and should art propose solutions – for example, point out alternatives on a small scale – and if so, how do we evaluate these? Or does this kind of green art risk being dogmatic, subservient, docile and, as a consequence, uninteresting art? Should art instead remain critical and ultimately avoid serving other agendas than its own, even if the future of the planet is at stake?

We believe it is more important at this moment to *raise* such questions in order to challenge traditional worldviews than to find quick answers. The discussions surrounding climate change do in fact have artistic, cultural and aesthetic dimensions, and to fully realize this, art is needed. As it was argued at the final panel at Transmediale's Deep North conference, the issue is not *whether* climate change is coming – because it is already here and will continue even if we managed to immediately stop all CO_2 emissions – but *how* we can learn to live with it, prevent the most catastrophic developments and handle the altered relationship between us and our surroundings which climate change signifies.[2] If we accept this line of reasoning, the next question could be whether *digital* climate art has a specific role to play, even though its dependency on technology and electricity entails that it is not immediately 'natural' or green.

The astrophysicist Roger Malina argues for the importance of digital climate art from the position of the growing distance between science, technology and our everyday experience of the world. The problem is that 'common science does not make common sense'. We use mobile phones, computers and many other results of modern techno-science, but we are generally unable to understand this technology, let alone its underlying cultural consequences. There is a need to work against the scientific world's institutional isolation and reach a deeper understanding of what takes place in the labs. We need a 'tool kit for survival' in which digital art is an important component: 'artists using new media and new technologies are not creating playthings for rich people but are part of the rapid cultural engineering we need to do to face the burning issues of our times'.[3]

As suggested by Jacob Wamberg, the evolution of the 'world picture' is connected to social and technical history.[4] Whereas especially critical, digital media art such as net art and software art has to some degree been occupied with a critical exploration of its own media, currently an environmental sensitivity is emerging in digital art. The environment and climate have increasingly been a theme at leading festivals for digital art such as Transmediale, Ars Electronica, FutureEverything and ISEA. Besides the climate crisis, this environmental turn is also related to the development and spread of ubiquitous and mobile computing. Whereas digital media were more or less localized and stationary until a few years ago, such as the game console in the living room and the computer in the office, now they are embedded, slip easily into the pocket and increasingly overlay all aspects of our everyday life. We are gradually getting used to networks, data and communication overlaying our sense perception and feeling of presence – for example, in the form of smartphones equipped with geo-localization, Internet access, music players, social media and cameras.[5] To an increasing degree, we are connected to data and the Net, even when walking in nature, and we are experiencing our surroundings mediated through mobile digital devices. As the environmental artist group HeHe phrases it, 'We are the Google Earth generation. We need to observe everything from above, from afar, through the glasses of a scientific model'.[6]

Consequently, digital climate art can be seen as part of the 'rapid cultural engineering' Roger Malina asks for, especially regarding our changing relations to the environment and how new technological paradigms such as ubiquitous and mobile computing increasingly become part of how we experience, relate to and interact with nature.

Imaginary Interfaces in the Deep Country

Entering a completely blacked-out room at Transmediale 2009, the audience receives night vision goggles and subsequently moves into the darkness. The work we are about to 'see' is called *Post Global Warming Survival Kit* and was made by the Bulgarian media artist Petko Dourmana. However, the darkness is thick, and our first experience is of being unable to see anything at all. Little by little we learn how the goggles work and discover an 'invisible' film featuring a coastal landscape projected onto a back wall. The film is shown via infrared light, which is invisible to the naked eye but can be seen through the night-vision goggles. We see a small caravan in the middle of the room from which infrared light emanates. On entering the caravan, we find a handwritten book, some communication equipment and various other instruments for sustaining life in a post-climate catastrophe world. With our night-vision goggles and a good deal of effort, it is possible to read in the book, which tells of life after the climate catastrophe, why it got dark and what the guard inhabiting the caravan is doing.

Slowly the layers of the installation are uncovered and its narratives pieced together, but still the difficulty of seeing is a central part of the experience. We move about uncertainly and awkwardly, meet other viewers that we cannot see properly, with their infrared goggles, and generally feel inhibited and alienated. The installation is a post-climate-catastrophic mystery that requires both intensive interpretation and night vision to be experienced. We leave with the experience of the inadequacy of our immediate senses, and a main point of the installation is the idea that the climate crisis demands a fresh perspective, our ordinary senses no longer being sufficient.

Digital media art like Petko Dourmana's installation offers the opportunity to experience a new, mediated perspective on nature. In addition to being a crisis for the globe and for humanity, the climate crisis is also an epistemological crisis, and we need to change our perception of the environment to better understand and deal with it. In other words, the climate crisis is also a cultural challenge.

The cultural challenge could be outlined as a breakdown – or perhaps more likely a reconfiguration – of the dichotomy between culture and nature. This breakdown leads to a semiotization of the weather, which we have previously regarded as something simply 'out there', beyond our reach, as nature in opposition to culture; however, the weather has now turned into yet another structure of signs to be read and interpreted. Furthermore, we have to act based on our interpretations: we cannot perceive the greenhouse gasses or their effects on the climate directly with our senses, so our understanding of climate challenges is largely based on complex climate models, and we must learn to act on this background. Epistemologically speaking, the climate crisis introduces us to the fact that our immediate surroundings are now being mediated by complex visualizations, statistics and carbon quotas. We have to rely on complex mediations or interfaces both in our interpretations and in our actions. In other words, an imaginary interface lurks in the blue sky, even deep in the country with no computers in sight. Or perhaps more precisely, we are increasingly experiencing the weather as an interface which shows the results of processes that we cannot immediately see with our eyes, but only through science; however, we know that we are influencing these processes, and what we experience might be signs of our influence. Just as in front of an interface, seeing through the mediation is a complex matter, but we become part of a cybernetic feedback loop by way of our interactions. In this way, the climate becomes a question related to interface criticism.[7]

Although this climate-as-interface might sound complicated, we are increasingly experiencing it concretely. For instance, as nagging doubts – perhaps a pang of bad conscience – when we are enjoying the unusually warm weather that perhaps, perhaps not, is a sign of climate change. Or as an extra worry when storms hit: is this only bad weather or a portent of the future climate? While we can feel the weather, the climate is more abstract, or to quote an old meteorological saying, 'climate is what you expect, weather is what you get'. In this sense, weather becomes the perceivable 'front-end' to an interface of which climate is the running system or the 'back-end'. It would probably be easier to change our behaviour if the changes were easier to see – if the greenhouse gasses were visible as black smoke, smelled foul or were otherwise directly perceptible.[8] We are used to acting on what we can perceive directly (the weather), not on scientific predictions as to what will happen in the course of the next centuries (the climate). In addition to being a subject of uncomplicated small talk, the weather is increasingly discussed as a potential sign of climate change and as such a subject for political debate, something we should do something about.

Recognizing this is a rather big challenge that demands a cultural process of change. What should we strive for now that it can no longer be growth, exotic holidays and bigger cars? How can we learn to acknowledge the invisible balances in a nature that is no longer absolute and unspoilt? To assist us in this process, art can confront us with new ways of seeing, as in the case of Petko Dourmana's installation.

The Things of Climate Change

If acknowledging the climate problem is difficult, it is even harder to learn how to address it and act politically. How do you act, what does it mean to act, what can you act upon and what will the results of your actions be? Identifying the actors is also a challenge: who are the actors capable of making real change and how can we judge their commitment?

Besides taking to the streets, as many did with varying success during COP15, one way of exploring how to act is through media activism. Net art activists like *The Yes Men* expose hypocrites and frauds of all kinds by kidnapping companies' brands, events, websites and press releases, and worthy opponents include Exxon Mobil, Halliburton and former President Bush's (lack of) climate ambitions. These activists lay bare the greenwashing and empty rhetoric of some of the people who, behind the façade, are trying to prevent binding climate goals. At COP15 they exposed the Canadian government in particular and the political leaders' lack of action in general by setting up a fake press studio, sending out fake press releases and giving people access to the press studio to make their own announcements.[9]

But what if criticism is not enough? Bruno Latour has raised the question whether criticism, as we know and cherish it in the humanities and social sciences, has 'run out of steam' and even risks serving lobbyists in undermining scientific authorities such as the United Nations Intergovernmental Panel on Climate Change (IPCC) by establishing a default critical attitude or scepticism.[10] Latour proceeds to discuss whether science studies have contributed to destabilizing scientific facts by doing analyses that show how scientific facts are produced by complex organizations and technological mediations and by directing 'attention toward the conditions that made them possible'.[11]

In this manner, Latour critically examines his own method and the impact and role of both the humanities and the social sciences in the climate crisis. He questions whether our critical methods are at risk of turning into a default scepticism towards the findings of science at a time when it might be vital for our survival to actually trust in science. And he answers this question by arguing for a new understanding of the world, according to which we recognize that matters of fact are only very partial renderings of matters of concern, and that the world is much more complicated and things much more intertwined with social dimensions and cultural perceptions than the bare facts suggest. If the critical mind is to renew itself and be relevant again, he argues, it demands 'the cultivation of a *stubbornly realist attitude* – to speak like William James – but a realism dealing with what I will call *matters of concern*, not *matters of fact*'.[12] The concept of facts being too limited for most contemporary phenomena, Latour instead makes a distinction between objects and Things, with the latter containing both the object out there and 'an *issue* very much *in* there, at any rate, a *gathering*'.[13] The concept of Thing is etymologically related to the Germanic and Scandinavian *tinge*, meaning an assembly such as the Norwegian Storting, the Icelandic Althing or the Danish Folketing.[14] In general, Latour suggests a new epistemology based on the concept of Things:

> things have become Things again, objects have reentered the arena, the Thing, in which they have to be gathered first in order to exist later as what *stands apart*. The parenthesis that we can call the modern parenthesis during which we had, on the one hand, a world of objects, *Gegenstand*, out there, unconcerned by any sort of parliament, forum, agora, congress, court and, on the other, a whole set of forums, meeting places, town halls where people debated, has come to a close. What the etymology of the world *thing* – *chose, causa, res, aitia* – had conserved for us mysteriously as a sort of fabulous and mythical past has now become, for all to see, our most ordinary present. Things are gathered again.[15]

Latour's argument is complicated and relates back to his book on the myth of modernity, *We Have Never Been Modern,* in particular its final chapter on 'The Parliament of Things', but it also rather precisely points out the complexity of the climate change debate.[16] First, as discussed above, it points towards an epistemological problem: what do we see when looking at nature and the world, what is behind what we see, and how can we interpret and relate to this complexity? What we see when looking at the world becomes a 'gathering' that relates objects, scientific interpretations, technologies, political and cultural discussions. A Thing, such as the IPCC, the Kyoto Protocol and COP15, is related to scientific observations, cultural frameworks for how we view nature and, finally, what we can observe as objects in nature, such as sea levels and the weather. The climate changes and observations are consequently not facts – and will not be facts before after the fact, which is too late – but this does not mean that the IPCC climate scientists are lying, suppressing evidence or succumbing to political goals, as they have been accused of in the so-called climate-gate, but that things (and Things) are more complicated. However, if understanding this is difficult, and implies an epistemological change requiring facts, objectivity and nature to be reconsidered as Things, finding out how to act upon this – which we need to do faced with the climate crisis – is even harder. If Petko Dourmana's installation shows us that the perception of our immediate surroundings is mediated and relies on interfaces, then we also need more sophisticated models of understanding to engage ourselves politically.

Planetary Pledge Pyramid – A Design for Things

With such complexities in mind, *Planetary Pledge Pyramid (PPP)* was designed as an artistic project aimed at discussing the issues of climate politics, through collaboration between the London-based artist group The People Speak, the Digital Aesthetics Research Center and the Center for Digital Urban Living[17] at Aarhus University. The project created an interface for change through collective idea generation, fundraising and online demonstration using popular media formats such as Facebook, talk shows and game shows. Whereas social media have often been used to quickly demonstrate how many of us support a specific cause, this project takes the next step and uses Facebook to recruit and connect people to generate, discuss and vote on ideas, also encouraging them to pledge a minimum of $10 for the best idea to be carried out. At the same time, groups could debate in round-table talk shows (Talkaokes) at the City Hall Square in Copenhagen, and the final decision on the winning idea was taken by the audience in the final game show, 'Who Wants to Be?', which took place at Klimaforum09–People's Climate Summit at the end of COP15 in Copenhagen.

The project tested whether people can use online and offline communities to negotiate solutions and raise the money to carry them out. The vision – built on Buckminster Fuller's idea of a 'World Game' from 1960[18] – was to recruit people from all over the globe to engage in discussions on how to solve the climate problems when the climate conference was launched in Copenhagen in December 2009. *Planetary Pledge Pyramid* was thus a global, planetary project designed based on the utopian idea that if people are given the appropriate tools in the form of an interface for collective activism on climate change, they will take on the task, including collecting the money through pledges. As such, it is a pyramid scheme for planetary activism designed to use Facebook's network of 'friends' for recruitment and dissemination.

Of course, the utopia did not fully succeed, and the project encountered problems staging serious debate and commitment in Facebook's attention economy. Still, the project managed to let the people speak and negotiate through various gatherings or Things (in Latour's sense). In fact, for the crowd of engaged users, it did create a community and open up discussions during the weeks leading up to COP15. It combined popular formats and serious discussions as a strategy to engage people but also to encourage them to reflectively explore the contemporary state of democracy. It aimed at simultaneously protesting against the ineffective negotiations of the serious politicians at the COP15 summit, supporting the hopes behind the summit and generating new ideas and taking responsibility. As such, *PPP* contained many paradoxes and dichotomies, which can be seen as weaknesses compared to more clear-cut protest strategies; and in fact the project was difficult to communicate, especially on Facebook, due to its complexity. On the other hand, given the situation, Things and explorations of how to create gatherings, make decisions and take action are necessary if we are to develop political action in the face of the climate crisis. The climate crisis is not clear-cut and based only on scientific facts, but rather is a new kind of gathering of fields within nature, science, technology, society and culture. Not only does this complexity make political action challenging, but it is also why the climate debate needs new Things to develop. By managing to engage more than a thousand people in Facebook and the Talkaokes and raise $3,000 for the winning idea of planting 2,500 trees in Colombia, *PPP* offered a way to discuss and explore climate politics and alternatives to the COP15 failure.[19]

The Aesthetics of Digital Climate Art – A Starting Point for Exploration

Although climate change seems to be an obvious stage for digital art, the role art plays while onstage is essential. To avoid being dogmatic, climate art should accept the challenge as an artistic challenge – by not 'merely' servicing the climate discussion and communicating the results of science but also discussing how the climate crisis challenges our culture, art and experience. It can do this by addressing and letting us experience the epistemological challenges, as in Petko Dourmana's installation; by criticizing and exposing empty rhetoric and greenwashing, as do *The Yes Men*; or by delivering interfaces for change, as in *Planetary Pledge Pyramid*. As we have argued above and exemplified with Dourmana's *Post Global Warming Survival Kit* and *Planetary Pledge Pyramid*, the field ranges from the epistemological to the political, and of course also includes combinations, and this observation should merely be seen as a starting point for exploring the aesthetics and poetics of digital climate art. Not just serving as green embellishments for the actions of serious politicians and companies, nor as mere illustrations and diversions for the circus surrounding the climate conferences, digital climate art is challenging explorations of what it means to live after the climate crisis.

Acknowledgements

This research has been funded by Center for Digital Urban Living (the Danish Council for Strategic Research, grant 09-063245) and by Aarhus University's interdisciplinary research center Participatory IT, PIT. *PPP* was furthermore supported by TEKNE Productions. Thanks to the many collaborators on *PPP*, including The People Speak, and to Stacey Cozart for proofing this article.

Notes

1 Most of this debate took place in the Danish press; see 'Kunstrektor: Klimakunst er politisk bestillingsarbejde', *Politiken,* November 6, 2009, http://politiken.dk/kultur/kunst/article827236.ece. The Rethink exhibitions took place during COP15 at various locations in Denmark, mainly in Copenhagen. See http://www.rethinkclimate.org, and Witzke and Hede, 'Rethink – Contemporary Art and Climate Change' (Aarhus, Denmark: Alexandra Institute, 2009).

2 The panel was called 'Environment 2.0' and took place at Transmediale 09, Deep North, January 31, 2009. The argument was introduced by Dr Jochen Richerts as an answer to a question about the role of climate change art.

3 Roger Malina, 'What Is a Climate Artist?', in *Deep North,* ed. Stephen Kovatz and Thomas Munz (Berlin: Revolver Publishing, 2009), p. 99.

4 Jacob Wamberg, *Landscape as World Picture: Tracing Cultural Evolution in Images* (Aarhus, Denmark: Aarhus University Press, 2009), p. 23.

5 Currently, in the early 2010s, social media platforms such as Facebook and Twitter are increasingly including geo-localization features, and we are seeing the first (still sketchy) augmented reality interfaces that use geo-localization, the camera and the Net to present what we see overlaid with data (Layar, Wikitude, Nearest Wiki). Besides, websites and Web services such as Google increasingly include geo-localization features, but just being able to access the Net, bring along our data and music and instantly connect everywhere might nevertheless be the most important development.

6 Alessandro Ludovico, 'Hehe Interview', *Neural* (2009): p. 40.

7 Cf. Christian Ulrik Andersen and Søren Pold: *Interface Criticism – Aesthetics Beyond Buttons* (Aarhus, Denmark: Aarhus University Press, 2011).

8 In relation to pollution and smoke, the work of the artist group HeHe is particularly relevant with their ways of high-lighting smoke emissions from power plants and cars (see http://hehe.org.free.fr/ [accessed July 30, 2010] and their 'Disappearing Clouds', in *Deep North,* ed. Kovatz and Munz (Berlin: Revolver Publishing, 2009), pp. 126–27.

9 Cf. http://good-cop15.org/ (a copy of the official COP15 website, now defunct), http://theyesmen.org and http://www.babelgum.com/yesmen (accessed July 30, 2010).

10 Bruno Latour, 'Why Has Critique Run Out of Steam? From Matters of Fact to Matters of Concern', *Critical Inquiry* 30/2 (Winter 2004): pp. 226–27.

11 Ibid., p. 231

12 Ibid.

13 Ibid., p. 233.

14 Cf. also Bruno Latour and Peter Weibel, *Making Things Public* (Cambridge, MA: MIT Press, 2005), p. 22ff.

15 Bruno Latour: 'Why Has Critique Run Out of Steam?', p. 236.

16 Bruno Latour, *We Have Never Been Modern* (New York and London: Harvester Wheatsheaf, 1993).

17 The authors are affiliated with the two research centres. The project was supported by TEKNE Production and part of Rethink, Hopenhagen and Klimaforum09.

18 Medard Gabel, 'Buckminster Fuller and the Game of the World', in *Buckminster Fuller: Anthology for the New Millennium*, ed. Thomas T.K. Zung (New York: St Martin's Press, 2001).

19 We estimate that more than 1,000 people experienced the Talkaoke table and among these 100 participated actively as speakers. On Facebook almost 500 users participated and the *Who Wants to Be …* show attracted a local audience of around 50. We participated ourselves in the events, observed participants and collected material such as questionnaires and extensive video material. Thorough analysis of this material is beyond the scope and space limitations of this article, but we are preparing further publications on this. See also http://darc.imv.au.dk/?tag=planetary-pledge-pyramid.

PART III

Localization/Exposure: Art and Technology Revealing Paradigms of Nature

10

A Perfectly Nebulous Experiment: C.T.R. Wilson's Cloud Chamber

Kristine Nielsen

The Romantic poet John Keats describes in a letter to his brother how drifting clouds on the mountaintop of Ben Nevis seemed to open and close like 'dome curtains'.[1] Inspired with awe, he wrote a sonnet, 'Read Me a Lesson Muse, and Speak It Loud' (1818), while sitting on the stony edge of a Ben Nevis cliff with the valley below him. A wondrous and sombre reflection on human limits marks his last three lines: 'all my eye doth meet is mist and crag, not only on this height, But in the world of thought and mental might!'[2] The desire for revelation is readily apparent in Keats's evocation that mist hides secrets to be illuminated. That clouds would elicit the imagination resonates with a long history of myths. In the Renaissance the shapes of clouds were feared to be the work of demons finding embodiment in fog and mist and transmuting themselves into ominous appearances in the sky.[3] Half a millennium later, the nineteenth-century writer and art critic John Ruskin described a billowing storm cloud as the hundred-headed dragon of the Hesperides and another as 'a black monster cloud – half dolphin, half tiger' that 'rolled and rose, and finally toppled and tumbled' like an avalanche.[4] In like manner, during the filming of *The Avenging Conscience* (1914), the director D.W. Griffith requested a series of cloud studies in a sequence featuring the apparitions of angels and demons.[5] The late nineteenth and early twentieth centuries seem to inherit an early modern fascination with creatures and spirits continuously saturating the air with clouds. Yet, it was also in the late nineteenth century that meteorology was transformed into a science proper. The weather was increasingly explained using science, thus progressively pushing aside the role of mythology and other less-than-materialist accounts. One of the significant contributors to the scientific understanding of cloud formation was the Scotsman Charles Thomson Rees Wilson (1869–1959), a meteorologist who discovered a means to render the path of electrons visible, thereby visually proving their material existence. He received the Nobel Prize in physics in 1927 for this work, which entailed photographing the tracks of charged particles in his so-called cloud chamber soon designated 'the most original and wonderful instrument in scientific history'.[6] The visualization of subatomic matter that this instrument promised initiated several generations of research into microphysics.

C.T.R. Wilson famously began his Nobel Prize lecture by attributing his success with the cloud chamber to his visual experience at the meteorological observatory on the Scottish mountaintop of Ben Nevis. Sounding more like an artist than a scientist, Wilson sought to recreate his experiences and so made a miniature world using a small laboratory chamber. Wilson's aesthetic interest in the cloud chamber is examined in detail by the historian of science Peter Galison.[7] When Galison questions why Wilson eventually purified the air in his chamber when dust is naturally present in the atmosphere, he concludes that Wilson's goal had turned from imitating to dissecting nature. But despite Galison's admirable cultural-historical analysis of the invention, it remains too safely rooted in the realm of present-day science. From its very beginning, the achievements of Wilson's cloud chamber reveal no clear distinction between what we might conceive as 'matters of fact' and 'matters of concern',[8] or between the realms of science, art and metaphysics. This essay argues that Wilson's earliest cloud chamber project sought to idealize the place he identified as home and recreate an exemplary nature from his experience on Ben Nevis. By reproducing local nature as perfectly pure, however, his early cloud chamber experiments relied on a metaphysically oriented scientific practice dominant before the mid-nineteenth century, one that had turned quite outmoded by the end of the century when Wilson was thinking through his project. As a result, in the 1890s, Wilson shared more artistic and environmental concerns with John Ruskin – including his lecture 'The Storm-Cloud of the Nineteenth Century' (1884) – than with John Constable and his cloud studies (1821–22). This paper's proposed shift in viewpoint for Wilson's cloud chamber highlights, then, how the seeds of a progressive science of meteorology and, later, particle physics were set in motion by one man's at times scientifically regressive ideas about art, science and nature. Thus, the scientific research on meteorology of the late 1880s and 1890s and his own metaphysical worldview confirmed the possibility of dust-free air on the mountaintop. Ben Nevis represented his pride in his local region and, with a geocentric picture of the world, Wilson imagined a purer realm above the earth and closer to heaven. It is this perfectly pure experience that Wilson wished to reproduce in his early cloud chamber through the recreation of optical phenomena.

Producing Clouds and Optical Phenomena in the Laboratory

Wilson's earliest predecessor for the cloud chamber was Robert Boyle's experiments with an air pump in the late 1650s in London, which spurred the idea of artificial experiments existing as matters of fact; that an experiment *as* human artifice could be just like nature. Wilson's desire to produce a second, mediated world was related to one already theorized in the realm of philosophy and aesthetics on the Continent in the nineteenth century. In the 1820s and 1830s, Hegel was lecturing on the necessity of understanding artistic activity as an act of reproduction, seeing art as an 'offspring' in which one planted one's own thought and nature. Alien nature was made familiar, inviting one to see an aspect of one's self in one's own doing: 'This human need runs through the most varied phenomena up to that particular form of self-reproduction in the external fact which is presented us in human art'.[9] The fabrication of Wilson's early cloud chamber in 1895 served the purpose of orchestrating and realizing an event as a matter of fact by means of an artificial, visual and tangible object that mimicked the object of nature sufficiently well to stand in for it:

> In September 1894 I spent a few weeks in the Observatory which then existed on the summit of Ben Nevis, the highest of the Scottish hills. The wonderful optical phenomena shown when the sun shone on the clouds surrounding the hill-top, and especially the coloured rings surrounding the sun (coronas) or surrounding the shadow cast by the hill-top or observer on mist or cloud (glories), greatly excited my interest and made me wish to imitate them in the laboratory.[10]

Though Wilson's initial trip to Ben Nevis made him 'fall in love with clouds'[11] and inspired him to create his first cloud chamber, he made many versions of it, each time modifying the apparatus according to its newly given purpose. While recreating fogs, mists and optical phenomena in 1895, for example, he also used his cloud chamber to measure levels of expansion at which vapour condensed or to observe moments at which colours were most brilliant. Soon after his creation of optical effects, his focus turned to an investigation of the possibilities for condensation. It is through his investigation with purified air in the cloud chamber that Wilson demonstrates his attitudes towards and pride in local nature. Examining a 1911 version of Wilson's cloud chamber, however, will aid in designating the various parts of the cloud chamber and how they operate (Figure 10.1). The top, flat glass chamber of Wilson's apparatus is where he would observe his experiments. The chamber sits on a brass cylinder inside of which a movable piston floats in a shallow trough of water keeping the air moist inside the cylinder. The large glass ball on the right is attached to a valve underneath the cylinder. The air in the glass ball is then evacuated using a pump. When the valve under the cylinder is opened, the piston inside the cylinder drops quickly, causing filtered (dust-free) air from the chamber to rush into the glass ball. The drop of pressure in the chamber makes the air expand, cooling the temperature and causing supersaturation in the glass chamber. Wilson would observe the top, flat glass chamber in darkness with a direct light source and make notes of the optical phenomena and cloud condensation created.

The introduction of photography into Wilson's later experiments changed the course of the project towards an investigation of subatomic matter not seen

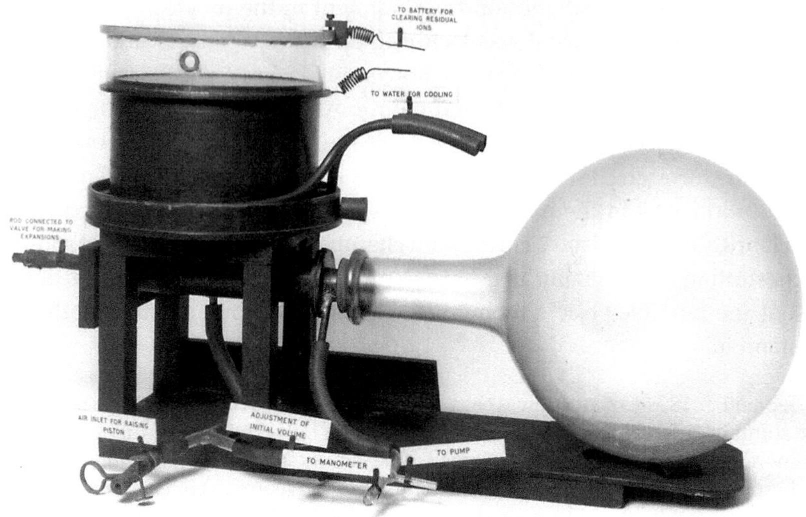

10.1 Wilson's Cloud Chamber (1911). Courtesy Cavendish Laboratory, Physics Department, University of Cambridge

by the naked eye. Thus, after approximately 10 years (1900–10) of focusing on thunderstorms and atmospheric electricity, Wilson returned to the cloud chamber with the intent of photographing rather than merely observing falling drops of liquid in the chamber. Once Wilson noticed cloud tracks in the chamber that looked like tiny, barely visible trails, presumably left behind by invisible subatomic particles, he managed to visually document the tracks through photographs showing 'little wisps and threads of clouds' excited by radiation. This new finding relied not only on the authoritative role of the photographic technology as an equivalent to sight, but, more importantly, on an image (almost) making possible an 'extension of vision' into the submicroscopic world.[12] As one physicist writes about the photographs of clouds tracks, '[b]ecause of the immediacy of the pictures obtained, the existence of the elementary particles is forced upon even the most skeptical'.[13] While Wilson's documentation of the tracks of charged particles is an exciting turn for his cloud chamber project, it is his early experiments, prior to his use of photography, at a time when he focused on the imitation of optical phenomena, that is of primary concern here. One of Wilson's notebook entries in 1895 poses the question, 'When [are the optical effects] best formed (In dusty or tolerably pure air …) [?]' The notebook entries that followed show that he proceeded by completely purifying the air.[14] The question of the inclusion or exclusion of dust in his cloud chamber foregrounds the extraordinary puzzle of Wilson's intentions with his initial experiments.

Wilson's colleague, the meteorologist John Aitken, had observed and measured dust particles in nature and in artificially created, supersaturated environments.[15] Aitken had also spent time at the observatory at Ben Nevis during the years 1889–93 and constructed a miniature dust chamber so as to count the dust particles in the atmosphere at the meteorological observatory.[16] Aitken found that the resulting condensation in his chamber was due to the presence of dust and concluded, as a result, that no clouds formed when dust particles had been filtered from the air. But Wilson knew that it was in fact possible to produce clouds through condensation of purified air *if* a specific 'expansion limit' was reached.[17] In fact, the environment created in the cloud chamber caused condensation in the chamber if one or all of the following circumstances were met: if dust particles were present in the chamber; if the expansion rate was sufficiently high (that is, if the drop in pressure in the chamber was sufficiently sudden and rapid); or, as he later discovered, and some had already theorized in the 1890s, if ions were present.[18] Through his experiments with the creation of optical effects – such as coronae (coloured concentric rings caused by light from the sun or moon hitting water droplets in clouds or fogs) and glories (coloured rings around the observer's shadow cast on a fogbank) (Colour plate 13) – Wilson had found limits at which expanded, moist and filtered air would indeed produce fog, mist or rain-like condensation. As he stated in his Nobel lecture several years later, '[w]ith expansions exceeding this limit, dense clouds were formed in dust-free air, the number of drops in the cloud increasing with great rapidity as the expansion was increased beyond it and giving rise on account of their small and uniform size to very beautiful colour phenomena'.[19] Wilson was clearly seeking to recreate magnificent optical phenomena, yet it seems less obvious why he chose to artificially filter all the dust from his chamber to reproduce optical effects when dust is naturally present in the atmosphere.

Peter Galison interprets Wilson's attempt to purify the air as a divergence from his previous intentions to make a mimetic experiment. 'This artificial aspect of his experiments', says Galison, 'signals a significant departure from the mimetic tradition in which his work was initially embedded'. According to this hypothesis, Wilson was no longer working with a matter of fact that was just like nature. Consequently, Galison concludes that, unlike Aitken who had wanted to recreate nature, 'Wilson created artificial laboratory conditions in order to "dissect nature"'.[20] But there are several problems with Galison's conclusion that Wilson was knowingly diverting from a mimetic experiment when he purified the air in the chamber. First of all, between the years 1904–10, Wilson put aside his cloud chamber, and did so because he worried that his use of X-rays, and the resulting condensation, was an artifice of his laboratory.[21] Considering this later and abrupt rejection of his project at a moment of marked doubt regarding its link with reality, it seems only plausible that Wilson's faith in his creation of a mimetic experiment was still intact during the time preceding his use of X-rays. Secondly, Galison seems to rely too comfortably on contemporary science when he draws his conclusion that Wilson was creating an artificial laboratory environment. For at the time of Wilson's experimentations, it was generally understood among meteorologists and laymen alike that the summit of Ben Nevis and the surrounding highlands provided the purest air, at times even exhibiting completely dust-free air. William Thomas Kilgour, a long-time staff member at the Ben Nevis Observatory, states that

> the summit of Ben Nevis ... enjoys the unique distinction of being the only station of the many at which observations have been taken where, at times, it has been found impossible to detect the presence of any dust. Sometimes, of course (particularly when the wind is from the direction of dense population), Ben Nevis air shows a large number of dust particles per cubic centimeter; but, on the mean, the West Highlands of Scotland, whether at a low or a high level, furnish the purest air as yet discovered in Europe.[22]

John Aitken was the authority on the measurement of dust, and in a 1902 retrospective paper on his work at Ben Nevis, he states that 'on a few occasions' he had observed no dust on the Ben. Meteorological observations reveal, adds Aitken, that the 'Western Highlands of Scotland is by far the least dusty in the world', noting soon again that 'the air of the Highlands is purer than at any other stations tested'.[23] Wilson's laboratory experiment with pure air cannot be divorced from a turn-of-the-century science of meteorology that conceived of the actuality of dust-free air under exceptional conditions.[24] Wilson was not only still working in the mimetic tradition, he had a specific experience and place in mind. Though rare but feasible according to Aitken's calculations, the atmosphere and upper air surrounding Ben Nevis could exhibit dust-free air.

Ben Nevis and Local Pride

Galison's account places insufficient emphasis on the locale that Wilson had in mind. That local meteorology mattered is evident in the descriptions by Kilgour and Aitken quoted above representing the quality of the air at the summit of Ben Nevis as well above the norm. This focus on locale seems more indicative of an eighteenth-century meteorology that sought to link the weather to the pride and identity of a region, a practice especially popular in the British provinces. Provincial weather phenomena were understood as part of the regional natural history and consequentially managed to signify the cultural importance of a local community.[25] Wilson's personal investment in the region near the Scottish hills appears in his propensity to describe his climbs on Ben Nevis and the surrounding hills alongside his laboratory notes. He was born and raised in the 'pastoral countryside', as he says, near Glencourse in the Pentland Hills and, upon retirement, returned to this place where his forebears had been farmers for generations. He acknowledged his debt to the locale in his Nobel Prize banquet speech when he declared, 'To those few weeks spent on the highest point of my native land I owe many happy years of work in the laboratory'.[26] And the constant emphasis on 'home' in his 'Reminiscences' (1960) confirms the extent to which he not only identified with the region but also understood his scientific work as bound to those experiences.[27] Wilson's meteorological practice was domestically centred and fundamentally a remnant of a bygone era.

The emphasis on personal experiences on the summit of Ben Nevis raises the question of Wilson's preferred imitation of the atmosphere at the mountaintop. For when Wilson purified the air in his cloud chamber, he was imitating what he believed to be a very specific and favourable condition right after condensation resulting in the purest experience possible on the summit. Dust-free air was not the norm; it was the exception, according to scientific research at the time. It was an extraordinary condition happening only rarely, as per Aitken's observations. The glories and coronae that Wilson sought to recreate were singular, specific to high altitudes and barely understood at the time.[28] Why was Wilson seeking to mimic the purest and most perfect experience possible, rather than what would have been conceived as an ordinary atmospheric condition on the top of Ben Nevis?

Constable or Ruskin as Artistic Source?

Late nineteenth-century meteorologists had laboured hard to dispel metaphysical and mythological speculations about the weather still lingering in the cultural mainstream.[29] In fact, meteorology as a science had long suffered from its own regressive theories, based as they were on Aristotle's *Meteorologica* as late as the 1600s. Even Luke Howard's 'Modifications of Clouds' (1803) was in principle merely descriptive of the various kinds of visual shapes formed by clouds. By the mid-1800s, meteorology was finally drawing on physics and chemistry for its practice, relying on experimental science rather than description or conjecture.[30] To now propose the possibility that Wilson's intentions might extend beyond scientific fact, beyond aesthetics and into the realm of metaphysics would be to suggest a major setback for the science of meteorology and, by extension, for particle physics. Perhaps it is for this reason that Galison chooses to omit the great Victorian writer, art critic, social critic, artist, scientist, poet and environmentalist John Ruskin from his cultural history of Wilson's cloud chamber, and this despite Ruskin's extensive writings on clouds. As a result, Galison provides a markedly different historical take on the Victorian imagination by including John Constable's cloud studies (1821–22), for example, while omitting John Ruskin's lecture 'The Storm-Cloud of the Nineteenth Century' (1884). These two Victorian men illustrate two very different, but contemporaneous ways of perceiving nature and truth. To Constable the idea of a 'truth of nature' would be nature itself made manifest, whereas to Ruskin 'truth of nature' would mean the revelation of a spiritual presence.[31]

In his lecture on the storm cloud, Ruskin described the existence of an unusual white cloud appearing in pure air, the cloud's 'prismatic colours' being 'impossible to paint'. He contrasted the pure, white cloud with the 'sulphurous chimney-pot vomit' of a 'diabolic' cloud.[32] In *Modern Painters*

(1888), Ruskin drew directly on biblical references when he claimed that clouds were God's chariot and His dwelling place, stating that it was precisely the colourful phenomena of light on the firmament of clouds that showed the divine most immediately.[33] Ruskin's emphasis on purity brings to mind an earlier scientific and artistic endeavour exemplified by Goethe's idea of the pure phenomenon (1798). He argued for a scientific experiment of a higher kind seen by the soul more than the eyes. After observing nature and subtracting the 'impure' and accidental, the pure phenomenon would emerge in the union of 'all experiences and experiments' combined.[34] In Goethe's mimetic science, man placed himself in nature (rather than against it), and through science and art he perfected it with the creation of an archetypal phenomenon. Wilson's early cloud chamber comprised a personal collection of thoughts externalized in a chamber placed in and beside nature. It served as an 'offspring', both like nature and more perfectly human. 'The nearest cloud maker is one's own self. All men are created cloud builders', says one meteorologist reflecting on the nature of clouds and the possibility of producing and controlling them. '[W]hen we cease cloud making we cease to live'.[35] In his cloud chamber Wilson planted his wonder and concerns, the experiences and experiments that for him truly mattered.

Acknowledgements

This is a revised and shorter version of an article with the same title published in *Chicago Art Journal* 16 (2006): pp. 32–45. My thanks to Lisa Zaher, Patrick J. La Riviere and Barbara Maria Stafford for reading and providing comments to earlier versions of this paper.

Notes

1 Gary Hawkins, notes to *The 64 Sonnets* by John Keats, ed. Edward Hirsch (Philadelphia: Paul Dry Books, 2004), p. 108 n.

2 Ibid. Charles Brown copied Keats's poem, probably the same year that Keats wrote it, and gives this account in his *Life of John Keats* (London: Oxford University Press, 1987).

3 Michael Cole, 'The Demonic Arts and the Origin of the Medium', *Art Bulletin* 84 (December 2002): p. 623; Hubert Damisch, *A Theory of /Cloud/: Toward a History of Painting*, trans. Janet Lloyd (Stanford: Stanford University Press, 2002), p. 23; Carlo Ginzburg, *The Night Battles: Witchcraft and Agrarian Cults in the Sixteenth and Seventeenth Centuries,* trans. Anne Tedeschi and John Tedeschi (London: Routledge and Kegan Paul, 1983).

4 John Ruskin, 'The Storm-Cloud of the Nineteenth Century' (1884), in *The Works of John Ruskin*, ed. E.T. Cook and Alexander Wedderburn (London: George Allen, 1908), p. 24, n. 3.

5 L.R. Phillips, *D.W. Griffith: Titan of the Film Art* (New York: Gordon Press, 1976), p. 109.

6 Lord Rutherford, 'C.T.R. Wilson – Bibliography', *Nobel Lectures, Physics 1922–1941* (Amsterdam: Elsevier Publishing, 1965), http://nobelprize.org/nobel_prizes/physics/laureates/1927/wilson-bio.html.

7 Peter Galison, 'Cloud Chambers: The Peculiar Genius of British Physics', in Galison, *Image and Logic: A Material Culture of Microphysics* (Chicago: University of Chicago Press, 1997).

8 Bruno Latour, 'Why Has Critique Run Out of Steam? From Matters of Fact to Matters of Concern', *Critical Inquiry* 30 (Winter 2004): p. 232.

9 G.W.F. Hegel, introduction to 'The Philosophy of Fine Art', in *Philosophies of Art and Beauty: Selected Readings in Aesthetics from Plato to Heidegger*, ed. Albert Hofstadter and Richard Kuhns (Chicago: University of Chicago Press, 1964), pp. 400–401.

10 C.T.R. Wilson, 'On the Cloud Method of Making Visible Ions and the Tracks of Ionizing Particles', Nobel Lecture (December 12, 1927), p. 194, in *Nobel Lectures, Physics 1922–1941*, http://www.nobel.se/physics/laureates/1927/wilson-lecture.pdf.

11 C.T.R. Wilson, 'Banquet Speech', Nobel Banquet in Stockholm (December 10, 1927), in *Les Prix Nobel en 1927*, ed. Carl Gustaf Santesson (Stockholm: Nobel Foundation, 1928), http://nobelprize.org/nobel_prizes/physics/laureates/1927/wilson-speech.html.

12 For a discussion of the problems inherent in the idea of an extension of vision through the use of instruments, see Joel Snyder, 'Visualization and Visibility', in *Picturing Science, Producing Art*, ed. Caroline A. Jones and Peter Galison (New York: Routledge, 1998), pp. 379–97.

13 Cyril Henderson, *Cloud and Bubble Chambers* (London: Methuen, 1970), p. vii.

14 C.T.R. Wilson's laboratory notes, A1 (March 26–August 15, 1895), Library of the Royal Society of London, quoted in Galison, *Image and Logic*, p. 96.

15 Boyle's air pump was made by his assistant Robert Hooke, who was to become the father of modern meteorology for inventions like the thermometer. Howard Frisinger, *The History of Meteorology to 1800* (New York: Science History Publications, 1977). Other important forerunners to Wilson's earliest work with the cloud chamber are Robert von Helmholtz, Carl Barus, Karl Kiessling and Jean Paul Coulier. See C.T.R. Wilson, 'Condensation of Water Vapour in the Presence of Dust-Free Air and Other Gases', *Philosophical Transactions of the Royal Society of London, A* 189 (1897).

16 John Aitken, 'Report on Atmospheric Dust', Meteorology of the Ben Nevis Observatory, *Transactions of the Royal Society of Edinburgh* 42/2 (1902): pp. 479–89; John Aitken, 'On the Formation of Small Clear Spaces in Dusty Air' (1883–84), *Transactions of the Royal Society of Edinburgh* 32/2 (1887): pp. 239–72.

17 Wilson, 'Condensation of Water Vapour', p. 266.

18 Galison, *Image and Logic*, pp. 94–96.

19 Wilson, Nobel Lecture, p. 195.

20 Galison, *Image and Logic*, pp. 95–96.

21 P.M.S. Blackett, 'Charles Thomson Rees Wilson', *Biographical Memoirs of Fellows of the Royal Society* 6 (November 1960): p. 287; Galison, *Image and Logic*, pp. 106–7.

22 William Thomas Kilgour, *Twenty Years on Ben Nevis* (London: Alexander Gardner, 1906), pp. 166–67.

23 Aitken, 'Report on Atmospheric Dust', p. 482.

24 Wilson himself justifies his use of dust-free air in his laboratory experiments by noting the difference in pressure between the air at the lower cloud layer and the air that rises above it where condensation becomes again possible. He adds, however, that any increasing, rapid expansion in his chamber beyond this re-emerging condensation would have little meteorological significance, as it would be unlikely to occur in nature. Wilson, 'Condensation of Water Vapour', p. 286.

25 See Vladimir Janković, *Reading the Skies: A Cultural History of English Weather, 1650–1820* (Chicago: University of Chicago Press, 2000), p. 166.

26 Wilson, Nobel Banquet Speech.

27 C.T.R. Wilson, 'Reminiscences of My Early Years', *Notes and Records of the Royal Society of London* 14/2 (1960): pp. 163–73.

28 Alexander Buchan, 'The Meteorology of Ben Nevis', *Transactions of the Royal Society of Edinburgh* 34 (1890): p. xlix.

29 Thunderstorms, for example, were thought to cause living organisms to rot. John Aitken, 'On the Number of Dust Particles in the Atmosphere of Certain Places in Great Britain and on the Continent with Remarks on the Relation between the Amount of Dust and Meteorological Phenomena' (1892), *Transactions of the Royal Society of Edinburgh* 37 (1892–95): p. 21.

30 Aristotle, *Meteorologica*, trans. H.D.P. Lee (Cambridge: Harvard University Press, 1952); Luke Howard, 'On the Modifications of Clouds, and on the Principles of their Production, Suspension, and Destruction: Being the Substance of an Essay Read before the Askesian Society in the Session 1802–1803', pts 1 and 2, *Philosophical Magazine* 16 (1803): pp. 97–107; 344–57; 17 (1803): pp. 5–11.

31 Ann Bermingham, *Landscape and Ideology: The English Rustic Tradition, 1740–1860* (Berkeley: University of California Press, 1986), p. 178.

32 Ruskin 'The Storm-Cloud', pp. 38–40.

33 Wilson was married to a reverend's daughter. John Ruskin, *Modern Painters*, vol. 4, *Of Mountain Beauty* (Orpington, Kent: George Allen, 1888), pp. 85–89.

34 Johann Wolfgang Goethe, 'The Pure Phenomenon' ('Das Reine Phänomen') (January 15, 1798), in *Goethe's Color Theory*, American edition, trans. and ed. Herb Aach (New York: Van Nostrand Reinhold, 1971), pp. 65–66.

35 Alexander McAdie, *A Cloud Atlas* (Chicago: Rand McNally, 1923), p. 27.

11

Images of Rain between Representation, Technology and Nature

Hanna Johansson

In Finland where I live, the amount of rainfall has increased in recent years. Meteorologists predict that, as average temperatures rise within the next 50 years, the amount of rainfall will increase another 30 per cent, just as the seasonal snow cover will move drastically further north. These current forecasts of climate change, especially the increased amount of precipitation, have led me to look at rain from a new, more intense angle. I have also started to pay attention to rain in artistic representations and to ask what role rain has played in landscape art.

In this essay I approach images of rain in two periods of time. The first is the early nineteenth century during the so-called second scientific revolution, when systematic observations of the sky gained in popularity. The second is the later part of the twentieth century, when landscape again became an interesting topos in art, after having been more or less neglected during the early part of the century.[1] Landscape reappeared in land and environmental art, as well as in experimental films and installations.

Although there is a continuation between early nineteenth-century landscape paintings and contemporary art in subject matter and a kind of realistic endeavour, I will argue that there are also interruptions and dramatic changes. These changes concern how landscapes, or more precisely in this case, phenomena in the sky, are understood in relation to human beings and technology. I suggest that this shift in landscape art can be compared to arguments made by the sociologist of science Bruno Latour, who talks about a move from first to second empiricism and a shift from matters of fact to matters of concern.

This new, realistic attitude searches for significant connections between human and non-human things and wants to explore the relationships between what is known and what is given by experience. Its aim is to uncover connections between different things instead of ascertaining facts. Latour explains this attitude by using the theatre as an example: a matter of concern is what happens to a matter of fact when you add to it the scenography as a whole, shifting your attention from the stage to the whole machinery of the theatre.[2] In other words, Latour argues that the second empiricism asks us to approach things not as they are in themselves but rather as they act together in an open network.

I will elaborate on the changes and transformations of images of rain through works of three artists. The first is John Constable's cloud paintings from the early 1820s. Secondly, I analyze the cinematic works and installations from the 1970s to the 1990s by the English filmmaker Chris Welsby. And thirdly, I look at atmospheric works by the Danish-Icelandic artist Olafur Eliasson. I believe that all the examples are related to so-called realistic art, because the artists have not sought to 'express', but rather to record, environmental phenomena as objectively as possible by drawing on contemporary science and technology. However, Constable belongs to the first empiricism – matters of fact – whereas the others belong to the second empiricism and its emphasis on matters of concern.

Interestingly, all of these works use technology in an experimental manner, that is, in a 'wrong' or unexpected way in relation to science. For Constable, technology must be understood as a painting technique, which he renewed by using 'too much' impasto, too roughly applied. Welsby uses the film camera's mechanisms unconventionally to highlight aspects of nature and weather that would otherwise remain unnoticed, while Eliasson experiments with sophisticated technology to produce seemingly natural phenomena inside museums. The misuse of technology has a subversive potential, which in the visual arts focuses on the manner in which we see the world. Donna Haraway's notion of 'situated knowledge' reminds us of the embodied nature of all vision, even if the eyes are those of technological science. Haraway draws a parallel between technological and human eyes by pointing out that human and non-human optical devices alike do not reproduce images passively, but rather produce knowledge.[3]

Apart from some shared aspects, the examples I raise represent highly different relations to nature and culture. Constable lived in the empirical atmosphere of early nineteenth-century England, in which different fields of sciences began to emerge and nature was studied as an object distinct from human culture. Welsby and Eliasson both live in the age of environmental concern. For them, nature is no longer separate from culture; technology is intimately intertwined with natural processes and, in Eliasson's case, nature as such is called into question.

The Science of Meteorology, John Constable's Cloud Sketches and the Effect of Rain

Although human beings have always studied and observed the sky, in the early years of the nineteenth century a certain change in attitude took place in

European culture. The sky was to be considered a natural phenomenon, which could be explored as a scientific object. The sky and weather phenomena played a substantial role in early landscape illustrations and depictions. They were seen as a symbol of the invisible and of transcendental religious emotions. In Western art, the sky has been regarded as a sacred and a symbolic place, its quick variations corresponding to the depths of inner life. The sky was also de-sacralized, and in Northern Europe in particular we find accurate images of different weather phenomena already in the late medieval period.[4]

Eventually, when laboratory sciences emerged, the atmosphere too was seen to be a vast chemical laboratory.[5] In England the dominant figures who formed the new science of meteorology were Luke Howard and John Foerster. Howard's idea of climate was based on two principles: that climate works according to laws and that it could be understood historically. He expressed a desire to found 'a history of climate',[6] which could be possible because people had acquired knowledge of the peculiar features of their own climate and its facts, 'which, properly arranged, would form its History'.[7]

It was this first stage in the scientific development of meteorology, which remained within the scope of simple, first-hand observation, that became important for Constable.[8] He thought that painting was comparable to science and that it could be considered a branch of natural philosophy.[9] Constable's idea of scientific thinking was situated on the threshold between the old, comprehensive natural philosophy and the new independent laboratory sciences.

Particularly interesting from a scientific viewpoint are Constable's cloud studies, painted in Hampstead in 1821 and 1822 (Figure 11.1). Although Constable was by no means the earliest landscape painter to give close attention to clouds, his cloud sketches occupy a special position among sky paintings. Sky painting had already developed on the continent, as Constable knew very well.[10] However, many earlier paintings of the sky and the heavens were laden with religious and apocalyptic meanings. Constable's sketches consist of dozens of small, accurate oil studies of clouds and the sky made *en plein air*. An annotation of the precise time the painting was made and the weather conditions accompanies every study. Some of the annotations even include a description of the weather before and after the moment of sketching. Art historian Ann Bermingham states that, during these years, the change of weather and the transitory aerial effects became more important to Constable than the familiar sites and places that he had previously favoured so keenly in his paintings. Constable's sun-filled naturalism was transformed into a meteorological expressionism.[11]

Sky sketching was considered a cultivated activity among Constable's contemporaries, both on the continent as well as in England. Artists were 'in the service of clouds', as John Ruskin defined the pictorial production of the early nineteenth century. In England William Turner, who complained about the inconvenience of painting outdoors, still filled a sketchbook with sky pictures and argued that his contemporaries surpassed their predecessors in

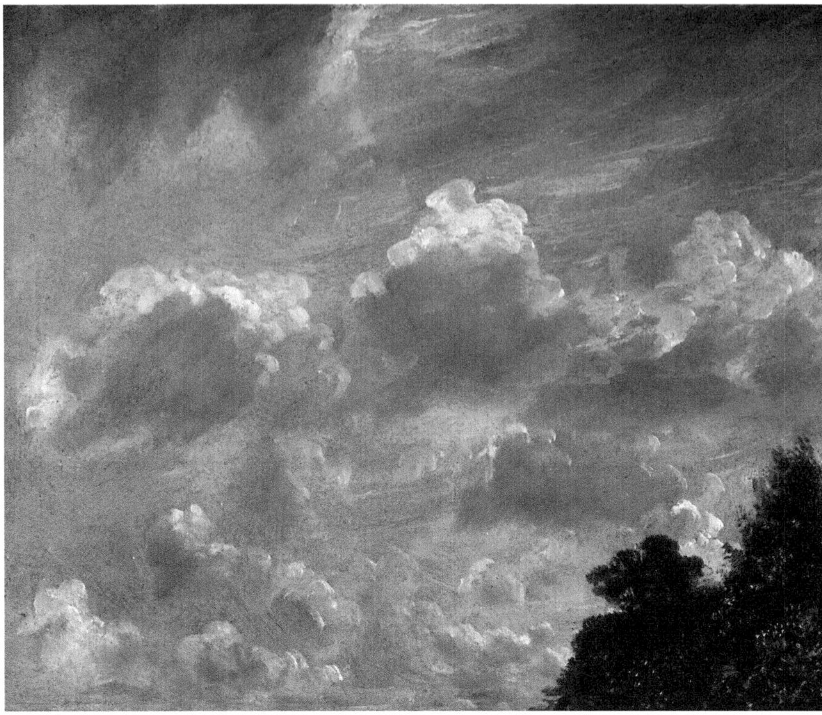

11.1 John Constable, *Cloud Study, Hampstead, Trees at Right 11 September 1821,* oil on paper laid on board, red ground. National Gallery of Australia. © Royal Academy of Arts, London/ Photo credit: John Hammond

landscape art because of the 'truth' of their skies. Constable's contemporaries, David Cox, Copley Fielding and James Harding, also painted swift changes of the sky, with Cox focusing especially on depicting the effects of the wind on landscape. But Constable's position among these sky painters is extraordinary not only because his brushwork was unconventional but also because of the abundance of his cloud paintings over a period of several years. In addition, he is arguably the first painter who systematically observed the sky from a scientific perspective. The annotations on the reverse sides of these sketches are records of these scientific observations.[12]

Already in the 1930s, the English meteorologist J.W. Bonacina stressed the importance of the scientific as well as the artistic approach to the atmosphere in Constable's works.[13] This discussion was amplified in the 1950s after the publication of Kurt Badt's *John Constable's Clouds*. Badt examines the dialogue between science and art at the turn of the eighteenth and nineteenth centuries. He charts the influence of the new science of meteorology on painting and poetry in particular. Badt's main concern is how the influence of science, with its search for certainty and regularity, can be traced in Constable's cloud paintings.[14]

In the last 50 years there has also been a dispute about Constable: was he a Realist painter who recorded the sky according to scientific knowledge or was he a pioneer of Abstraction and a prophet of European High Modernism up to

Abstract Expressionism? Curiously, his cloud studies occupy a key position in both of these arguments. Sky and clouds both provided evidence of the new science of meteorology.[15] Constable was accused of not having any subjects in his landscapes. But one can argue that his sky paintings turn the attention to that part of the landscape, the sky itself, which is the source of light in nature, and which governs the entire landscape. The lack of a horizontal landscape composition with figures, however, tilted his works towards Abstraction.

One characteristic of Realism in painting was a disregard for the artist's inner life and the apparent withdrawal of the author, a concentration on both an outer reality and a technique of representation that highlights the artificial qualities of the image. Thus, realistic painting paradoxically emphasizes the artificiality of the means of representation. The thick impasto, often left rough, makes the viewer conscious that she is faced with a painted canvas, and not looking at a landscape. Constable is often mentioned as a predecessor of Realism. Although he emphasized the materiality of painting and led it towards Abstraction, the paintings do awaken emotions, emotions that might have been there for Constable himself.[16] In recent years the realistic understanding of Constable's cloud sketches as weather records has again been taken up.[17] What seems to be the key issue here is understanding the link between Constable's cloud studies and the 'spirit' of the work of the early meteorologists. In the 1970s John Thornes compared the weather statistics in southern England to Constable's cloud paintings on the same places and days, and he was convinced that the paintings are rather precise weather records.[18] According to Constable's diary, the artist preferred to paint in the west wind, which involved the visually impressive, rain-carrying cumulo-clouds, as well as the so-called cumulo-nimbus clouds, which cause thunderstorms. However, in none of his cloud studies is it actually raining, although in one annotation he writes, 'September 3rd. Noon. Very Sultry with large drops of rain falling on my palette'.[19] However, we do see sudden rain showers and distant rainbows in his other larger and more 'substantial' works. According to his diary, he liked to paint landscapes right after a night of rain. 'Nature is never seen, in this climate at least, to greater perfection than at about nine o'clock in the mornings of July and August / and it is still more delightful if vegetation has been refreshed with a shower during the night'.[20]

In effect, there seems to be a temporal aspect to the way Constable painted rain. He either anticipated rain or painted a landscape just after a shower. Several writers have explained that Constable's intention was to look at a landscape as a development in time wherein simultaneous natural phenomena might be comprehended in a single image. This may also be the reason that Constable added remarks about the wind and temperature to his cloud studies, because previous weather changes explained the different forms of the clouds. The same goes for the paintings: an expert can judge the previous state of the weather or forecast what is to come from the details of the painted scenery.[21]

Cooperation with Weather and Technology

The idea of systematic descriptions of weather in his cloud sketches gave Constable an understanding of weather as a temporal phenomenon that told the 'story' of moisture and heat, rain showers and wind. His paintings regard nature as an ever-changing history of the weather, where rain has an effect on the changing image of the landscape. Today one can say, as Gillen Wood has suggested, that owing presumably to climate change, these paintings have become apocalyptical, because they represent the gradually disappearing climate of southern England, regardless of the fact that climatologists predict precipitation in Britain will increase slightly in the future. Somewhat surprisingly, in that part of southern England where Constable painted his cloud sketches, the annual rainfall will be about 30 per cent less.[22]

In 1980 the art theorist Peter Wollen stated that the films of the English experimental filmmaker Chris Welsby (b. 1948) should be seen in the tradition of John Constable's landscape art.[23] Wollen acknowledges the meteorological narration in Constable's and Welsby's art, and also, more importantly, the scientific and technological aspects of both artists' methods.

Wollen draws a parallel between Constable's meteorological paintings of landscape and the nearly contemporary invention of photography. Photography and later film offered artists new ways of integrating their observations of landscapes and climate into a visual representation, for instance, by using serialization. This allows the artist to capture the sequence of instantaneities, also found in Welsby's manner of mediating meteorological investigations.[24]

Although there is an obvious affinity between the two artists, as Wollen points out, the works of each differ dramatically in the way observations are made, how the technical devices are used and how the works represent the aerial subject for the viewer, none of which Wollen takes into account. The differences appear when the works are approached with matters of concern in mind.

Chris Welsby is one of the main figures among English experimental filmmakers. He has been making films since 1971 and, more recently, film installations as well. Throughout his career he has had a special focus on landscape, and his films belong to avant-garde British landscape films, as defined by Deke Dusinberre.[25] According to Dusinberre, these experimental landscape films made an important contribution to the development of avant-garde filmmaking because they succeeded in asserting both the illusionism of cinema through the sensuality of landscape imagery and the material nature of the representational process.[26]

Yet according to the film theorist Laura Mulvey, a better classification for Welsby than 'landscape artist' would be 'weather artist', because weather plays such a crucial role in Welsby's films. Mulvey emphasizes the difference between landscape and weather; for her, 'landscape stands as an object for exterior observation and contemplation, whereas the weather is a force for change'.[27] The difference between landscape and weather is definitely not so clear. As Luke Howard argues in his writing[28] and as Constable and Welsby point out in their works, weather is, in fact, what makes a landscape.

Welsby's films are based on the interaction between nature and technology. Weather and other kinds of natural forces, such as the rotation of the Earth, the tides, the sun and wind, act as a director, although the artist adjusts the camera and its mechanism. The artist is not necessarily observing the sky at all while the camera is shooting. In Welsby's practice, natural processes are no longer simply recorded from the outside, as objects of observation, but are made to participate in the scheme of observation itself. The observation is made by the automatic procedures of technology as directed by nature, not by the artist's rendering of it. The promise at the heart of Welsby's work is a new type of equal relationship between technology and nature and between the subject and the object of observation.[29]

In some of Welsby's films we see rain, as in a 20-minute film called *Seven Days* (1974), one of Welsby's early mature works and filmed in Wales. According to Fred Camper, Welsby offers a tour de force such as cinema had never seen before.[30] The camera was mounted on an equatorial stand that was aligned with the Earth's axis and that rotated on its own axis once every 24 hours. The image selection was controlled by the cloud coverage. When the sun was shining, the camera was turned towards its own shadow; when the weather was cloudy, the camera was turned towards the sun. Consequently, the shape of the film is a product of the interaction between the predictable mechanistic technology and the chance-like qualities of the natural world. The rigid system that structures the film nevertheless depends upon an aleatory element – the weather – for its realization.[31]

In the later part of the film there is a long period depicting the rain. The image is of raindrops hitting the lens, and the landscape behind is blurred. Welsby's technology and the structure of film mechanics allow rain to be depicted. Yet at the same time the rain is destroying the visual qualities and representation of the landscape; the wet lens of the camera becomes the only observable thing.

In the film installation *Rainfall* (1983), the screen is a horizontal plane suspended two feet above the gallery floor (Figure 11.2). Rain has a double function in this installation: it is both the performer and the subject of the work. The projector beam is a vertical cone of light from which rain appears to be falling onto the horizontal screen below, transforming the screen into a surface of water pitted by raindrops. The recorded sound of heavy rain falling in a forest fills the acoustic space.

Besides falling rain, Welsby's installation emphasizes the situation itself wherein the viewer's participation becomes significant. The installation focuses the viewer's attention on the process of perceiving.[32] A continuous and constant image and the sound of falling raindrops in a closed gallery space creates a strong atmosphere, an ambiance of artificial timelessness, but it also disturbs the demarcation between the natural phenomenon of rain

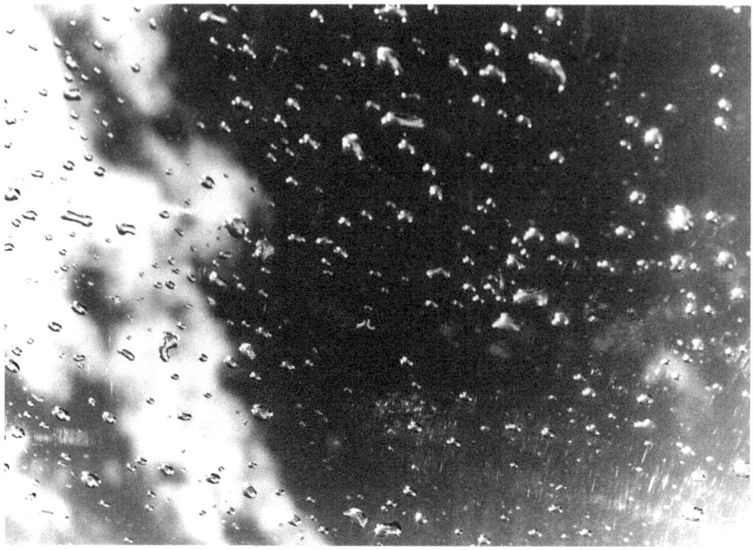

11.2 Chris Welsby, film still from the installation *Rainfall* (1983), 16 mm.
© Chris Welsby/Photo credit: Chris Welsby

and its indoor representation. According to the film scholar Chrissie Iles, Welsby's *Rainfall* marks the beginning of a shift away from earlier, structured films of the 1970s, towards more emotionally charged work.[33] However, this emotionality is not a result of the method of the work, but derives instead from the spatial properties of the installation.

Even if nature and its representation intermingle in the viewer's perception, the border between real and artificial rain is still there. In fact, *Rainfall* can be seen as an upgrading of the classic examples of trompe l'oeil in cupolas in the fifteenth and sixteenth centuries, for instance, Andrea Mantegna's *Spouses' Chamber* in Mantua (1472–74) or Correggio's *The Vision of Saint John on Patmos* (1520–24), which are constructed so that the paintings on the cupolas seem to pierce the ceiling and open the view up into the sky.[34]

Mechanical Production of Rain and Its Affective Image

Among the abundant body of works of the Danish-Icelandic artist Olafur Eliasson (b. 1967), a group of installations deals directly with weather-based natural phenomena, such as fog or freezing and melting ice. Eliasson uses mechanical devices to simulate natural phenomena in installations, which he stages both within a museum context as well as in outdoor environments. In museums his works evoke ambiguous landscape situations, such as *The Mediated Motion* at Kunsthaus Bregenz (2001), in which three different conditions of nature – earth, an algae-covered pond and fog – were set up in an exhibition space.

Architectural theorists Ilka and Andreas Ruby call these situations pseudo-natural landscapes and compare the effect of the works to epic theatre and its double alienation. This means that when the museum context makes a natural phenomenon seem weird, the natural phenomenon in the museum accordingly makes the museum seems strange.[35]

This alienation effect continues in Eliasson's handling of nature. His natural indoor phenomena disturb the experience of nature and sharpen the viewers' perception of naturalness. Eliasson's works underscore the ways in which cultural institutions mediate the viewer's perception of seemingly pure processes. By leaving the mechanics of the works deliberately visible, he lays bare the artifice of illusions and further accentuates the elliptical relationship between reality, perception and representation, giving attention to matters of concern and to the scenographical connections, of which our environment and our world is made.

Among the works that reproduce natural processes and phenomena are two installations that deal with rain and related phenomena. I argued above that Welsby's *Rainfall* gives an atmospheric task to rain, which again engenders affective qualities in the perception of the work. Eliasson's installations have the same kind of affective and emotional range as *Rainfall,* but they carry the blurring or intermingling of real and fictional, or factual and artifactual, beyond Welsby. In *Your Strange Certainty Still Kept* (1996), Eliasson focuses the perceiver's attention on a mundane phenomenon – the falling of water drops, that is, rain in the gallery space (Figure 11.3). The work consists of a pierced hose from which droplets of water fall into a pool, and lights that go on and off quickly, causing a cinematic flickering effect. *Beauty* (1993) also consists of tiny drops of water sprinkling down from a perforated hose and light coming from above. Light splits the air and goes through the water curtain to produce a rainbow in the space. The rainbow, however, is perceivable only from a certain viewpoint; one step away and it disappears. Both works execute two possibilities that are almost irreconcilable. On the one hand, they work as an image or a representation; they allow one to look at the moving image of rain and the rainbow where electric light and falling drops create a continuously shifting protocinematic experience.[36] Yet on the other hand, it is also possible to touch this cinematic image – namely, water – to walk through it and to play with it.

As these two examples almost deliberately show, Eliasson suggests that rain is not necessarily caused by nature alone, but might also be constructed by human beings and culture. His installations call the viewers' attention to contemporary climate issues, showing that air conditioning and the global climate system cannot be regarded as completely separated entities. As Latour notes, Eliasson's will to control climate is not after all inspired by ambition for total mastery over the elements, but by a reasonable desire to ascertain what sort of breathing space is conducive to civilized life.[37]

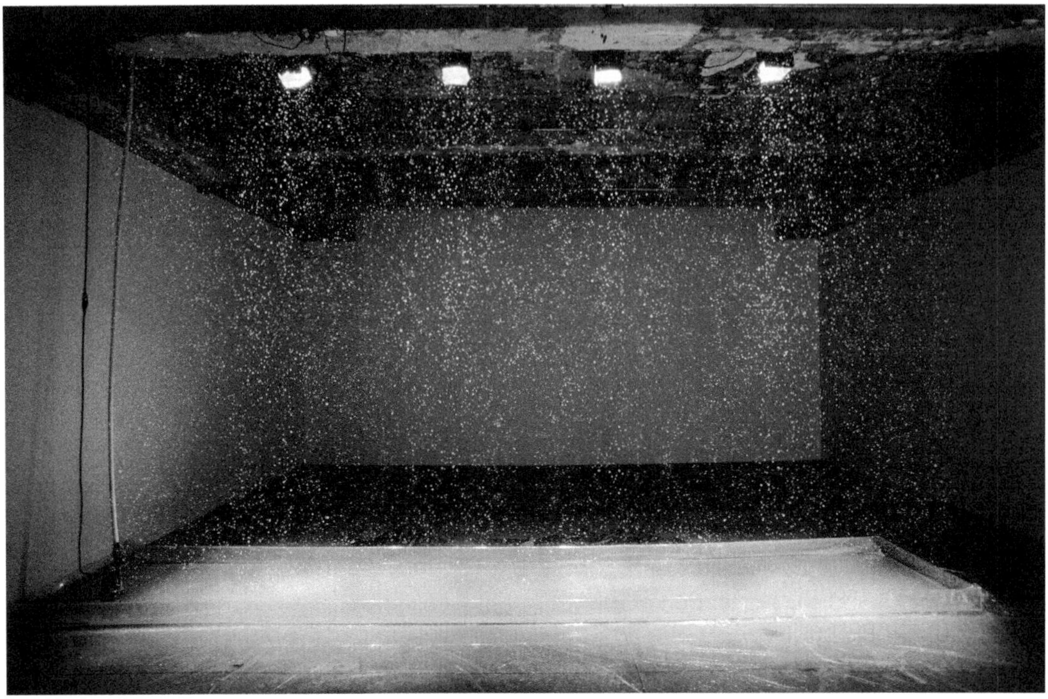

11.3 Olafur Eliasson, *Your Strange Certainty Still Kept* (1996), installation, water, strobe lights, acrylic, foil, wood, pump and hose. Dakis Joannou Collection, Athens. © 1996 Olafur Eliasson/ Photo credit: Olafur Eliasson

The Explicitation of Rain

The three artists discussed here use the modern scientific discoveries of their respective times. Although Constable's paintings have substantial personal traces that make them oscillate in a paradoxical space between illusion and truthful representation, all aspire to anti-subjective representation. The result in every case, however, is not only descriptive but also affective and emotional. These works erase the division between the objective scientific approach and the emotional, bodily feelings based on affects and spontaneous reactions. And still, all of these works take a distance from the subjective expression of the artist's innermost self.

They bring forth the changing conceptions of representation, from painted records of the historical development of climate to the technical recorded temporal representation of rain, and in the last case to rain itself as a material and affective image produced mechanically. In addition, the examples emphasize the changing paradigms in man's relation to the natural world. In Constable's case, nature is revealed through the budding science of meteorology and by paintings that both record the new science and give an emotional interpretation of it. In Welsby's and Eliasson's works, the separation between nature, technology and culture gradually disappears. In Welsby, nature interacts with technology, but the division has not yet totally vanished. In Welsby's works, the fusion between technology and nature takes place on the level of illusion or representation, but not physically.

In Eliasson's installations, rain is a natural but technologically domesticated phenomenon and an image artificially produced and controlled. The old divisions between wild and domesticated, naturally given and technologically made, are ignored and replaced by affective experimentations that reposition human beings in the environment by letting us feel and see things differently.[38]

It seems to me that by following the historical traces of the representation of rain, one can argue that rain has become more and more tangible and explicit; moving from the background to the foreground. This is the movement that resembles the German philosopher Peter Sloterdijk's incisive argument – namely, that our history has not been so much about modernization as it has been about another phenomenon that he calls 'explicitation'. According to Latour, Sloterdijk's argument means that as we have moved on through our scientific and technological inquiries, we have rendered more and more explicit the fragility of the life support systems that make our spheres of existence possible, until eventually we come to realize that the self-evident natural elements might disappear. Accordingly, I would like to suggest that rain has become explicit not only in the artistic representations I have discussed here but perhaps also in our everyday experience, because we have begun to realize that we will either get too much or too little of it. That means weather is not what it used to be – that is, reliable.

This confirms the change in attitude that Latour describes as a shift from matters of fact to matters of concern. Based on these representative examples, I argue that rain has become a matter of concern the more it has been revealed that the atmosphere is a chemical laboratory and that instead of being infinite, atmosphere has strict limits.

Notes

1 Hanna Johansson, 'The Revival of Landscape Art', in *Landscape Theory*, ed. Rachel deLue and James Elkins (London and New York: Routledge, 2007), pp. 221–28.

2 Bruno Latour, 'What Is the Style of Matters of Concern', Spinoza Lectures at the University of Amsterdam, April and May 2005 (Amsterdam: Van Gorcum, 2008), p. 39; Bruno Latour, 'Why Has Critique Run Out of Steam? From Matters of Fact to Matters of Concern', *Critical Inquiry* 30/2 (Winter 2004): pp. 225–48, p. 232.

3 Donna Haraway, 'Situated Knowledges: The Science Question in Feminism and the Privileged of Partial Perspective', *Feminist Studies* 14/3 (1988): pp. 575–99, p. 583.

4 For a detailed analysis about the meanings of the sky and weather in the history of Western landscape art, see Jacob Wamberg, *Landscape as World Picture: Tracing Cultural Evolution in Images* (Aarhus, Denmark: Aarhus University Press, 2009), vol. 1, chapter 4, and vol. 2, chapter 11.

5 Luke Howard, *The Climate of London*, vols 1–3 (London: Harvey and Darton, 1833), vol. 1, p. 1; Kurt Badt, *John Constable's Clouds* (London: Routledge and Kegan Paul, 1950), pp. 62–63.

6 Ibid., p. 64.

7 Howard, *Climate*, p. ii.

8 There were different apparatuses in use in Constable's time such as different kinds of barometers and thermometers, but for the most part the observations were made with the naked eye. See Edward Morris (ed.), *Constable's Clouds: Paintings and Cloud Studies by John Constable* (Edinburgh and Liverpool: National Galleries of Scotland, 2000), pp. 118–20.

9 Charles Robert Leslie, *Memoirs of the Life of John Constable* (London: Phaidon, 1995), p. 274.

10 Ibid., p. 270.

11 The shift of emphasis from sites and objects to varying climate conditions in Constable coincides with the more abstract style; Ann Bermingham, *Landscape and Ideology: The English Rustic Tradition, 1740–1860* (Berkeley: University of California Press, 1986), p. 148. See also Michael Rosenthal, *Constable: The Painter and His Landscape* (New Haven: Yale University Press, 1983), p. 166.

12 John Cage, 'Clouds over Europe', in *Constable's Clouds: Paintings and Cloud Studies by John Constable*, ed. Edward Morris (Edinburgh and Liverpool: National Galleries of Scotland, 2000), pp. 125–34; Anne Lyles, '"The immense canopy": Studies of Sky and Cloud by British Artists c. 1770–1860', in *Constable's Clouds*, ed. Edward Morris, pp. 143–48.

13 L.C. Bonacina, 'John Constable's Centenary: His Position as a Painter of Weather', *Quarterly Journal of the Royal Meteorological Society* 63/272 (1937): pp. 483–90.

14 Badt, *John Constable's Clouds*.

15 See Graham Reynolds, *Constable: The Natural Painter* (London: Cory, Adams and Mackay, 1965), p. 86; Robert Rosenblum, *Modern Painting and the Northern Romantic Tradition: Friedrich to Rothko* (London: Thames and Hudson, 1975).

16 Charles Rosen and Henri Zerner, *Romanticm and Realism: The Mythology of Nineteenth Century Art* (London and Boston: Faber and Faber, 1984), pp. 143–57.

17 See Gillen D'Arcy Wood, 'Constable, Clouds, Climate Change', *The Wordsworth Circle* 38/1–2 (2007): pp. 25–34.

18 John Thornes, *The Accurate Dating of Certain of John Constable's Cloud Studies 1821/22 Using Historical Weather Records*, Occasional Papers, 34 (London: Department of Geography, University College London, 1978).

19 Ibid., p. 13.

20 R.B. Beckett, ed., *John Constable's Correspondence*, vol. 3 (Ipswich: Suffolk Records Society, 1950), p. 96.

21 Badt, *John Constable's Clouds*, p. 63.

22 Wood, 'Constable, Clouds, Climate Change', pp. 31–33.

23 Peter Wollen, 'Chris Welsby – Films/Photograps/Writings' (London, 1980).

24 Ibid.

25 Like the Realist paintings, avant-garde films in the 1970s also emphasized the material basis of the film, which strove for illusionist representation. Deke Dusinberre, *Avant-garde British Landscape Films in the Tate Film Theatre, 3–21 March 1975* (London: Tate Britain, 1975). See also Deke Dusinberre, 'On British Avant-garde Landscape Films', *Undercut* 7/8 (1983): pp. 48–49.

26 Deke Dusinberre, 'On British Avant-garde Landscape Films', p. 49; see also Deke Dusinberre, 'St George in the Forest: The English Avant-Garde', *Afterimage* 6 (1976): pp. 4–19.

27 Laura Mulvey, 'Introductory Essay', in *Chris Welsby British Artists Films*, DVD Jacket (London: British Film Institute, 2005).

28 Luke Howard, *Seven Lectures on Meteorology* (London: Harvey and Darton, 1843), pp. 2–3.

29 Wollen, 'Chris Welsby', p. 2.

30 Fred Camper, 'Blowin' in the Wind Films by Chris Welsby', http://www.fredcamper.com/Film/Welsby.html, 2001.

31 Dusinberre, 'On British Avant-garde', p. 50.

32 Chrissie Iles, *Signs of the Times: A Decade of Film, Video and Slide – Tape Installation in Britain 1980–1990* (Oxford: Museum of Modern Art, 1990), p. 53.

33 Ibid.

34 Hubert Damisch, *Theory of /Cloud/: Toward a History of Painting*, trans. Janet Lloyd (Stanford: Stanford University Press, 2002), pp. 1–2.

35 Ilka and Andreas Ruby, 'Take-away Landscape', in *Your Black Horizon Art Pavilion: Olafur Eliasson and David Adjaye* (Cologne: Verlag der Buchhandlung Walther König, 2007), pp. 142–43.

36 The notion of the protocinematic in relation to Eliasson's art comes from curators Klaus Biesenbach and Roxana Marcoci. They suggest that Eliasson's works are descendants of Lissitzky and Moholy-Nagy in two ways: firstly, because Eliasson uses simple and unconventional optical techniques and experiments with light sources, and secondly, because of the social potential these experimental projections achieve in asking people to take part in the works thus allowing them to create a temporal community. The two works discussed here provide good examples of these features. See Klaus Biesenbach and Roxana Marcoci, 'Toward the Sun: Olafur Eliasson's Protocinematic Vision', in *Take Your Time: Olafur Eliasson*, ed. Madeleine Grynsztejn (New York and London: San Francisco Museum of Modern Art with Thames and Hudson, 2007).

37 Bruno Latour, 'Atmosphère, Atmosphère', in *Olafur Eliasson: The Weather Project*, ed. Susan May (London: Tate Modern, 2003), p. 30.

38 Mieke Bal, 'Light Politics', in *Take Your Time: Olafur Eliasson*, ed. Madeleine Grynsztejn (New York and London: San Francisco Museum of Modern Art with Thames and Hudson, 2007), pp. 158–58, 176–77.

12

Crossovers: The Art of Rodney Graham, Roni Horn and Diana Thater between Technology and Nature

Hans Dickel

Before the early modern age when art and technology began to diverge, both endeavours were jointly juxtaposed to nature as intellectual achievements; Leonardo da Vinci, for instance, was still active in both fields. The more recent history of art and technology, however, is characterized by mutual aversion: the decided emphasis on the social function of technology and the autonomy of art has – generally speaking – prevented their alliance. Meanwhile, following the culmination of abstract painting, the autonomy of modern art is now in danger of hitting a dead end. This may explain why artists are seeking connections to science and to service industries; using image media based on technology, they are once again introducing Representationalism into their art.[1] But civilization, which rests on the exploitation of nature, seems to have reached the 'limits to growth'.[2] Even scientists are aware of the inherent limits of their positivist field in its advancement and its dependence on nature.[3] The harbingers of a possible climatic disaster make us all too clearly aware that the 'second nature' man has established could, one day, be overtaken again by another one. In Western consumer societies, which were the first to develop an awareness of the ecological crisis, numerous artists have attempted, especially since the 1960s, to convey a holistic, expanded perception of nature. There was a move away from easel painting and a detached view of nature in favour of presenting natural materials beyond the patterns of 'landscape'. In environments or installations, the latter were to be experienced with all the qualities that affect the senses. Yet it must be considered that art, like technology, is essentially at odds with nature, because to have natural materials appear in art as authentic substances, an artificial framework is always necessary. Why then should not contemporary technologies also be used to expand the aesthetic perception of nature? However, the precondition for such a revision of artistic traditions is also a new understanding of technical equipment, how it functions and its media use and modality. Though technology-based image media are generally used only as means to an end, in contemporary artwork they are themselves a topic. I explore the use in art of our culture's key media – namely, photography, film and video – with regard to nature by looking at selected works by Rodney Graham, Roni Horn and Diana Thater. All three artists have liberated available technology from its customary use and wielded it for their own, independently defined functions and modalities in investigating nature. These artists have overcome or enlarged the customary ways of using photography and film, such as those established by, say, tourism or the culture industry.

Rodney Graham

In exploring the mediated perception of nature, Rodney Graham first focused on photography and film. *75 Polaroids* (1976) – the title of an early work – is exhibited on a wall like a frieze at eye level in a darkened room.[4] Graham snapped these shots while walking through a forest at night, with only his camera flashlight to show him the way. He moved clumsily in the dark, carefully groping his way forward, and, by contrast, the Polaroid photos are the regular square, standard shape produced by the camera. Thus apart from trees and bushes, the viewer also sees the technically conveyed portrayal – in other words, the Polaroid images produced by chance as he moved through the dense undergrowth. Wilderness and technology, nature and civilization, are juxtaposed with one another: they are identified as two systems that are alien to each other. That the Polaroid camera is not used for its normal purpose but instead serves as a way for the artist to find his way in the dark forest reveals the artistic statement of this work, presenting the dichotomy between nature and culture, which is based on technology.

We encounter this very constellation in later works by Graham. For example, in the happening *Illuminated Ravine* (1979), he led participants along the course of a river on the campus of the Simon Fraser University in Buranby, British Columbia, at night. And for the later *Two Generators: A river illuminated by means of two self-sufficient commercial lighting systems for a duration of time fixed by the length of a single role of film* (1984, 35 mm), he filmed the same river, illuminated by two generators. At first the viewer of the film sees only the darkness of night, in other words, nothing – but hears a roaring river in the distance. Then one after the other, the generators start up with much noise and gradually begin to cast what is initially a flickering light onto the river. Later their light grows stronger, and then they shut off, disappearing into the darkness – and finally the loop begins afresh. The flickering light of the generators and the frequency of the film images allow the viewer to detect differences in the ways both technologies work. By contrast, the course of the river, which is lit up, illustrates the continuity of real time. Another measurement of time is that provided by the film's show (one hour), after which the emergency lighting is switched on in the projection room. In other words, Graham makes visible the media equipment with all attendant circumstances. As such, his examination of the phenomenon of film/cinema also demonstrates the limits of that enlightenment which only knows the nature it has illuminated or questioned according to its own laws as Kant defined it.[5]

The climax of this topical body of work by Graham is the work *Edge of a Wood* (two-channel video installation, recorded on a DVD, 8.05 min., 1999), which is projected onto two mega-screens installed obliquely to each other, with accompanying stereo sound (Figure 12.1).[6] Graham filmed the perimeter of a forest using two film cameras, and the images are projected separately onto two large screens. By presenting two separate screens, Graham is evidently referring to human perception relying on two eyes, and emphasizes the difference between two-dimensional and three-dimensional, retinal and mental perception, which the human brain is capable of compensating for, while the technical medium fails in the attempt.

We see the edge of a forest at night – in other words, initially we see nothing. As indicated by ever-louder sounds, we hear an approaching helicopter which cannot be seen but whose bright search lights soon light up the forest on the two mega-screens. However, these lights are incapable of revealing the forest's interior. Neither the light nor the helicopter can surmount the border drawn by the trees' foliage. What is nature remains hidden to technology, but art is able to visualize both components of our life, man's origins from the darkness of matter and his or her orientation towards light and the overcoming of gravity or the movement of the intellect. As in the paintings by Max Ernst, there is

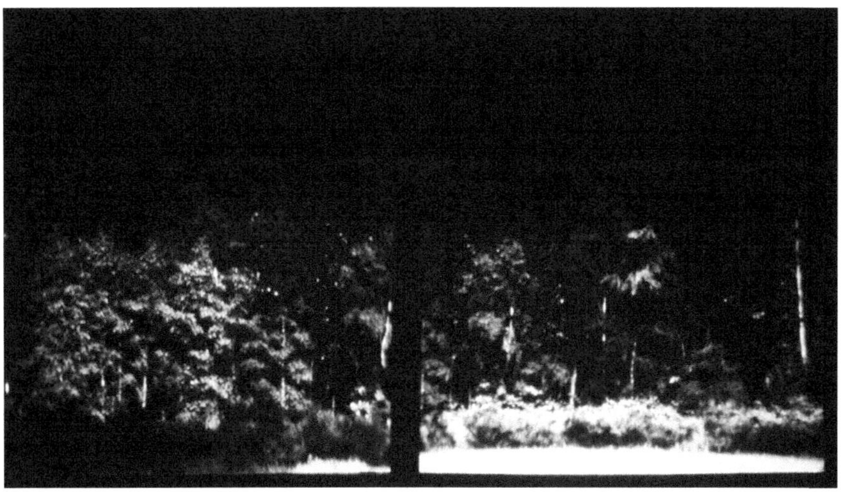

12.1 Rodney Graham, *Edge of a Wood* (1999), still from video/sound installation, transferred to DVD. Loop 8.05 min. From: *Beyond the Cinema. Films, Videos and Installations from 1965–2005*. Museum für Gegenwart – Hamburger Bahnhof, Berlin 2006. © Rodney Graham/Photo credit: Rodney Graham

something threatening about the forest in this video projection, something eerie, yet simultaneously reassuring. Nature is seen as Other, something our culture cannot grasp and it almost seems as if this loud roaring of the helicopter epitomizes our vain attempts to understand nature using technical means. By contrast, the helicopter's light rises up from the density of the trees as if in free flight, as if this motif were also a reference to the intellect in the dialectic of its separation from nature and its search for nature.

Graham's series of experiments explore in artistic form the principal otherness of natural phenomena and the cultural phenomena of their perception which depends on technical means. A similar interest in overcrossing these areas can also be detected in the artworks of Roni Horn and Diana Thater.

Roni Horn

Like Graham, Roni Horn explores 'relations to something that exceeds us'.[7] Photography and sculpture are her favoured means of dealing with the topic of nature beyond its trivial treatment in the media. In photographic series, she transfers phenomena of time, preferably the weather or the course of the weather, into the timelessness and spacelessness of the image. A clear distinction is made between what is depicted and its portrayal. Presenting photographs in series and combining them with texts mitigates the isolation of the photographic image and refers to the continuity in nature, which the individual image can elude.

For *Another Water, The River Thames For Example* (2000), Horn had the Thames photographed from a boat, the camera pointing down towards the river to maintain an even sharpness in depicting the water and to avoid creating a horizon.[8] The photographed moments of the moving surface of the Thames reveal a startling diversity; as a viewer, one could not see the river since the water moves too quickly for one's eyes to follow. And an iconic inversion takes place: in the series, the index-like images become iconic images, which show how differently the water moves, how differently the changing environment is reflected. A frieze of images is created by this composition of the photographs, a virtual film, while the presentation in a printed book implies a novel. Horn printed texts on the Thames on a narrow white strip beneath each image. The attractive appearance of the water shining in delicate shades of brown, green, ochre and grey becomes dubious when you read in the text placed as footnotes about not only the river's pollution but also the high suicide rate in London. Phenomenological descriptions relating to the behaviour of the current and changes in the water caused by climate change, dictionary definitions and personal thoughts and reflections are also to be found here – for example, about how water reflects the emotional state: 'Have you ever noticed how light camouflages water?' Beneath the 830 texts there are passages from songs, poems, narrations and films by, for example, Robert Altman,

Michelangelo Antonioni, Robert Bresson, Charles Dickens, Emily Dickinson, William Faulkner, Ira Gershwin, Al Green, Alfred Hitchcock, James Joyce, Stanley Kubrick, Flannery O'Connor, Edgar Allan Poe, Bruce Springsteen, Wallace Stevens and Neil Young. These references serve to remind us of what is not visible in what is visible, making us aware, above all, of the horrors the Thames conceals.[9] Apart from the literary projections, the texts also note the hard facts of everyday life in London; for instance, on separate, white pages several 'Dead Body Reports' are printed, naming the cause of death, and the dates the suicides jumped into the river. In thus portraying the Thames, Horn deviates from the visual tradition of the mythical 'lifeblood of Britain', which since Joseph Mallord William Turner's *Rivers of England* delivered views of the river in its entire course.[10] Instead, she directs the gaze into the water and has the horizon in the photos sink. In the shots, which let the viewer look into the dark depths of the water, the strong fascination the Thames exerts on suicides becomes evident. But in some of the images, taken in sunlight, the viewer can imagine the water's heavenly origins, although in London the river has long since become a waste-water outflow. In other words, Horn uses photography not only to depict the visible but also to unfold a wide thematic spectrum on the Thames. Her book documents the many manifestations of a river whose surface on the one hand reflects climate change, and on the other conceals the residue of urban civilization. Horn shows a river in which traces of living nature might still be detectable but appear as alien in her images as the nature of death has always been to those still living.

Horn created a highly contrasting piece to *Another Water*, devoted entirely to a life with and beside the 'nature' of Iceland, in a library in the small coastal town of Stykkishólmur, the *Library of Water* (2007) (Colour plate 14).[11] Her installation is located in a low-slung library building from the 1950s, which sits on a rock above a wide bay. With its semicircular front window, it seems to look out as an eye on Iceland's nature. It is difficult for art to garner attention in such a special place as this, so Horn installed in the large reading room 24 clear glass columns without bases or capitals. They are filled with melt water from 24 glaciers. The water in them stands upright at variance with nature; the water is fixed vertically behind glass so that its consistency and colouring is determined by not only the various minerals but also the diverse reflections of visitors and the surroundings. Visitors move through the columns as they would through a museum collection, and can read in the colour nuances the origins of the water and experience the mood created by the light. As such, the colourless glass and water become media, which make environmental phenomena easier to detect. Horn's glass columns link the space, people and nature: they have a spatial dimension, the vertical form of a person and displayed water as the dominating element of the natural surroundings.

Visitors do not experience the constant change in the weather in a directly sensual manner but in a sublimated form, that is, visually and emotionally. As the climatic situation is somewhat removed owing to the glass, visitors feel moved to reflect, and the influence the weather has on one's mood becomes apparent. Horn's typographically styled words in Icelandic and English, displayed on the ochre floor, help visitors understand how changes in weather affect human spirits. Used as a museum, the room might separate visitors from the surrounding nature but through the columns filled with water, the room is subjected to something like visual dilution. Thanks to their reflections and various colouring, the columns seem to absorb the surroundings and transcend phenomena to create something aesthetic. Though the viewer is not really affected by the weather conditions, he or she can reflect upon and 'read' them, as the title *Library of Water* suggests. Adorno's poetic metaphor 'What nature strives for in vain, artworks fulfil: They open our eyes'[12] precisely describes the quality of this library. Visitors can experience Iceland's nature as the Other, as something alien, effectively at eye level, and can simultaneously recognize how and why it affects them. By experiencing something atmospheric, they also feel themselves as nature and yet remain as a describing viewer, non-nature, a recognizing eye. Horn's work of art also fulfils another purpose: it is a meeting place for the locals, a communal place where a collection of personal weather reports ('Weather reports you'), recalls especially intensive experiences. Horn visualizes natural phenomena that touch people but only become intelligible through the artistic use of photography or in the 'water library'. The exhibition room in Iceland is also infused with a vanitas motif, painfully recalling in glacier melt water the first symptoms of the climatic disaster.

Diana Thater

When Diana Thater began to work as an artist, video technology was already firmly established. But she avoided using video in the conventional manner in order to discover new options for the medium, particularly with regard to the perception of nature. For example, rather than project onto screens, Thater projected her images directly onto walls, ceilings and floors. In doing so, she accomplished an important 'stepping out of the picture' for video art, because the visual material is used in a sculptural manner. This is why Thater says of herself, 'I think about nature and sculpt with images of it in space'.[13]

By way of illustration let us look at *Dolphins* (1999) and *Broken Circle* (1997). It was during Thater's studies in Giverny, enabled by a grant, that she developed her expansive form of video art. Thater was inspired by Claude Monet's landscape painting in which the horizon falls, which paved the way for abstraction in painting. Thater uses computer technology for her spatial and temporal extension of the moving image, but does not use visual opulence to create a narrative illusion. Rather, she uses the technical components of an operating system such that the phenomena of vision themselves become visible: that of the technically conveyed (long-distance) vision of the camera, that of the human eye and that of the filmed animals. The images are reduced to a few motifs (above all from the animal world) and are projected in various directions and layers, they run around corners and incorporate visitors into the current field of vision, because it is possible to walk freely around the equipment. Similarly, the genesis of the images remains visible, with either the camera operators or the technical equipment of a shot clearly captured. In the installation *Dolphins* (1999), the viewer looks at projected images of the animals in the lower level and sometimes sees the scuba-diving flippers worn by the film crew (Figure 12.2). By comparing perspectives, the viewer realizes that the habits of vision are culturally determined, and that there are other modalities of perception in nature such as the echo system used by dolphins to find their way around in water. Unlike, say, the television series *Flipper* with its narrative plot, here the modalities of the technology itself generate meaning. These modalities connect the difference between the film and video shots (i.e. divers wearing snorkels filmed with a Super 8 camera from above, while deep-sea divers using digital video cameras recorded the animals from below) to the difference between the two zones the dolphins inhabit, air and water.[14] As the media foil of perception itself comes into focus, it becomes a tool of reflection for the beholder.

Thater arranged the projections using filters in the colour components of the video spectrum – of the three cathode ray tubes for blue, green and red, two are eliminated, and she also produced the complementary colours yellow, cyan and magenta. In a similar way the daylight was given different facets by filtering it through coloured foils in front of the windows. The monochrome visual fields and light zones combine to form a spatial mosaic, which illustrates diverging forms of perception.

In place of the mono-perspective programming of vision using technical equipment, what is created here is a visual form that makes it possible to view not only the depicted phenomena but also the phenomena of the depiction itself. In becoming visible, the medium is lent the function of referring to specific types of animal behaviour such as that of dolphins at various levels of the ocean. In this sense, Thater's works of art are not so much the subject as they are means of observation.

In opposition to what Paul Virilio referred to as the 'aesthetics of disappearance', by which mass media engulf everything depicted in the depiction, Thater succeeds in anchoring her images in space such that for the viewer a dimension of aesthetic experience is regained. And to counter the permanent reiteration of mono-perspective vision that the camera ritually performs, she uses the medium in an experimental, artistic manner to multiply the perspectives. Viewers are motivated to move about in the installations and explore alternative angles. Vision is not restricted to the surface of a depiction, rather it is multiplied and activated through the spatial dramaturgy of the images.

Similarly, her installation *Broken Circle* (1997), shown initially in the tower Buddenturm in Münster during the *SkulpturProjekte* (Sculpture Projects) exhibition and since 2001 in the Siegener Museum für Gegenwartskunst, also

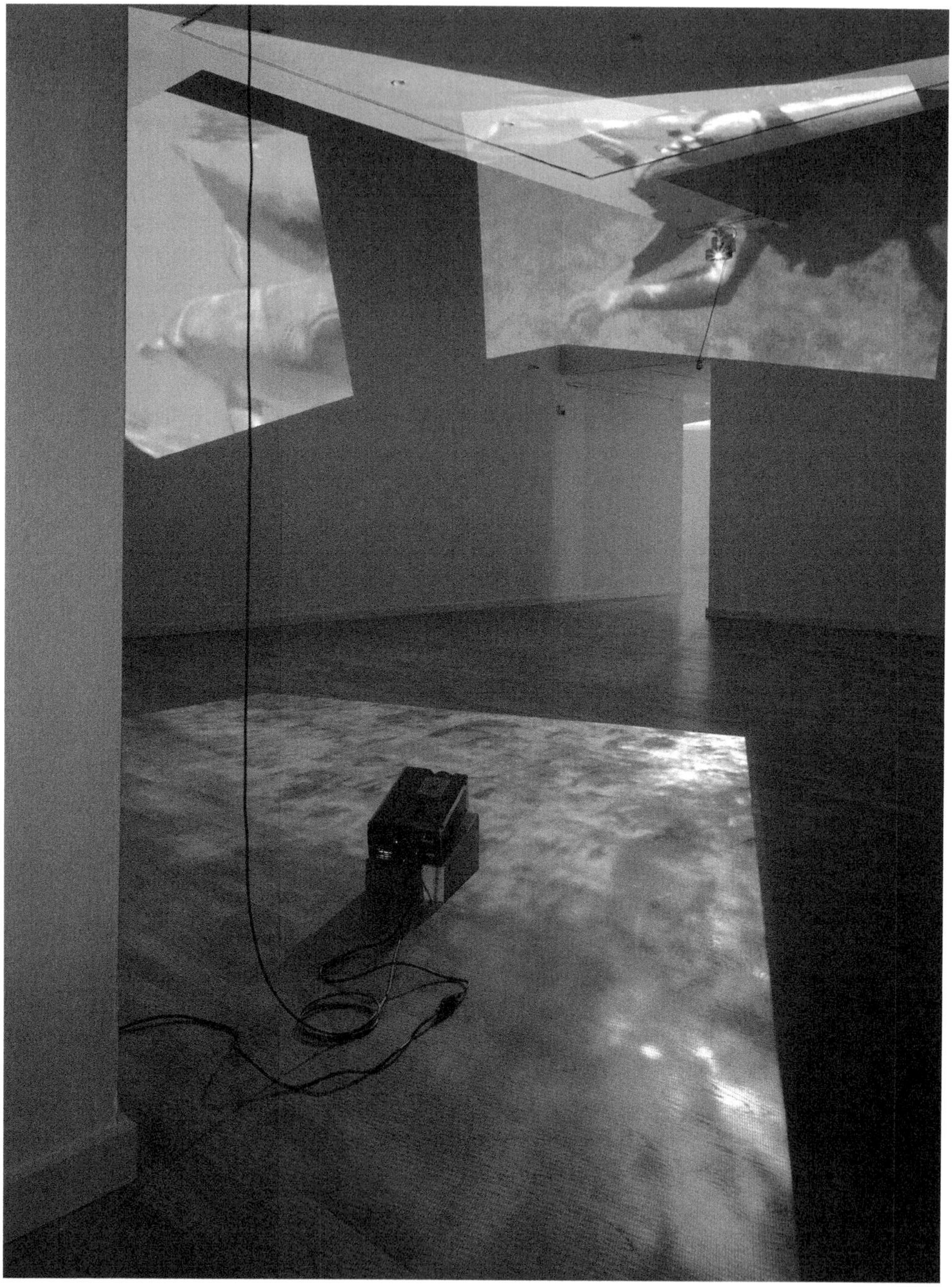

12.2 Diana Thater, *Dolphins* (1999), four video projectors, nine video wall monitors, five DVD players, five DVDs, one synchronizer, one video wall processor, Lee Filter. From: Diana Thater, *Delphine* (Vienna: Wiener Secession, 2000). © Diana Thater/Photo credit: Diana Thater in a tower, shakes up the habits of perception conditioned by the media.[15] It consists of images of horses that were trained for Westerns in Hollywood, but here were shot moving freely by six video cameras set up in a circle. The images do not, as is customary for Westerns, merge into a suggestive complete view but rather unfold in sequences. As the installation is located in a narrow staircase, visitors can only move in the vertical; on each of the four floors they only get to see one segment of the film (*Broken Circle*), and then only in one colour. The towers in Münster and Siegen, evoking as they do watchtowers and medieval knights' tournaments, do not direct the gaze outside as their windows are covered in coloured foil. What we have is a reversal of cinematographic vision because the cameras were not – as was customary since the shots by Eadweard Muybridge – directed at the horses, but captured their movements casually. And in the large projection of the images, it is essentially their intrinsic movement the viewer finds oppressive, as the projection on the curving walls (*Broken Circle*) in a semi-circle envelops the viewer such that the image space largely encloses the viewing space.

> The piece that I made for the Münster Skulptur Projekte, *Broken Circle*, is all about that very specific shot in Westerns where the galloping herd of horses splits around the camera set up. This piece is really an essay on how to shoot that shot, how to get it and what it looks like from every angle. But of course I never get the shot, which is the whole point. I don't want to propose the singular eye of the filmmaker as the be-all and end-all of the film. So there are multiple points of view and from any point of view you see the heard of horses splitting around the camera, but every time they are about to do this, we go off to another perspective where we see an empty field or camera people working or a horse in the corner eating grass. You're always distanced from 'the shot' and you look back at it and realise that it can't be what it appears.[16]

In the scenic differentiation, the images follow the behaviour of the animals, the rhythm of their movements. Thater's visual spaces not only show the subjects but also the images themselves, which are presented next to each other as in an experiment on perception. The artist is very careful in using her technical equipment to arouse our awareness of the otherness of nature in opposition to culturally defined vision. She dismantles the equipment and reassembles it differently to gain new perspectives in the sense of philosopher Vilém Flusser (1920–91) by using technical equipment in an experimental fashion.[17] The medium is used and its conditions and means questioned rather than remaining invisible for the purposes of illusion. It is perhaps possible to transfer this to a general questioning: namely, if one understands the medium as analogous to human reason, which in the sense of Kant's *Critique of Pure Reason* can only recognize nature using the criteria of reason, then it is no longer reason or rather the medium that recognizes nature according to its laws but, on the contrary, it is nature that is presented in a work of art precisely in its unavailability. Diana Thater also succeeds thanks to her artistic form in opening up new perspectives on technology and nature.

Summary

In their use of modern technology and emphasis on sensory perception, Rodney Graham, Roni Horn and Diana Thater render viewers of their works aware of their own physical nature, their human nature, which underlies all culture.

'Art possesses its beauty thanks to the power of unadulterated views of the mysterious foundations of body and soul'.[18] Here Martin Seel refers to a general problem of contemporary art in its relationship to nature, because the beauty of nature, which is per se a visual manifestation of nature, cannot be portrayed in media or art without being concretized as a result. Insofar as 'nature, as something beautiful, cannot be represented', because 'as something that appears it is itself an image', as Adorno noted, every visual surrogate is inappropriate because 'by objectifying what appears, it eliminates it'.[19]

But this concretization is performed by technical picture media, whose images are generally taken for what they show but are not the beauty of nature. Without the artistic components, which were described here for the works of Graham, Horn and Thater, technical media are generally lacking the very artistic beauty that Kant defined as having a mimetic relationship to nature: 'Art can only be called beautiful if we are conscious of it as art while yet it looks like nature'.[20] As technically based image media that favour vision, photography, film and video might seem to let us see everything, but precisely in doing so they distance what is seen and tend to reduce it to its visible surfaces and oust sensual and thus also moral aspects in the encounter with nature. The beauty of nature, fleeting and transient as it is, is captured in this manner, concretized, rather than being shown in its unavailability. By contrast, Graham, Horn and Thater are intent on communicating a new investigation of nature, an expanded perception of its phenomena. And numerous other contemporary artists also use technical image media for similar purposes. Why is that? Man's relationship to nature is so determined by technology and structured by industry, from the car to the camera, from the refrigerator to the heating system, that painting as a pre-industrial form of art can fail to achieve these structural preconditions of an industrialized society when it naively relies on visualization alone. But art can be genuine when it incorporates these conditions into the field of vision. As a cultural production located alongside nature but also alongside technical civilization, art can credibly comment on their current constellation, and by securing new modalities from technology it can also expand the view of nature.

Owing precisely to image technology and its automated programs, many artists are attracted to the idea of trying out other modalities, and in so doing, not depicting the beauty of nature, but demonstrating its unavailability. Caught up between technology and nature, the works of art considered here direct special attention to the physical nature of the observer. Apart from the image media favouring vision, they are also directed at other modes of perception. In the case of Horn's *Library of Water*, this also includes references to context and the perspective of a possible function of the artwork for visitors. By being juxtaposed to nature and technology, art can help revise our relationship to both, and prompt us to question our habits and values.

Acknowledgements

This essay was first presented as a lecture during the conference 'The Recognition of Nature – the Construction of Landscape in the 21st Century' at Ritsumeikan University, Kyoto. I would like to thank Professor Yuko Nakama for the invitation and opportunity to publish the text in advance in English.

Notes

1 Marga Bijvoet, *Art as Inquiry: Toward New Collaborations between Art, Science and Technology* (New York: Peter Lang Publishing Inc., 1997).

2 Dennis L. Meadows, *The Limits to Growth* (New York: Universe Books, 1972).

3 Stephen Toulmin, *The Return to Cosmology: Postmodern Science and the Theology of Nature* (Berkeley and Los Angeles: University of California Press, 1982).

4 Rodney Graham, exh. cat. (Vancouver: Vancouver Art Gallery, 1988), catalogue essay by Jeff Wall; Rodney Graham, exh. cat. (London: Whitechapel Art Gallery, 2002), catalogue essay by Carolyn Christov-Bakargiev; Rodney Graham, exh. cat. (Toronto: Art Gallery of Ontario, 2004); Rodney Graham, *Through the Forest*, exh. cat., ed. Friedrich Meschede, Museu d'Art Contemporani, Barcelona (Ostfildern: Hatje Cantz, 2010).

5 (Die Naturforscher) 'begriffen, daß die Vernunft nur das einsieht, was sie selbst nach ihrem Entwurfe hervorbringt, daß sie mit Prinzipien ihrer Urteile nach beständigen Gesetzen vorangehen und die Natur nötigen müsse auf ihre Fragen zu antworten'. Immanuel Kant, *Vorrede zur Kritik Der Reinen Vernunft*, in Immanuel Kant, *Gesammelte Schriften, Preußische Akademie der Wissenschaften* (Berlin: Walter de Gruyter, 1902), vol. 3, p. 10 (XIII).

6 Rodney Graham, exh. cat., ed. Susanne Gaensheimer, *Kunstverein München/ Westfälischer Kunstverein Münster* (Cologne: Oktagon, 2000); Dorothea Zwirner, *Rodney Graham* (Cologne: DuMont, 2004).

7 Roni Horn, 'Interview with Lynne Cooke', in Louise Neri, Lynne Cooke, Thierry de Duve and Roni Horn, *Roni Horn* (London: Phaidon, 2000), pp. 6–25, p. 6.

8 Roni Horn, *Another Water* (Zürich: Scalo; London: Thames and Hudson, 2000); Hans Dickel, *Künstlerbücher mit Photographie* (Hamburg: Maximilian-Gesellschaft, 2008), pp. 227–33.

9 'Die Auseinandersetzung – nicht mit dem Unsichtbaren – sondern mit dem Nicht-Sichtbaren war in meiner Arbeit schon immer etwas ganz Wichtiges. Eigentlich ist es da und man könnte es sehen. Das Nicht-Sichtbare ist vollkommen konsistent mit dem, was man sehen kann'. Roni Horn, in interview, *Kunst-Bulletin* 10 (October 1995): pp. 10–18, p. 12.

10 Cf. David Hill, *Turner in the Thames* (New Haven and London: Yale University Press, 1993).

11 Roni Horn, *Vatnasafn/Library of Water* (London: Artangel, 2007); Roni Horn, *Vatnasafn/Library of Water* (Göttingen: Steidl, 2009); Mark Godfrey, 'Roni Horn's Icelandic Encyclopedia', *Art History* 32/5 (2009): pp. 932–53.

12 Theodor W. Adorno, *Aesthetic Theory* (London and New York: Continuum, 1997 [1970]), p. 86.

13 'I think about nature and sculpt with images of it in space'. Diana Thater in interview with Carol Reese, in Diana Thater, *Delphine*, exh. cat. (Vienna: Wiener Secession, 2000), pp. 8–14, p. 8, p. 9.

14 Ibid; Diana Thater, in *Parkett* 60 (2000), pp. 76–116; Diana Thater, *Keep the Faith: A Survey Exhibition*, exh. cat., ed. Barbara Engelbach and Wulf Herzogenrath. Kunsthalle Bremen, Museum für Gegenwartskunst Siegen (Cologne: Salon, 2004).

15 Skulptur Projekte in *Münster 1997*, exh. cat., ed. Klaus Bussmann, Westfälisches Landesmuseum Münster (Ostfildern-Ruit: Hatje, 1997), pp. 406–11; Anne Hoormann, *Diana Thater: Broken Circle* (Siegen: Museum für Gegenwartskunst Siegen, 2004).

16 Douglas Folge, 'Diana Thater: Being Inside of a Work of Art', *Flash Art* 198 (January/February 1998): pp. 86–89, p. 87, p. 88.

17 'Freiheit ist, gegen den Apparat zu spielen'. Vilém Flusser, *Für eine Philosophie der Fotografie* (Göttingen: European Photography, 1983), p. 73.

18 Martin Seel, 'Das Unsichtbare sichtbar machen. Zur aktuellen Lage des Austauschs von Kunst und Natur', in *Ferne Nähe. 'Natur' in der Kunst der Gegenwart*, exh. cat., ed. Volker Adolphs, Kunstmuseum Bonn (Cologne: Wienand, 2009), pp. 161–81, p. 163.

19 Adorno, *Aesthetic Theory*, p. 87.

20 Immanuel Kant, *Critique of Judgment*, pt 1, § 45, trans. J.H. Bernard (New York: Hafner Press, 1951 [1790]), p. 149.

13

Haacke, Systems and 'Nature' around 1970: An Art of Systems / Systematic Art

Caroline A. Jones

The artist Hans Haacke is best known for socially engaged work that inaugurated what is known as 'institutional critique' in the 1970s, dominated by research and documentation, crucial to the postmodern turn to come. But in the less familiar phase of Systems Art that immediately preceded Haacke's social turn, a different aesthetic prevailed. In works from 1961 to 1968, the artist took off-the-shelf, medium-scale technologies and materials (fans, fabrics, Plexiglas cases, weather balloons, refrigerator units, earth, water) and arranged them in different configurations so they would operate with minimal intervention: ribbons of fabric blowing in the air, balloons suspended in wind, tables developing a coating of ice, cases weeping with internal condensation, grass growing. The argument in this essay will be that these works, evolving from kinetic art into an explicit embrace of 'systems', constituted an antihumanist, techno-utopian staging of Nature. As Haacke put it at the time, his systems works were 'entirely independent' of the human viewer and allowed Nature's processes to unfurl with autonomy.[1] The aim was not to bridge the distance between technology, Nature and the still-human subject, but to stage the absolute alterity of non-human systems, full stop. Although ultimately backing away from the rather extraordinary implications of these pieces, Haacke and key systems theorist Jack Burnham forged an aesthetic almost clinically designed to eliminate the human from the affective-perceptive cycle. 'Ultimately', Burnham would later ruminate, 'systems theory may be another attempt by science to resist the emotional pain and ambiguity that remain an unavoidable aspect of life'.[2] Systems was a way to get the human out of Nature's path.

Modernism has a robust tradition of such positivist and empiricist modes, what we might call 'Systematic Art'. As I have argued elsewhere, even Claude Monet's late paintings (haystacks, cathedrals, Japanese bridges, Nymphéas), which are usually seen as relaxing invitations to the individual gaze, were produced in systematic series by converting the artist into a registration device, the extreme regularity of his brushstrokes functioning like a human clock, recording 'petites sensations' over time with such chilling automaticity that they had to be tempered by icons of religious, rural, or Oriental retreat.[3] Humanity becomes a regulated part of the machine world, in order to allow Nature's more indolent temporality to appear.

Systems Art took this strain of modernism to an extreme in the late 1960s, peaking at the Massachusetts Institute of Technology in a brief moment extending from the fall of 1967 to the spring of 1968. In that span, an exhibition was mounted of Haacke's air and water works at the Institute's Hayden Gallery. Just across campus, Haacke's mentors, Zero Group artist Otto Piene and Systems Art theorist Jack Burnham, were named in the first group of fellows at MIT's Center for Advanced Visual Studies (CAVS), under the direction of New Bauhaüsler Gyorgy Kepes. Haacke's show was covered in numerous articles in the student paper at MIT; Burnham and Piene had more impact outside the university, as Burnham's articles 'Systems Esthetics' (1968) and 'Real Time Systems' (1969) came out in *Artforum*, and Piene's 'Sky Art' events occupied Central Park in New York (1967).

Haacke's exhibition had no thematic title. Organized by Department of Architecture professor Wayne Andersen, it consisted of the kinds of objects described above, unfamiliar in their art gallery setting – bubbles sliding through a suspended water level, immiscible liquids sloshing between Plexiglas sheets, 'rain' percolating through small holes in transparent plastic acrylic, silk chiffon flowing in ribbons and waves, a parachute suspended in a column of air, a pile of dirt germinating grass seed, a 'weather cube'. In the context of MIT's architecture program, this made sense, giving form to a century-long aspiration to transform academic Beaux-Arts traditions (rendering complete with chiaroscuro and perspective cues) through protocols of engineering (structural analysis, material innovation, calculation and mechanical drawing) tempered by science (physics, chemistry, biology). MIT had always aimed at an entirely new relation to Nature, one that would leave conventional pictures behind to investigate the deeper orders of matter and force.[4]

Haacke was clearly stimulated by the opportunity to exhibit in this pedagogical/technical/military-industrial context, and the 'systems' championed in his discourse and exhibition changed the terms of engagement. In place of Kepes's use of technologies of the image (X-rays, microscopic photographs, etc.), which still aimed to ameliorate relations between art and science (or nature and engineering) with humanist readings of 'new images', Haacke's 'systems' posited a radically antihumanist alternative.[5] In the month before the MIT exhibition, Haacke wrote many statements that reveal his thinking. Once he arrived, the new agenda was confirmed in his accounts to reporters for the student newspaper:

> Haacke rejects the name 'sculpture' for his works. He calls them 'systems', noting that they 'have been produced with the explicit intention of having their components physically communicate with each other, and the whole communicate physically with the environment. … Changes are desired and are part of the program – they are not due to the shifting experience of the viewer'.[6]

The systems Haacke was proposing featured dynamic equilibrium. 'Nature' was clearly conceived as a set of physical forces comprising energy, temperature, humidity, plant growth – all of which could be simulated and modelled by Haacke's electrical fans, refrigerator units, earth/seed piles and condensation devices. Notably, Nature was not to include *human* nature.

Haacke's exhibition made it clear: Nature should be known and emulated in her processes, not her forms. Such 'Systems Art' upended modernist tropisms to style or 'abstraction' from Nature, and went after the *core protocols* governing matter and its transformations. Such a systematic modernism can be seen as both enduring (stemming from the Enlightenment and maturing in the Industrial Age), and as a new development that occupied and gave form to the 'great hiatus between standard modernism and postmodernism', in the retrospective account of Systems Art's great prophet and theorist, Jack Burnham.[7] Systems referenced feedback, recursive loops and autopoiesis. Most remarkably, even if they sometimes needed human agency to set them in motion, Haacke's systems artworks in 1967 were positioned explicitly as being outside standard aesthetic discourses involving emotion, interpretation, culture and memory. Systems worked to eliminate 'human factors' altogether.

An 'urban mentality'

Systems pervaded MIT by 1967, and it is thus significant that it was at this moment the idea became explicit in Haacke's art. Fueled by Claude Shannon's information theory, systems and cybernetics stretched at the Institute from Jay W. Forrester's applications in social science to the major contributions of Norbert Wiener in mathematics and Vannevar Bush in computational engineering. Wiener, a galvanic presence at MIT until 1964, had first published *Cybernetics* in 1948, with a revised second edition put out by MIT Press in 1961; once Ludwig von Bertalanffy's *General System Theory* was published in English by the New York literature and art firm Braziller in 1968, the art world had its bookend bibles for a systems revolution.[8]

It seems clear that the MIT context stimulated the transfer from mere influence from systems to an explicit aesthetic. Haacke's earliest work in this direction dates from 1961–62 (with, for example, a tall Plexiglas box containing water which drips through its many perforated sections, called *Rain Tower*), but he did not use 'Systems' as a descriptor until 1967.[9] By that year, he could write of the ice and fabric pieces he was then showing at the

Howard Wise Gallery in New York (*Ice Stick, Ice Table, Blue Sail*) and preparing for the Hayden at MIT in terms of a 'system', still in scare quotes:

> A 'sculpture' that physically reacts to its environment is no longer to be regarded as an object. The range of outside factors affecting it, as well as its own radius of action, reaches beyond the space it materially occupies. It thus merges with the environment in a relationship that is better understood as a 'system' of interdependent processes. These processes evolve without the viewer's empathy. He becomes a witness. A system is not imagined, it is real.[10]

Haacke's adoption of systems theory can be seen as the artist's way of avoiding the traditional Germanic trope of empathy in art. As Haacke recalls his mood as an art student in the late 1950s, 'I was intrigued by non-compositional developments'.[11] In place of 'expression' the artist sought the appearance of an algorithm that would replace intention with iteration, delectable composition with non-composition. He became exposed to systems theory after meeting Jack Burnham during his U.S. Fulbright grant in 1961; shortly thereafter he began to challenge himself to explore the implications of Wiener and Bertalanffy for art. His condensation pieces (cubes, walls, low-slung rectangles) were an early result. Here is Haacke, again in 1967, reflecting on the autonomy of the systems he was then interrogating (the scare quotes are gone):

> In the past, a sculpture or painting had meaning only at the grace of the viewer .... Without his emotional and intellectual reactions, the material remained nothing but stone and fabric. The system's program, on the other hand, is absolutely independent of the viewer's mental participation. It remains autonomous – aloof from the viewer. As a tree's program is not touched by the emotions of lovers in its shadow, so the system's program is untouched by the viewer's feelings and thoughts. The viewer becomes a witness rather than a resounding instrument striving for empathy.[12]

Haacke thus rejects centuries of the humanist 'pathetic fallacy' in art, wherein representations of the natural climate are deployed (in paint, in poetry) to stir the empathic response of the reader. Burnham, in his 1968 *Artforum* manifesto for 'Systems Esthetics', recognized the rather extreme position Haacke was taking in eliminating empathy and humanism: 'Haacke's systems have a limited life as an art experience ... . He insists that the need for empathy does not make his work function as with older art'. In support of his conclusion that Haacke's 'Systems exist as ongoing independent entities away from the viewer', Burnham quoted the Haacke statement cited above, that 'These processes [of the "system"] evolve without the viewer's empathy'.[13] Philosophically, I would argue that some notion of 'experience' and empathy within a viewer/reader/watcher is inextricable from talk about aesthetics – but historically it should interest us that Haacke was determined to eliminate it.

This is the paradox – that mere months before his turn to the social, Haacke was capable of arguing for a systems art that was wholly independent of the humans perceiving it. Rather than prosthetic 'extensions of man' as in the 1960s theories of Canadian English professor-turned-media-guru Marshall McLuhan, Haacke's technologies anticipated the 1980s work of German literary-theorist-turned-media-guru Friedrich Kittler – less enhancements of a coherent human body than propulsions to the posthuman.[14] Burnham, who had proposed to Haacke that his art could be understood as offering something like Thoreau's 'organic rapport' with Nature, was 'shocked' when he received the artist's abrupt response:

> I hate the nineteenth century idyllic nature loving act. I'm for what the large cities have to offer, the possibilities of technology and the urban mentality. Plexiglas, on the other hand, is artificial and strongly resists either tactile sensuality or the 'personal touch'. Plexiglas, mass-production – Thoreau – they don't really fit together.[15]

Haacke's rejections, as in Kittler's later attempt at 'driving spirit [Geist] out of the humanities',[16] were certainly coloured by the recent German history in which fascism had appropriated these 'Nature' tropes (nature-loving, blood, soil and spirit) for the purposes of genocide. Haacke's favouring of the 'urban mentality' was both a cancellation of this poisonous past and an embrace of his new context in New York City. As Haacke recently recalled his position, 'I rejected the traditional thinking of the romantic, and rejected the psychological, which exudes the magic of all art criticism'. In place of Nature and its romantic magic would be systems.

Climate in a Box

The canonical *Condensation Cube* (first conceived in 1963, and executed in its smaller form in 1965) (Colour plate 15) reveals the barometric operation of the systems Haacke wanted to employ. That a larger version was exhibited at MIT in 1967 under the title *Weather Cube* is significant. At the least, it troubles some interpretations that place this work either with the 1960s cubic objects of Minimal Art (courting what Michael Fried called 'objecthood'), or along the path to full-blown institutional critique (as theorized by Benjamin H.D. Buchloh).[17] Like *Weather Cube*, I argue, it instantiates precisely the relation to Nature that Haacke was after: *The human is black-boxed, Nature is revealed in full transparency, with allusions to planetary climate systems loaded into the title.* Constructed of the synthetic glass substitute, thermoplastic polyacrylic (developed in wartime research on airplane windshields and marketed in the U.S. under trade names such as Plexiglas, Lucite or Perspex), the cube's flawless clarity extends to its chemically fused edges. A tiny hole drilled in a top corner allows the introduction of water to a depth of about an inch in the bottom of the box. Distilled water forms the substrate for the cube's theatrics of *Weather*.

Crucially, the box becomes a device for staging events and processes. As in the cloud chambers made by Victorian physicists for mimetic experiments, the water in Haacke's *Weather Cube* forms a microclimate system.[18] Although Haacke now demurs that 'weather' is no different from 'condensation' ('they are the same thing'), I would argue for their difference.[19] Condensation is commonly experienced on the surfaces of a nearby object (at a minimalist scale); weather happens atmospherically (at a systems scale). To model condensation is a modest aspiration; to model weather aims at the orders of the world.

In 1967 Burnham could already see how these 'weather' boxes were related to Haacke's wind devices, with global implications: 'The Earth itself could be looked upon as a great wind making device forming patterns of evaporation, rain and humidity over its surface as a kind of enormous condensation container'.[20] Unaware of anthropogenic climate change, Burnham shared Haacke's view of the *impersonality* of 'weather' in 1967. *Weather Cube* allowed water to circulate through several of its available states: beginning as liquid, evaporating into mist, and slowly condensing again into liquid droplets that form orderly yet subtly random patterns, eventually 'raining' down the sides of the cube.

Nature here is not named as such, but in his constant invocation of forces, Haacke is calling upon Nature in kind if not in name:

> The original use of the term ['systems'] in the natural sciences is valuable for understanding the behavior of physically interdependent processes. It explained phenomena of constant change, recycling, and equilibrium. ... there are sound reasons for reserving the term 'system' for certain non-static 'sculptures', since only in this category does a transfer of energy, material, or information occur.[21]

Now that human-driven changes to the global climate system are all too clear, other questions arise about Haacke's 1967 *Weather Cube*. Around which inputs and outputs is this system defined? What are its boundaries – does the system include us? The artworld? The larger climate? *Where does Nature end and the techno-human begin?* We may venture answers to these questions, but for Haacke in 1967, the boundaries of the system in question still excluded the human. The miniature model, and the closed universe it implied, called up the magic of autonomous art: 'in spite of all my environmental and monumental thinking, I am still fascinated by the nearly magic, self-contained quality of objects. My water levels, waves and condensation boxes are unthinkable without this physical separation from their surroundings'.[22]

Details reveal the artist's efforts to maintain that 'physical separation'. The distilled water introduced into the various Plexi structures must be treated with copper sulphate to prevent unwanted biotic 'systems' from blooming. Similarly, the tiny hole drilled at the top must be managed – either left open, or partially covered by clear tape (depending on the temperature of the gallery) for the condensation 'system' to function. Art history's canonization of *Weather Cube* as *Condensation Cube* is largely unreflective about these aspects of the work, celebrating the work as an object, or at most as the container of processes 'completely independent of the viewer's perception' (as Buchloh correctly relays Haacke's intentions).[23]

In complicating these received views (and in the spirit of systems theorizing), I suggest that what is now known as *Condensation Cube* would be unlikely to display its internal weather theatre without two variable inputs from outside the box – the ambient temperature of the gallery and the heat generated by the bodies of visitors. That Haacke was uninterested in both of these as 'human' aspects of the system is consistent even today. When discussing his late 1960s contribution to a conceptual art show in which he put a gallery humidity detector on display, Haacke insisted there was no input from the human – although it is precisely the sweat, heat and vaporous breath of crowds that the device is ordinarily used to monitor.[24]

Haacke was unusual in viewing the human as uninvolved with these systems, refusing even the Duchampian role of designator (with its roots in *design*) – although later Edward Fry would insist that Haacke's works were 'readymades'.[25] Burnham saw the artist's role much more broadly, reclaiming a humanism that could include the entire planet:

> Rather than a novel way of rearranging surfaces and spaces, ['Systems Esthetics'] is fundamentally concerned with the implementation of the art impulse in an advanced technological society. ... [Man] assumes a new and more critical function [as] that of man the maker of esthetic decisions [that] control the quality of all future life on the Earth.

Ultimately pushing hard on the very systems aesthetic that he had done so much to put in place, by 1969 Burnham hovered on the edge of post-structuralism. As he wrote in 'Real-Time Systems',

> Both in the sciences and in the humanities something is rapidly happening: we are beginning to see scientific 'objectivity' as an illusion, as are the notions of independent scientific 'disciplines', of isolating subjects of scientific inquiry from their settings, and of the possibility of making unobtrusive measurements.[26]

Segueing to the computer-controlled real-time systems coordinating contemporary economic and military domains, Burnham acknowledged the 'instinctive antipathy' that most humanists have towards 'these immensely complex computer systems. Their Orwellian overtones far overshadow their conceivable use as artists' tools. But practically, it is imperative that artists do understand them – both technically and philosophically'.[27]

But if Burnham was beginning to understand in 1969 that the artist and scientist were themselves a part of any system they might model, Haacke was not yet willing to take that logical leap. True, Haacke is quoted by Burnham admitting in 1968 (after the MIT exhibition) that 'An artist is not an isolated system' – and yet his work often contradicts that intellectual insight.[28] In Haacke's ongoing practice, *neither Nature nor artwork include the artist.* The plants and animals in what the artist humorously calls his 'Franciscan' phase operate without reference to human experience or expectation (even if the St. Francis allusion already complicates that position in a metaphor of divinely inspired ministration to the biome). Haacke's stated expectation of radical separation between the work of Systems Art, the Nature it reveals and its human maker or viewer is more enduring than his pledge to see the artist as part of the system, and reinforces the most interesting *philosophical* problematic of systems theory – namely, *around what nexus of entities is the 'system' boundary to be drawn?*

From Systems to the Social

To return to one of Haacke's statements of 1967 is instructive. Here Haacke reveals the inframince of an elite aesthetic that he still wants to preserve as a boundary around the art-world system, separate from both his agency and the world, in order to avoid the complete collapse of art into life:

> A system is most generally defined as a grouping of elements subject to a common plan and purpose. These elements or components interact so as to arrive at a joint goal. To separate the elements would be to destroy the system.
> ...
> The physical self-sufficiency of such a system has a decisive effect on the viewer's relationship to the work, due to its hitherto unknown independence from his mental involvement. ... there are systems which function properly even when the viewer is not present at all, i.e., their program operates *absolutely independently of any contribution on the part of the viewer.*[29]

Haacke's commitment to the 'hitherto unknown independence' from the mental involvement of the viewer is a dramatic circumscription of the boundary of the proposed system. If Burnham would polemicize that 'in an advanced technological culture the most important artist best succeeds by liquidating his position as artist vis-à-vis society',[30] Haacke not only refused to go this far, but in projects he proposed for the Guggenheim (and in response, presumably, to the trauma of the show's cancellation in 1971),[31] he would make a decisive turn towards preserving the autonomy of the artist in works that would now interrogate *social* systems. This autonomy was desperately necessary, of course – as the Guggenheim cancellation showed, Haacke was operating in a radically contested cultural sphere where freedom of expression was put under extreme pressure. Post-MIT, the social turn that Haacke's work was taking forced the art system into the epoch of its own institutional critique. But the artist's principled insistence on an avant-garde unimpeachably outside the society it proposed to criticize derailed the postmodern potential of this move for his own work, while ensuring Haacke's role as progenitor for the very postmodern generation to come.

Exemplary of this sequestration of the artist from the social system, and the preservation of the art museum as a public site for the interrogation of that system, is a unique project that Haacke produced in Germany in 1972, one of the last that both referenced Systems and took 'Nature' as its unstated theme. Installed at the Krefeld Museum Haus Lange, the *Rheinwasseraufbereitungsanlage*, or *Rhine Water Purification Plant* (Colour plate 16), is usually represented in a single photograph showing a large Plexiglas tank, flat on the ground of the gallery's parquet floor and filled with water and fish. On the table in the back of the room is a set of glass jars filled with murky water, together with an apparatus consisting of a much smaller tank and what seem to be charcoal filters, all hooked up to a pump and connected by a single tube to the tank on the floor. From the other side of the fish tank comes another tube that disappears into the floor; outside the window can be seen the greenery of a garden.

Declaring a new focus on conceptual and land art in 1969, Krefeld's municipal museum (opened in 1955 to show contemporary art in a former Mies van der Rohe house) doubtless took Haacke's exhibition in stride.[32] Welcomed back to exhibit in Germany after the scandal of his exhibition cancellation in New York, Haacke chose a title for the exhibition that explicitly reflected both natural and now social systems: *Hans Haacke: Demonstrationen der physikalischen Welt: biologische und gesellschaftliche Systeme*. He chose the extreme pollution of the Rhine near Krefeld as the 'biological and community system' he would address. Towards this purpose, Haacke brought into the gallery gallons of polluted water discharged by the city sewage system (this is what filled the glass jars in the back of the room). The water was purified and oxygenated by being pumped through a set of filters, after which it was then piped in to sustain the fish in the tank. From there, it exited under the floor and flowed out into the garden, where it seeped into the ground.

Not visible in the usual installation view is a large colour photograph that Haacke installed on a nearby wall, showing the bank of the Rhine in Krefeld from which the water was being brought into the gallery. (The influence of Robert Smithson's 'site/non-site' works is obvious.) Everything in the horizonless photograph is stained bright orange with pollutants: rocks, plastic bottles, cardboard, sticks, contracting debris – all unified by the extreme colour of chemical sludge. There was also a discursive component to the installation. In the catalogue, Haacke had printed an extensive, fold-out schematic showing the Rhine dotted with all of the plants that were pouring out effluent from its banks on a daily basis. As he now recalls it, the town's sewage treatment plant was completed shortly after the exhibition – but his art had made an invisible infrastructural upgrade *the focus of a visibility*. In this now-visible system, Nature was contrasted with the social, the irruptions of pollution a problem for society to solve.

Haacke, following the social turn of his systems work after MIT, was clearly grappling with the interface between the human and the natural in this Krefeld installation. But I suggest that Nature, as such, was still an autonomous realm that could be 'purified' of human desecration. Smithson, of course, treated human despoliation very differently, preferring the postmodern ambiguities of the technological sublime to modernist hygiene (proposing an earthwork in a tailing pit, for example, in the *Bingham Copper Mining Pit – Utah / Reclamation Project*, 1973), forcing contemplation of human-driven entropy rather than plumping for the idealism of a Nature made whole. Haacke would never take up Smithson's cynicism, nor would he follow the trajectory of a younger generation marked by postmodernism and embracing the post-structuralist turn. These would include Olafur Eliasson, who revisited Haacke's 1960s *Water in Wind* (where mist from a nozzle produces a rainbow in the dispassionate eye of the camera recording the experiment) and converted it into *Beauty* in 1993 (thereby neatly acknowledging all the romantic appurtenances of aesthetics with a post-structural nod to the viewer, without whom the rainbow could not be seen). Eliasson's more recent title structure announces, 'Your' (... *Black Horizon*, for example), enforcing the imbrication of maker and viewer in the hermeneutic of systems; Haacke has kept aloof from this development.

But if Haacke remained a modernist and an idealist, we can see how the systems with which he was concerned nonetheless made their mark on contemporary artists working between Nature and the social. Take the Center for Land Use Interpretation (CLUI). Founded in 1994 by Matthew Coolidge, it has been praised for a relation to Nature that avoids either the romantic or technological and achieves 'the Administrative Sublime'.[33]

CLUI is unabashedly systems oriented, following Haacke in using systems' neutral and technocratic voice in service of a critical agenda. (The center rightfully claims that merely visualizing the appalling traditions of land use in America is a critical act.) Staging exhibitions such as 'Loop Feedback Loop: The Big Picture of Traffic Control in Los Angeles', CLUI's exquisitely accurate photographs and reporting clarify a 'site's mechanisms' for a wide set of users.[34] The center's 'voice', its images and its 'tour guide' relation to Nature were spawned somewhere between Haacke and Smithson. Although indebted to Haacke for its systems orientation and documentary thrust, CLUI may ultimately reinforce Smithson's postmodern deadpan rather than Haacke's modernist dreams of purification where wasteland sites are concerned. For Smithson, the human was already everywhere (except perhaps the Palaeolithic past and the unforeseeable future). For Haacke, a romantic after all, it was just possible to imagine natural forces without humanity, water without organo-polymers and wilderness without people. Systems, in Haacke's practice, performed the necessary removal of the human, imagining a world in which Nature could still be itself.

Notes

1 My thinking on Haacke's systems art grows out of an exhibition and publication produced for the MIT List Visual Art Center in 2011. See Caroline A. Jones, *Hans Haacke 1967* (Cambridge, MA: List Visual Art Center, 2011). In the present essay, I specifically attend to the thematic of Nature as it can be traced in Haacke's

thinking from this period. The reader should note that there are many routes to Systems Art (as in the London Independent Group with its exploration of structures in the life sciences) and other sources that influenced Haacke's path. Among these, for instance, would be his study with Stanley William Hayter in Paris in 1960–61, where he was on a DAAD (Deutscher Akademischer Austausch Dienst / German Academic Exchange Service) scholarship at a time when Hayter was exploring mathematics as applied to nature's systems in waveforms; in Paris he was also exposed to the systematic investigations of the GRAV group (Groupe de Recherche d'Art Visuel / Visual Art Research Group). I am grateful to the anonymous reviewer for pressing these points.

2 Jack Burnham, 'Introduction', *Great Western Salt Works: Essays on the Meaning of Post-Formalist Art* (New York: Braziller, 1974), p. 11.

3 Jones, 'Rendering Time', in *Einstein for the 21st Century*, ed. Peter Galison et al. (Princeton: Princeton University Press, 2008), pp. 130–49.

4 Nature is capitalized here and elsewhere to keep it strange, and to enforce it as an epistemic category. For more on this MIT history, see Arindam Dutta et al. (eds.), *A Second Modernism: MIT, Architecture, and the 'Techno–Social' Moment* (Cambridge, MA: MIT Press, 2013).

5 For an indication of what the image of 'Nature' was at MIT, see Gyorgy Kepes, *The New Landscape in Art and Science* (Chicago: P. Theobald, 1956). Kepes was of course deeply interested in the systematic, and anthologized both Weiner and Bertalanffy in his publications in the *Vision and Value* series, but I am arguing here that his was a deeply humanist project in distinction to Haacke's.

6 Hans Haacke speaking with Peter Mechsler, 'Haacke to Exhibit Kinetic Art', *The Tech* (Cambridge, MA, October 17, 1967). After this article, and Haacke's gentle corrections, the phrase 'kinetic art' was no longer used by student reporters.

7 Jack Burnham, Personal correspondence with the author Edward Shanken, April 23, 1998, as cited by Shanken in 'The House That Jack Built: Jack Burnham's Concept of "Software" as a Metaphor for Art', *Leonardo Electronic Almanac* 6/10 (November 1998), online at http://www.artexetra.com/House.html, accessed October 2012.

8 Norbert Wiener, *Cybernetics: Or, Control and Communication in the Animal and the Machine*, 2nd ed. (Cambridge, MA: MIT Press, 1961); and Ludwig von Bertalanffy, *General System Theory: Foundations, Development, Applications* (New York: Braziller, 1968).

9 As proof of this periodization, see the early sketches of an unrealized 'Seagull Sculpture' for the Zero at Sea project in 1965–66, and the realization of this project as 'Live Airborne System' in New York in 1968. The year 1967 and the fall MIT exhibition provide the pivot for this shift. Documentation in Jones, *Hans Haacke 1967* (Cambridge, MA: List Visual Art Center, 2011).

10 Hans Haacke, statement written in English in September 1967, New York; see Jones, *Hans Haacke 1967* (op. cit.), p. 48.

11 Hans Haacke, interview with the author, New York City, March 29, 2009. Henceforth, quotes without citations are taken from this interview.

12 Hans Haacke, statement written in English in September 1967, New York; see Jones, *Hans Haacke 1967* (op. cit.), p. 50.

13 Jack Burnham, 'Systems Esthetics' (*Artforum*, September 1968), as anthologized in *Great Western Salt Works* (op. cit.), p. 22.

14 Kittler's best-known work in English is the 1999 translation (by Geoffrey Winthrop-Young and Michael Wutz) of his 1986 study *Gramophone, Film, Typewriter* (Stanford: Stanford University Press); McLuhan's comparable treatise is the 1964 *Understanding Media: The Extensions of Man*, reissued by MIT Press in 1994.

15 Hans Haacke to Jack Burnham, letter sent prior to April 1967, as quoted by Burnham in his monograph 'Hans Haacke Wind and Water Sculpture', Northwestern University, *Tri-Quarterly Supplement 1* (Spring 1967): p. 13. Although Haacke begins with an ameliorating 'Good old Thoreau' and acknowledges 'there's some [romanticism] in me', his rejection of Burnham's transcendentalism is complete.

16 Friedrich Kittler (ed.), *Austreibung des Geistes aus den Geisteswissenschaften: Programme des Poststrukturalismus* (Paderborn, Munich, Vienna, Zürich: Schöningh, 1980).

17 Michael Fried, 'Art and Objecthood' (1967), anthologized in his *Art and Objecthood* (Chicago: University of Chicago Press, 1998), pp. 148–72; Benjamin H.D. Buchloh, 'Hans Haacke: Memory and Instrumental Reason' (1988), as anthologized in Buchloh, *Neo-Avantgarde and Culture Industry: Essays on European and American Art from 1955 to 1975* (Cambridge, MA: MIT Press, 2000), pp. 202–41.

18 On the experimental cloud chamber and mimetic experiments, see Peter Galison, *Image and Logic: A Material Culture of Microphysics* (Chicago: University of Chicago Press, 1997). On the connection of Haacke's *Condensation Cube* to contemporaneous climate-control systems in architecture, see Mark Jarzombek, 'Haacke's *Condensation Cube*: The Machine in the Box and the Travails of Architecture', *Thresholds* 30 (Summer 2005): pp. 99–103.

19 Discussion with the author, May 24, 2011, New York City.

20 Jack Burnham, 'Hans Haacke Wind and Water Sculpture' (op. cit.), p. 11.

21 Hans Haacke, statement written in New York, September 1967, just prior to the MIT exhibition. Jones, *Hans Haacke 1967* (op. cit.), p. 50.

22 Haacke, letter to Jack Burnham written before April 1967, as cited in Burnham, 'Hans Haacke Wind and Water Sculpture' (op. cit.), p. 14.

23 Benjamin Buchloh, 'Hans Haacke' (op. cit.), p. 218. Buchloch's thoughtful piece on Haacke has certainly been the most influential since the writings of Jack Burnham and Edward Fry in the late 1960s. Fry's longest essay on Haacke, which remained unpublished in English after the cancellation of the Guggenheim exhibition in 1971, has now been published: see Jones, *Hans Haacke 1967* (op. cit.), pp. 29–45.

24 In my March 26, 2009, interview with Haacke (ts 13), he stated that the hygrothermograph piece was intended for a conceptual art show as an ironic commentary on the institution, not as an actual measure of the human/HVAC systems interacting in the galleries:

CAJ: 'By '69 you're looking at hygrothermographs in an art gallery … so you know that when bodies come into the gallery, the temperature rises'.

HH: 'No, that was not the idea behind it. I would suspect that the visitors, unless it's really packed, have no effect on the humidity'.

25 Edward Fry's Duchampian reading of Haacke is not well known in the U.S., since it was first published in German in 1974. For a recent publication of the essay in its original English, see Jones, *Hans Haacke 1967*, op. cit.

26 Burnham, 'Systems Esthetics', *Artforum* (1968), and 'Real-Time Systems', anthologized in his *Great Western Salt Works* (New York: Braziller, 1974) pp. 25, 29.

27 Ibid., p. 30, illustrated with a picture of Haacke's *Chickens Hatching*, 1969.

28 Haacke, 'from a talk … at the annual meeting of the Intersocietal Color Council, April 1968', as cited by Jack Burnham, 'Real-Time Systems', in *Salt Works* (op. cit.), p. 30.

29 Hans Haacke, statement written in English from September 1967, New York; in Jones, *Hans Haacke 1967* (op. cit.), p. 50.

30 Burnham, 'Systems Esthetics' (1968), in *Salt Works* (op. cit.), p. 16.

31 Haacke proposed a series of systems works for his one-person exhibition at the Guggenheim, set to open in 1971. Famously, one of them – *Shapolsky et al., Manhattan Real Estate Holdings, a Real-Time Social System, as of May 1, 1971* – was viewed as unexhibitable by the museum's director owing to its trenchant politics regarding Manhattan slumlords and real estate speculation. The curator protested and was fired, the show was cancelled, and Haacke's reputation for 'institutional critique' was cemented. My point here is not to rehearse this well-documented and shameful episode; it is to assert the importance and anti-humanism of Haacke's previous systems art.

32 See the current website of the 'Kunstmuseen Krefeld' detailing this history, http://www.nrw-museum.de/en/#/en/more/museums/detailview/details/musea////kunstmuseen-krefeld-kaiser-wilhelm-museum-haus-lange-haus-esters.html, accessed October 2012.

33 Michael Ned Holte, 'The Administrative Sublime, or The Center for Land Use Interpretation', *Afterall* 13 (Spring/Summer 2006), online at http://www.afterall.org/journal/issue.13/administrative.sublime.or.center.land.use.interpre. Accessed October 2012.

34 Ibid.

14

It Is the City That Makes the Walking What It Is: Interview with Olafur Eliasson, in His Studio in Hellerup, Copenhagen, November 7, 2012

Jacob Wamberg

Since the 1990s Danish-Icelandic artist Olafur Eliasson has won international acclaim for his art projects, which typically engage the corporeal foundation of our senses and the ways in which we reflect upon them. His works often transgress inherited disciplinary boundaries – exploring the interaction between art, architecture and design, between art, technology and science, and indeed between art, technology and nature (from landscape to physics).

JW: Often in your work, natural phenomena turn out, on closer inspection, to be demonstratively technologically produced – rainbows, for instance, appearing in artificially generated fog (*Beauty*, 1993), or waterfalls falling from scaffolds (*The New York City Waterfalls*, 2008; Figure 14.1).

OE: My thinking on this has evolved over the years. By staging natural phenomena through a mechanical setup and making this discoverable for the audience, I allow the viewer to become co-author of the experience. I am now particularly interested in how the viewer's co-production of the work can build shared confidence in the art situation, which, although typically found in a museum or an exhibition, can also be in an environment where art is not announced or institutionalized. Natural phenomena seem to form a good exercising field for investigating this type of contract. They allow for an interesting democratic cocktail, where you are both plural and singular – unlike other systems, which seem to demand that you be either altruistic or egoistic.

Natural phenomena sustain polarized situations in which we share an experience but agree to understand it as highly personal or subjective. This means that we are conscious of being collective – we are out in the sunlight or in the rain, we walk through the landscape together – but there is no doubt

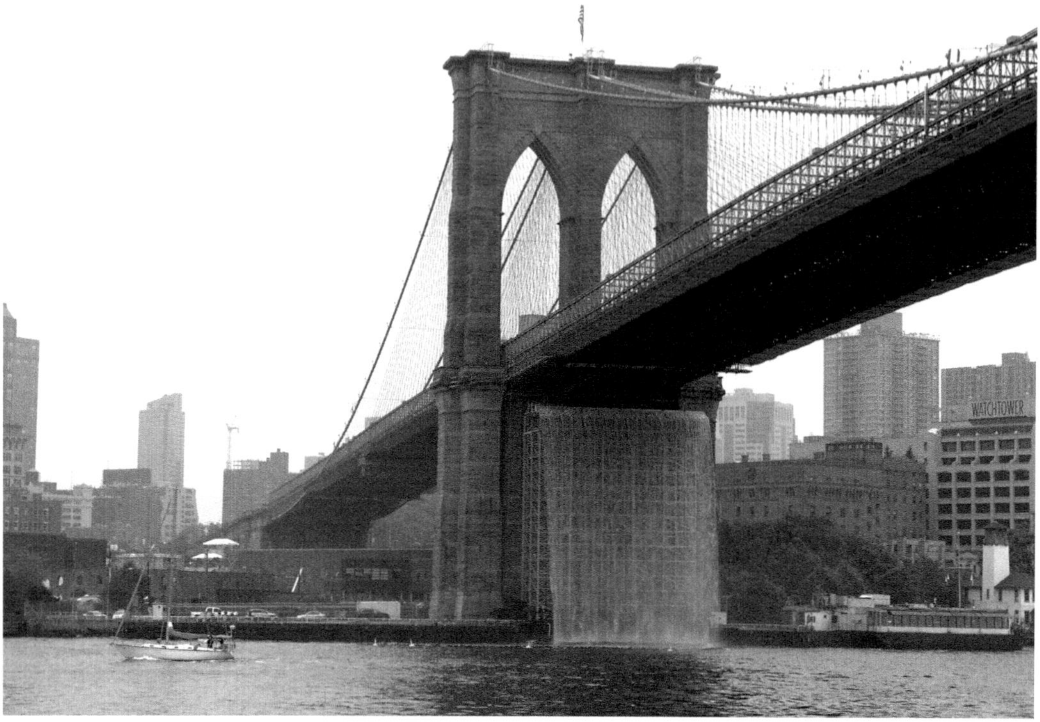

14.1 Olafur Eliasson, *The New York City Waterfalls* (2008), installation. East River, New York. Julienne Schaer/ Courtesy Public Art Fund. © 2008 Olafur Eliasson/ Photo credit: Olafur Eliasson

that we are also having a highly personal experience at the same time. So in various ways, my interest in landscape has often been rooted in how landscape enables social principles and systems to be re-evaluated.

JW: So you think natural phenomena are more easily situated in a kind of collective feeling than cultural ones?

OE: I think that natural phenomena and the landscape and the range of experiential potential found in nature offer a very strong resource, a model, for investigating the generalization of our senses that takes place in urban life. But I don't mean this in a romanticizing, animistic sense. Typically, we organize our production of reality within a very optic frame of reference – meaning that we have a clear understanding of being in a space with an object, and between me and the object, there's nothing. Nature, on the other hand, allows for a model of experience in which the object plays a less important role.

The notion that we build up knowledge based on two-dimensionality is very strong in our society. Two-dimensionality is representational space. It could be a rendering for selling a house. It could be an advertisement on a billboard. It has significantly weaker, but clearly defined, power structures. Even when we stand in front of a landscape, we can physically read it as a two-dimensional phenomenon. What is particularly interesting for me personally is that the landscape in Iceland has a relatively low degree of recognizability, meaning that when I look at it, I feel alienated from it to the extent that I have a hard time understanding it as a space; I refer to it as a picture. 'This is an excellent view', I say.

What makes a space three-dimensional for me, rather than two-dimensional, is the labour I put into engaging with it. Looking, for instance, at a waterfall from a distance, I can say, 'Oh, the water's falling very slowly. It must be a very high waterfall. But from where I'm standing, it is very small, which means it must be very far away'. The waterfall is a temporal indicator, which allows me to judge scale. Walking through a landscape, I see one mountain move faster than I expect it to, while another stays fastened to the horizon – this means that the former is a small mountain close by, and the other one is a big mountain far away. Once I start to walk, I physically build a different relationship through temporality; the space expands from two-dimensionality into three-dimensionality, and everything gains scale, from the smaller rocks to the bigger rocks, to the mountains and rivers and so on.

The whole point, though, is not about making landscape three-dimensional; the point is that we also gain our three-dimensional body back. When we look at a two-dimensional landscape, we tend to read our own bodies as representations. Once you start walking, you gain your physical confidence, and once you are physical, you are also more likely to meet another person. You are not going to meet anybody on a two-dimensional plane.

JW: Couldn't you say that the degree of alienation in the landscape, the two-dimensionality that makes it difficult to relate to your own body, is the Romantic factor? And hasn't it very much to do with the virginity, the untouched aspect, of nature? Are you interested in exposing and then deconstructing Romantic nature in your works?

OE: It's funny, but I think the notion of virginity is a cultural heritage rather than a social-scientific phenomenon. Virginity refers to entering a field in which you have little or no skills. You could also refer to it as a state of numbness: you are insensitive in a certain way. In my work I have often referred to the lack of sensitivity, either spatially or socially, in the urban context. In Western society, people are incredibly clumsy when it comes to simple empathic practices, how to execute public values in public spaces – so much so that one could call them the new virgins. When we enter a landscape, there is a significant difference between how we theoretically act, based on what we know, and how we actually recompose our posture physically and change our agendas when we engage with the landscape. It seems to me, from the time I have shared in landscapes with other people, that they are not virgins, that there is not the Romanticism of art history and cultural history. On the contrary, we are interestingly less numb when we are moving through a landscape than when we are walking through a city. I don't find this very surprising, as landscapes are highly stimulating. We are suddenly confronted with a lot of natural light in a lot of different shades, which immediately stimulate a whole range of psychobiological processes in our bodies. Actually our whole evolutionary system has developed based on spending time in nature. In terms of evolution, it is only within the last five minutes of human history that we have started living in shelters or huts.

In some Romantic ideas – some of which I also use myself – there is a sense of disappearing, detachment and contemplation, of losing contact with one's sense of existence and entering into the sublime, which I do see as an interesting aspect of landscape and nature. But I also see it has a lot to do with vanitas and egoism as well, your desire to be forgotten or to forget, which I can really follow, considering the way the world looks today. Still, I think the greater potential in landscape and nature lies not in the beautiful view itself, but rather in the less spectacular, pragmatic activity of walking to the outlook point. The journey or the time it takes to get there, the trajectory or the sense of the path, the process in which you negotiate the scale, the depth, the quality of the line, the movement of the sun – these are the strongest tools for rediscovering our own bodies. Our respiration, our physicality, our muscular activity and our thinking in the actual act of travelling through the landscape are of incredible importance.

JW: Up to now we have mostly talked about the transformation of nature *out there* to culture *in here*: how you transform landscape into technology in some of your works. I'm also interested in another aspect of nature, which maybe moves from the opposite direction: starting with a physicist's nature, of light, colours, projections, refractions, shadows, trajectories, vibrations, waves – phenomena that are typically theorized in the hard sciences and that, as Ina Blom has interestingly observed, are also used in media technologies. At the same time as you keep this more abstract, scientific nature inside a technologically framed exhibition space, you also seemingly want to turn it into a corporeally tangible, bodily accessible space: wresting it out of the grip of physics and technology and into our everyday space. You even turn it into architecture, as for instance in the models of *Suncity* (2005) materializing the sun's route through the sky.

OE: I've long been interested in how technology can assist us in seeing things that we do not normally see, such as light. Normally, we can see the light cast on a wall, for example, but by using technology, we can see what the light does on the way from its source to the wall. That means you make something tangible out of something that was already there but was not detectable. *Suncity* comprises 52 rings representing the trajectory of the sun across the sky during one year, with one ring per week. The model is based on a stereographic chart of sun paths. Drawings demonstrate the development of the sun building from a diagram into an enclosed space and how the path of the sun would be perceived from within such a structure. So technology allows us to understand in different ways that the result has to be seen in relation to the way it was produced.

Still, a great challenge for me is that my rather obsessive interest in these ideas encourages a language that prioritizes subjectivity in a very Western, almost egoistic way. My obsession reminds me sometimes of phenomenology, which can be so obsessed with subjectivity that it tends to forget that an object, such as a machine or a technological device, can also have intentionality and a degree of performativity. My challenge is to maintain a robust, sustainable relativity in using such metaphors and language because the notion of working with some models is, of course, a very centralized one, or one of great classicism or classicist values.

JW: In what sense?

OE: There is always one source of light or one point of view. I think it is important to allow the single or subjective viewpoint to exist alongside other viewpoints, without one necessarily being better than the other.

When I touch the world with a walking stick, the stick is an extension of my hand. But there is a moment in which the quality of the stick becomes apparent to me. I'm using technology in this way: normally we use light to touch things which are otherwise in darkness and make them apparent to us. But we could also evaluate the construction, the culture, the model of the light that actually co-produces the things on which it falls. When I shine my lamp onto a white wall, and then increase the brightness of the lamp, the colour of the wall changes. One type of light has a different impact than another quality of light. Blue light makes you more active than red light, and so on. Technology generally allows us to shift from measuring the world to focusing on measuring the way we measure the world. That holds as a principle for a few potentials of critical nature: potential for self-evaluation and also the potential for evaluating your own constitution in the context of our environment.

There is a sub-answer to this as well. The art world has a very predictable and somewhat flattering, but counterproductive, notion of artistic potential. Calling a work of art a piece of technology can sometimes be liberating, as you can then shift away from the focus on objecthood and the modern notion of it as an inherent, built-in quality of the work, because the performativity of an object is based on its context and the agents involved with it. When I refer to an artwork as a machine, as I often like to do, I'm trying to demystify it and also to compensate for the somewhat conservative and over-obsessive push for a mystical notion of objecthood in art.

JW: It seems to me that, overall, there is a certain anxiety about technology inside the art institution. There shouldn't be too much technology, because it risks instigating a kind of technophilia. Have you encountered this anxiety in relation to your work, or does everything go just smoothly in relation to the art institution?

OE: There is certainly a degree of anxiety inside the art institution about technology. I think this primarily has to do with two things. One is simply the lack of competence regarding what technology is. The other is a general fear, based on the post-Benjaminian idea of technology as the enemy of authenticity. Technology, of course, is at the origin of mechanical reproduction, and the museum is still one of the last fortresses claiming authenticity. There is a general assumption that if you introduce technology, you open the floodgates to the mismanagement of authenticity. Fundamentally, this combination of a lack of competence and fear leads to a lack of self-confidence, which blocks the art institutions from asking how and why technology works.

It might sound strange, but even though I have received a lot of exposure over the years, I'm also still a marginalized artist. The work I do represents a relatively small portion of the art world today, painting and sculpture being, of course, the mainstream. In this respect, there is an overlap between the art market and the institutional system. There are a number of – typically private – collectors who will not buy something where it is only possible to own the work by certificate, and the parts, such as a transformer for a lamp or cables, are replaceable, as they do not hold any grade of authenticity. The museums have taken on a similar, conservative approach. They find it hard to own something that is based on fog or other ephemera. Museums are not even built for the degree of maintenance such works require.

Of course, this is a broad generalization, and we do see a much more complex, changing landscape of institutions as well. There is also a significant reevaluation going on of the question of what on earth museums are doing. This does allow for a shift of focus from the object itself to the performativity of the object as the impetus for collecting it.

JW: You have stated that you are not interested in the question of whether what you are doing should be categorized as art or not. At the same time you have been quite involved with architecture and design, for instance in your many geometrical constructions, which seem to come out of a utopian architectonic tradition harking back to Constant, Buckminster Fuller, Yona Friedman and Einar Thorsteinn, with whom you have collaborated since the 1990s – maybe in a slightly psychedelic version of Bauhaus functionalism. In your studio in Berlin, a team of about 70 people, including craftsmen, specialized technicians, architects, artists, archivists, art historians, cooks and administrators work with you to develop, produce and install your artworks and exhibitions. Are you dreaming of a new meeting of art, technology and science, letting us see again a Leonardo da Vinci–like universalism, or a Greek *techné*?

OE: I'm interested in peripheral languages, people who have been defined, according to general notions of success, as not fitting the definition of normality – to borrow from their vision is quite simply a great source of inspiration. What they all have in common – Constant, Buckminster Fuller, Yona Friedman, Einar

Thorsteinn and, we could also mention, Poincaré – is their confidence in the equation of an algebraic idea and a modelled idea. They all thought, or think, out loud using algebraic or metaphorically algebraic arguments – in the sense that you can prove a point in reality through representation. They evaluate the causality of the argument, the shape of the idea, within the context of the world. They work with pushing the field of thinking into the field of doing – with sculpting thought. That is what drives my interest in them: the idea that a certain phenomenon is not isolated; it changes or co-produces the world. So I'm not so interested in utopia as an idealized, modern final output; I'm more interested in the feeling of utopia, the feeling that things matter.

JW: So do you think it would be possible to reintegrate this more embodied science and knowledge with art?

OE: I am pretty much constantly in contact with scientists and have worked closely together with a number of scientists on projects such as *Look into the Box* (2002; with Luc Steels) or *Your Uncertainty of Colour Matching Experiment* (2006; with Boris Oichermann). There is a rather blurry landscape between art and science, which often takes a little bit of sorting out before it becomes productive. Typically, science is interested in science, whereas artists who are interested in science are interested in how they can use science to connect to the world. Many scientists interested in art – neuroaestheticians, for example – seek scientific generalizations in art, trying to find systematic rules for good composition in painting, for instance. Artists, of course, do not really find that particularly interesting.

Yet science as a discipline has sharpened a number of tools for working with multidimensional complexity, whereas art has fallen short in this respect. Art has failed to reexamine its own methodological practice; it has been unable to include other thought processes in the making of a work of art. In this respect, I think science can be highly productive as an assistant or a guide. It can give us a notion of how to work with things in a more complex way, trying to take everything into account and evaluating the construction of it and its output as part of the result.

As a systematic workplace, my studio has also become much more self-aware through working with scientists. The studio itself has become a work of art that can never be detached from the artworks produced there. A person in a museum might not be interested in this fact, but once you get deeper, the structure of my studio and its methodology constantly influence the performativity of the artwork. It also works the other way around: the museum that shows the work of art, or the system in which the artwork engages, also influences the studio.

Traditionally, it was generally accepted that there was *thinking* (the art school or the university), *doing* (the studio) and *communication* (the museum). If we want a sophisticated university today, it has to include all three – the thinking, the doing and the communicating. We have now acknowledged that thinking *is* doing. Thinking is not just a representation of doing; you can change the world by simply thinking about it. In my studio, we are not only involved in doing. We consider ourselves a thinking university, and I believe if one were to look at the studio as a museum, it might very well be a more interesting museum than a midsize Berlin museum in understanding what art can do to change reality. As a studio, we are perhaps not so utopian, because we are already thinking, doing and communicating, not least in relation to technology and social media and so on. One of the places we have learned this from is science. I think science has become methodologically more sophisticated, visionary and resourceful through the pressure to bring about results, whereas art is really a very conservative field.

Some people refer to my studio as a laboratory that uses the real world as a frame of reference. But I would prefer to say that the world is a laboratory, and my studio is a reality machine, the only real place in the world. Looking at it this way is a more responsible way of creating reality – we are not in the representational-alchemistic non-responsible setting of withdrawn-ness and philosophical detachment. We are in the reality-making part of the world. Once you step out onto the street, you have to work very hard to find what constitutes any sense of reality on the street.

JW: So in that sense your studio is in the forefront of breaking down borders between society and art. Would you consider this even in the old avant-garde sense of art coming closer to life?

OE: Well, the avant-garde might initially have been about bringing art closer to life, but of course it ended up being about its own marginalization, since it was so obsessed with not being mainstream. So I do not think we're trying to be avant-garde. My studio represents both memories and expectations. It has a history, and it also collects history internally in the form of an archive. The archive is not, however, only concerned with the past; it is also productive of the future. Normally, this notion of history and of future and experience is claimed by the museum: it collects art from the past and somehow points toward the future, making an exhibition and trying to give it some relevance.

For two years now, I've been investigating ways to reorganize my archive, so that it is not only a chronologically retroactive, systematic way of looking for something that is lost, but, rather, the looking itself should become a process of production. You are driven by expectations and not by memories. The studio has a very strong sense of the production of now, of ongoing daily life. We enjoy exposing that. When I show my work, I often show a picture of my staff at the studio having lunch as well. So there is that notion of the now.

JW: But turning back to your relation to the avant-gardes, you have stated that they have marginalized themselves in relation to society. And you seem to want to go into much closer dialogue with society. So might the difference between you and the historical avant-garde project be the question of negativity versus positivity? The avant-gardes always said no even when they said yes, whereas you want more to say yes. Is your art more positively edifying than critical?

OE: Yes, maybe it is. Or maybe this is more a reflection of my naivety. But I do think that my confidence in making art has to do with not polarizing myself into an oppositional, and therefore predictable, role. I'm interested in the de-materialization of art in general. I have a feeling that art has something to offer on many different levels. Art can function in both commercial and non-commercial contexts, it can be political or apolitical; I generally think that one should not be afraid to trust art to perform artistic activity in every possible environment.

Take, for example, my solar-powered lamp, *Little Sun*, which I developed together with the engineer Frederik Ottesen. It was prompted by my confidence in the great language of art and in the ability of its surrounding systems to actually have an effect outside the art system. The *Little Sun* presentation at Tate Modern last summer gave the lamp some leverage that can help it operate in other areas – like the commercial marketplace. Essentially, *Little Sun* is about getting clean and affordable light to the 20 per cent of the world's population with no access to the electrical grid, and a business approach is the best way to spread the lamps to the largest number of people.

Actually, I think it is great that there are more people visiting museums than ever before. Art has simply become less elitist. On the other side, art has lived so long by itself that it has gotten used to the belief that a painting on a wall in a museum can best be digested when you're alone in that space with that painting, that you have a stronger experience when you're not surrounded by hundreds of people in that space. I have tried to work in various ways with this challenge in my work, specifically by making work where being together is an amplifier. I worked in very big spaces, trying to create a theatrical sense where you feel included despite there being too many people around. Maybe we need to train our ability to have sophisticated and sensitive experiences when we are in a space with many people.

JW: Focusing on the question of negativity versus positivity, I think it is easier to be negative and critical when one makes an autonomous artwork, whereas when one makes something which is truly environmental, like design or architecture, there is a limit to how much criticism such environmental interventions can contain, if users should truly inhabit them. You yourself have remarked that design has a certain alliance with fashion and with the capitalist market, which in itself has a homogenizing effect. Now, Caroline Jones has said of your work that you counter spectacle with interpretation. If in design there is this risk of spectacle and you build in this hermeneutic dimension, it seems that you can get away with working in the middle of the market sphere and yet at the same time retain a critical stance. Do you really think it is possible for the single spectator or user to take up this kind of interpretative challenge and not just be swallowed up by the demands of the market?

OE: Well, I think we should at least have some confidence in the capacities of the user. I think there is a tendency, especially in the very elitist theory of the art world, to have no confidence in the user at all, which is a sign of the social arrogance that governs the art world. I think we can challenge the viewer; we can ask and make demands. Generally, I think we should not underestimate the individual's experience of the works of art. From the perspective of the arrogant art world, that experience is maybe not really optimal. But if we consider it from the perspective of a person who is intoxicated by the rationalistic, indoctrinating approach to experience created by the dominant power structures – and it is a little unfair that the market economy has to take the whole burden for that – we need to see the experience of a work of art as a seed in the sequence of things. Perhaps it is my naivety, but I have some trust in this seed's potential when it is inserted into the other systems.

JW: Is that seed a kind of experience that is outside the experience economy, then?

OE: I don't think you can have an experience outside the experience economy. I don't see any museums, at this point, which are not a part of the experience economy.

JW: But still you have stated that the artwork is in some sense sovereign in relation to its platform of communication. So is there something that reaches beyond the experience economy?

OE: The experience economy is a part of the way our society is put together, so although you, in theory, can refer to an inside and outside of the experience economy, you are, in reality, experiencing it all the time. There is no situation that is not, to some extent, infiltrated by other interests (such as optimizing profitability) than those that are apparent. I cannot think of a place outside the experience economy. The Danish political system is an experience economy, not to mention the American electoral system. The question is to what extent we actually see in the experience presented to us that there is a discrepancy between what it says it does and what it really does.

This is where museums, I think, are different. They hold a different contract with reality, based on how art is produced and communicated. You could say that art is both inside and outside the experience economy. For that reason, I think it is very important to have confidence in the museum visitor. The way to deal with the experience economy is to be honest – to include people in the premise, so that they can think for themselves and decide whether they find this productive or not, whether they find it socially resourceful or socially counterproductive.

JW: So you basically think that theoreticians like Horkheimer and Adorno were on the wrong track when they theorized about the paralyzing effect of the cultural industry and serious art being necessarily outside of it?

OE: No, that's not how I meant it. I think that had they not done what they did, we would not be having the discussions we are having here. And I still believe that the fundamental values in art lie in full – maybe not *transparency* (which somehow suggests that you can see through to some pure state of reality) – but at least *constructivism*, so you see that what you are engaging in is constructed, it's not real. That is, I guess, what Adorno shows: that we should be sceptical about taking reality for granted.

JW: So this is a kind of avant-garde alienation effect, in the Brechtian sense? This is the hermeneutic potential?

OE: You might say this, yes. I just don't know of any museum that exists outside the experience economy, where you could suddenly step into a different system from the market-driven reality outside. In my view, a resourceful museum would create awareness of presence by showing people the sequence of their bodies moving, from focusing on the floor, how you touch the floor, what the part of the limb that touches the floor feels like, and how that feels within the construction of your whole body and its connection with the brain; how essentially the painting, the wall, the floor, the footstep, the leg, the body and the brain are one system. If you succeed with that, you can achieve a degree of presence, which allows for a reflectivity that, in my view, contains great potential, not least because it doesn't exist in many other places in society. It might happen in school, in an artist's studio, in a museum. It is certainly manipulated to happen at church, in the consumer-driven environment, in some urban planning where diversity is an asset and not a threat. This might be a shift that has taken place over the last 15 years, where diversity is considered not counterproductive but rather a resource. For a museum to succeed in doing this, I think it would be a mistake to say that the market economy or the experience economy should stop at the front door of the museum. It would be a very modern type of naivety. Just as you do not forget the city in which you live when you are walking in the Icelandic landscape. It is the city that makes the walking what it is.

Epilogue

James Elkins

It was a surprise and a pleasure when Jacob Wamberg and Camilla Skovbjerg Paldam asked me to contribute an epilogue to this book. It was a surprise, because given the book's subject – the convergence of art, nature and technology – I would have expected the editors to choose someone else. There are a number of prominent theorists of their three themes; Bruno Latour would have been my first choice. But it was also a pleasure, because my own interests in the visual prompt me to think that visuality plays a central role in the ongoing reconceptualization of technology and nature as well as art.

This book is too rich, and too various, for me to do justice to all the essays; instead, I will concentrate on the articulation of the book's three guiding concepts, in order to see what kind of historical understanding they create. I'll also be limiting my remarks mainly to modern philosophy and art, because I want to try to get at an issue that has to do with the general conditions of writing intellectual history.

What Is 'Art' in This Book?

Each of the three fundamental terms of this book is broad and contentious. Of the three, 'art' is the least problematic, I think – not because the term is itself stable or well defined, but because the contributors use it fairly consistently to mean Western practices from the end of the Renaissance to the present. Without those three qualifiers the word would have a much wider range of meanings. For the editors, especially I think for Wamberg, the word 'art' is very broad, and that raises different issues.

As the editors say in their Introduction, it may seem unusual to pursue the theme of art, technology and nature 'all the way back to the beginnings of culture in the early Stone Age'. Wamberg has done this himself in his monumental study of landscape, but most authors would not want to trace histories past certain cultural divides. In effect, however, the conceptualization in this book does not begin with prehistory, or even with Pliny (mentioned in the Introduction) but with Kant. The editors succinctly set out the stakes, proposing that art and technology can be seen as parallel and complementary in the Kantian critique, differing only in *telos*. That permits them to speculate that art and technology were both originally meant to 'improve upon a nature that somehow did not seem sufficient in its naked state'. It is a strong, clear schematization, something missing from some of the recent literature on art and nature, including Jeffrey Kastner's otherwise excellent anthology *Nature*.

In the editors' schema, 'art' is at first imagined as 'imitation of natural genesis', then as a representation of natural forces and, finally, as a counterpoint to technology. In the end of this final stage, identified with postmodernity, 'art' and 'technology' are drawing nearer to one another. Between them might be the cyborg – a possibility that seems to be bolstered by Markus Rath's very unusual essay on Renaissance wooden dolls, except that his subject is four centuries before postmodernism.

Readers will have to decide if the word 'art' keeps sufficient coherence as it moves down the long sweep of history (and prehistory). For example the editors propose that European cave paintings 'traced the forces of the fertile netherworld onto the cave walls' – an example of the 'initiatory function of art' before the modern era. Whether or not it makes sense to say that cave paintings had this function, is the word 'art' apposite here? Likewise, it is easy to agree that *techné* and *ars* were once related, and that 'this common domain – called *techné* in Greek, *ars* in Latin – not only bears witness to the originally shared purpose of all artefacts, whether functional or mimetic, but its etymology highlights the separation that was to come in the High-Modern era'; but it is less easy to see how the word *ars,* in that context, is continuous with what most contributors to this book think of as art. For the contributors in the second and third sections of this book, art often wants to 'interrupt' technology by 'appropriating' it 'in extraordinary ways'. The most technophilic art, as the editors accurately observe, is 'to a large extent ghettoized in special technologically oriented institutions and festivals'. In what contexts is this situation best understood as a continuation of the themes of *natura, ars* and *techné* from past generations? In my reading, the claims of continuity are most persuasive in the Introduction and Part I, and they become less so as the essays focus more on the past two centuries.

What Is 'Technology' in This Book?

Perhaps this is mainly my own perspective, but I have often thought that the relation between science and technology works differently in art and in the art world, than it does in science studies or philosophy. The distinction is not between theory and utility, or epistemology and pragmatism, or natural philosophy and engineering, or between falsification and verification, but between two very different things. Science, in the art world, tends to be pictured as an institutional, political entity, aligned with power and ideology. Technology can sometimes be imagined as a tool in the ideology of progress, but it is much more often understood as whatever interesting tools and machines might be used to make art. I used to teach a course called 'Science and Art', but I stopped when I realized the students were mainly interested in cool devices: they were fascinated by subjects like Tesla, augmented reality and micro-implants, but they didn't care at all about the correspondence theory of truth, the concept of hypothesis or Karl Popper's theories about falsification. (The students were interested in Feyerabend, Kuhn, Hacking and Latour to the extent that they seemed to underwrite their anti-realist and relativist convictions about the constructed nature of science.) Over the years I have had many conversations with artists about such things as laser routers, muscle wire, miniature surveillance cameras, liquid photo emulsion, the wiring of EKGs, ultra-slow worm gears, spray insulation foam, the procedures of DNA synthesis, tissue culture and robotics, and very few conversations about the ideas of science or technology. (These are the current equivalents of what were once 'sciences', in the sense that Bjørn Okholm Skaarup documents in the Renaissance: that is, artists' tools.) I could have kept teaching a course called 'Technology and Art', but I imagine even that would have been more about how to build Tesla coils than how to critique Heidegger's theories of *Gestell* (discussed by the editors in the Introduction).

I mention this somewhat personal perspective to suggest that what counts as technology for the artists discussed in this book is different from what counts as technology for the philosophers who are cited by the contributors to this book. Pernille Leth-Espensen's essay is a good example of this. She discusses work by three artists, and places it in the context of work by several philosophers (mainly Latour and Don Ihde) and art historians (including Tom Mitchell). Latour and Ihde, she notes, 'interpret the role that instruments, inscriptions, technologies and metaphors play in the generation of scientific knowledge', emphasize the mediated nature of transcriptions and representations, and 'stress that technologies and scientific representations have ethical consequences as they affect our view of nature, our own body and society'. I agree with her summaries of the philosophers' interests, and I like her descriptions of the artists, but I don't think the artists are expressing the philosophers' ideas at all. The artist Anne Niemetz (who turns data from an atomic force microscope into music) is producing an arbitrary aesthetization of a certain technology. Paul Vanouse (who produces electrophoresis gels that look like maps) is not doing 'molecular biology in reverse' but rather producing a formalism with no scientific content. The work *MEART* by the SymbioticA Research Group (which produces mechanical drawings from neural activity) is producing another formalism, indecipherable as, or in relation to, its originating science or technology. These are good examples of contemporary artists' uses of technology: in them, technology is seen as a complex system of transcriptions and representations, and the artists produce aesthetic re-transcriptions and re-presentations in return. But the artworks do not contribute to the kinds of chains of signification that Latour studies, or the mediations that interest Ihde, for three reasons: they are not significant as science or technology; they are usually not fully explained to viewers, so there is no way to study exactly how they are doing their work (Vanouse is a partial exception here, but only partial); and their products are aesthetic objects, which produce meaning in an entirely different register from the original technologies. In a word, the art is formal, and the technology and science are informational. Or, as the editors put it in the Introduction, 'avant-garde art typically alienates technology from its already known utilitarian functions'.

Technology, in this book, is often divided in this particular way: philosophic analyses of technology are used to explore practices where technology is a black box, used for aesthetic purposes. The artworks may conjure many issues, but they do not 'interpret the role that instruments, inscriptions, technologies and metaphors play in the generation of scientific knowledge'. This way of interacting with science, in which art is done using techniques that are not themselves explained, resulting in works that are available primarily as aesthetic objects, is one of the principal reasons why it is so difficult to make art about environmental issues, as Søren Bro Pold and Christian Ulrik Andersen note. There are many others reasons: I might cite Jacques Rancière's observation that the most overtly political art is often in fact the least political (at least in his sense of 'political'), or I might note, in the mode of *October* magazine, that work about the Earth is seldom also about the conditions of its own medium or artistic lineage. Søren Bro Pold and Christian Ulrik Andersen have an interesting argument; my position is that if it is necessary to work on saving the Earth, then it is probably necessary to suspend talk about art. The authors ask, 'Should art instead remain critical and ultimately avoid serving other agendas than its own, even if the future of the planet is at stake?' If the Earth is the first thing we should worry about, then we need to acknowledge that what we make may not be art. As my colleague Frances Whitehead says, if someone comes along in the future and says, 'You made art', then that's fine, but for now, I have other things to worry about.

What Is 'Nature' in This Book?

Of this book's three leading terms, 'nature' is perhaps the most interesting. Do artworks represent nature? To this one might answer no, because nature is such a compromised term that the statement, *art represents nature*, could at best be an empty formula. As Latour says, nature is a 'jumble of Greek philosophy, French Cartesianism and American parks'. (Bruno, if you're out there: you deserve a separate Afterword. Your work is woven into most of these authors' essays, but it is seldom drawn out as it might be. If I can just put in a sentence on the subject: if the contributors, or the artists they report on, were thinking of 'nature' the way you do in *The Politics of Nature*, they wouldn't have written as they did: to me, that's a sign of an ongoing entanglement, one that might be resistant to any re-conceptualization.) But back to my subject: 'nature' was, of course, chaotic before Latour; the best essay remains Arthur Lovejoy's classic '"Nature" as an Aesthetic Norm', which divides the word into 60 different meanings.

Aside from this 'jumble' (if it is possible, even temporarily, to set aside the deep confusions about the word 'nature' in favour of an informal, vernacular sense of the word), the art world has long since given up the idea that 'nature' is the privileged object of artists' attention. Ever since M.H. Abrams's *The Mirror and the Lamp*, it has been clear to scholars that Romanticism fundamentally altered artists' ideas about their project. Before Romanticism, nature was universally, if problematically, the object of art; art was the mirror of nature. Since Romanticism, art has been conceived on the model of the 'lamp': it expresses the artist's inner life, her *Innerlichkeit*, her 'spirit', thoughts, emotions, affects or feelings. That sea change in culture emerges in various essays in this book. It is evident in the distance between the Spanish nature-printing practices Alisa Luxenberg describes, and contemporary representations of nature. Claire Farago notes the Council of Trent's opposition to any 'explicit display of artifice for its own sake, contrived composition and artistic ingenuity' – that is, anything that would detract or distract from the artist's only true goal, the representation of the sacred. Farago provides an excellent formulation of this now-forgotten sense of 'nature':

> To summarize, Leonardo's richly conceived discussions of painting as a scientific art that results from the artist's understanding of nature's causes through observation and ingenious mental discourses that translate nature's ornaments into painted ones, are informed by Scholastic classifications of knowledge, deeply engaged with humanist debates over the value of poetic artifice.

She adds that Leonardo's discussions are 'strongly reminiscent of contemporary issues regarding methodology in the natural sciences', but not, I would say, of art – the difference is one of the main points of separation between post-Romantic art and modern science. This is an enormous issue, and there are as many counterexamples as there are examples. This book necessarily emphasizes art practices in which 'nature' is something represented in, through or as art, but I think those practices are still in the minority. Most post-Romantic, modernist, postmodern and contemporary art is aimed at expression, and 'nature' is not directly part of the equation. Grammatically, 'nature' is an indirect object: art expresses affect and other experiential qualities, and it may do so by re-presenting something identified as nature. The rise of environmental art, and art projects aimed at public awareness of global climate change, water shortages, extinctions and other issues are, I think, the exceptions that prove the rule. 'Nature' may be 'the most complex word in the language' (as Raymond Williams has said), but its complexity is sometimes beside the point because nature is no longer the grammatical object of sentences that name what art addresses, or modifies, or represents or does.

These are two large problematics: that 'nature' means so many things and that art has, in the most general sense, turned away from representing nature. A third problematic also needs to be considered, and that is our lingering romanticism about nature. Hans Haacke's early anti-Romantic stance, described in Caroline Jones's essay, provides a useful contrast. If we were to reject both what Haacke calls 'the traditional thinking of the romantic' and 'the psychological, which exudes the magic of all art criticism', then we'd have very little of the contemporary sense of nature left. Eliasson, Wilson, Dion, Horn, Graham, Thater, Sugimoto, Viola ... any number of contemporary artists are deeply immersed in late-Romantic constructions of nature. Even environmental artists, whose work engages the mutual dependence of the human and the natural, mingle modern science (or, in this book's terms, technology) with Romantic representations. It is not a position from which we can disentangle ourselves.

These three problematics weave through the pages of this book. One essay, Kristine Nielsen's study of cloud chamber images, concerns a practice that was perfectly poised between the Romantic image of nature and an emerging mathematized image. Nielsen quotes Peter Galison's reading of the cloud chamber, which he places at the beginning of his book *Image and Logic:* to Galison, the cloud chamber was a moment when 'imitating' gave way to 'dissecting', when the unique representational image gave way to the statistical aggregate, when 'image' gave way to 'logic'. But Nielsen is right to say that Galison's interpretation is 'too safely rooted in the realm of present-day science', and no such clear distinction can be made. I would only add that in the twenty-first century we are still entangled in this issue. Galison's work in *Image and Logic,* like his work on objectivity, has the tremendous virtue of conceptual clarity, but the drawback of conceptual optimism: implicitly, the claim is that clear and distinct ideas can articulate historical understanding. Perhaps the harder route is to picture the inadequate conceptualizations and ensuing analytic entanglements that constitute the self-understanding of modernism in this regard.

When Hans Dickel says that the water in Roni Horn's *Library of Water* in Stykkishólmur 'stands upright at variance with nature', he means that water in glaciers doesn't stand upright. But when he says, a few sentences later, that 'Horn's glass columns link the space, people and nature', he is using the word in a very different sense, something like 'the world outside of culture', or even 'the world outside of representation'. When he says the room separates 'visitors from the surrounding nature' then 'nature' is more like 'whatever is outside architecture'. Dickel then cites Adorno to the effect that 'what nature strives for in vain, artworks fulfil: they open our eyes', but that sense of 'nature' is different once again, because Adorno is conjuring the pre-Enlightenment (or 'early modern', in Farago's sense) understandings of generative nature ('nature as producer', in Hansen and Paldam's terms), *natura naturans,* the spirit of the world as manifested in nature. There are many other examples even in these few pages of Dickel's essay. A few lines later he says Horn's installation lets visitors 'experience Iceland's nature as the Other, as something alien, effectively at eye level'; in that sense 'nature' is something that we are all part of, and the artist re-presents it to us by defamiliarizing it. And in the next sentence, Dickel argues that 'by experiencing something atmospheric', viewers 'also feel themselves as nature, and yet remain as a describing viewer, non-nature, a recognizing eye', implying that we are not all parts of nature, and we need art to re-present nature to ourselves so that we can feel part of it.

I don't at all mean to suggest that Dickel's essay suffers from conceptual incoherence. It only 'suffers' from the same ailment that we all feel: we are entangled and confused by the many things we want nature to be. And that confusion is only deepened by our ambivalence about what art expresses. Horn's art, like Rodney Graham's and Diana Thater's, is about nature but always also about the artist's sense of nature, and there is an enormous gulf between those two formulations.

Sometimes it is possible to be clear about the distinction between the 'mirror' and the 'lamp'. For instance, Hanna Johansson says that the art she studies is 'related to so-called Realistic art, because the artists have not sought to "express", but rather to record, environmental phenomena': a nice condensation of the 'lamp' and the 'mirror'. Each of the artists she considers aspires 'to anti-subjective representation. The result in every case, however, is not only descriptive but also affective and emotional … . And still, all of these works take a distance from the subjective expression of the artist's innermost self'. What matters here is the 'but also' and the 'and still': the qualifiers that link 'objective' or 'scientific' representation to inner expression. Towards the end of her essay, Johansson paraphrases Mieke Bal on the subject of Olafur Eliasson's work: 'the old divisions between wild and domesticated, naturally given and technologically made, are ignored and replaced by affective experimentations that re-position human beings in the environment by letting us feel and see things differently'. Here 'ignored' and 'replaced' may be a little too strong. I prefer expressions like 'but also … and still'. As Eliasson says in Wamberg's interview, art in which nature is both natural and artificial can 'allow for an interesting democratic cocktail, where you are both plural and singular'.

What Histories Have Purchase on the Present?

Historiography can perhaps be divided into two strands. Some scholars are interested in the histories of their disciplines, and their work uncovers past moments and practices, elucidates the lives of scholars and demonstrates the attention that was paid to certain practices, problems, economies and institutional configurations. That kind of research is archival, scholarly or archaeological in the sense that it contributes to a narrative of past cultural configurations. Another kind of historiography is archaeological in Foucault's sense: it aims to uncover the conditions of the possibility of present-day institutions. It is activist historiography in the sense that it wants to find a past practice that can be of use to the present. Activity historiography proposes a different sense of history than archival historiography: for an 'activist' scholar, only certain elements in the past of her discipline are of interest, because only they can help solve problems that affect the present. In my discipline, art history, some historiographic projects are aimed at past practices that are no longer connected to the living present. In the past few years, for example, I have seen research done on Willem Vogelsang and Johan Jakob Tikkanen, early twentieth-century art historians in Utrecht and Helsinki, respectively. Neither has immediate use for the present, because young art historians could not write like either author, or take ideas from them or use them as models. They are rather of interest for a richer, deeper sense of art history's past, and specifically its past in certain countries and universities. On the other hand, the recent interest in Alois Riegl and Aby Warburg is 'activist' in the sense that it aims to enrich current practice, and even to directly influence future practice. Georges Didi-Huberman's work on Warburg is a good case in point: his curiosity is not archival, but interventionist: he means to solve current problems in the conceptualization of art history and of the image.

Archival and activist historiography are sometimes blended, and no historiography can be purely archaeological in the non-Foucauldian sense – that is, no historiographer can think that the objects of her scholarly attention are entirely disconnected from the present. But the mixing of archival and activist historiography can result in conceptual puzzles. If the concept 'nature', for example, is taken in any of its past senses, then it will appear in our scholarship as an archival term, which is to be understood as something that once had a particular meaning, which it has since lost. But if 'nature' is taken, for example, in Latour's sense as a 'jumble of Greek philosophy, French Cartesianism and American parks', then it becomes a concept that does work for current contexts. Latour's project is decidedly activist and philosophical, but when he is cited in what I am calling archival historiographic essays, his concepts can appear as simply parts of the historiography of 'nature'.

Mixtures of archival and activist historiography occur in a number of fields. Another way to put this issue is to distinguish between histories that are chronicles of other people's pasts, and histories that are reflections of our own concerns. In this book that theme becomes visible in several ways. Marion Endt-Jones says her purpose is to trace the history of the cyborg to Surrealism and to 'an early modern "age of curiosity" in which the meanings of objects and the boundaries between different fields of knowledge and modes of creative production were flexible and permeable rather than set in stone'. I think she is correct to trace the idea of the cyborg to Surrealism and to the curiosity cabinet, but the two are different kinds of histories. Surrealism is still very much continuous with the present, in art theory, criticism, history and practice. Among others, Hal Foster has explored many connections between the practices of the 1930s and those of the 1990s and 2000s. Any study of Surrealism is necessarily an activist historiography, because it has to address current and future possibilities for art and scholarship. Rosalind Krauss and Yve-Alain Bois's *Formless: A User's Guide* remains the case in point, because the authors are explicit in their intention *both* to provide an exemplary scholarly treatment of Bataille's 'Dictionary', and *also* to augment, amend and criticize it, so as to produce 'a user's guide'. The history of the 'age of curiosity' is a significantly different matter. Revivals of the Wunderkammer have been continuous at least since the 1986 Venice Biennale, and they continue in work by Mark Dion, David Wilson and many other contemporary artists. But before the 1980s, curiosity cabinets were only intermittently noticed: Breton was interested in them, and they were studied by antiquarians and historians of science. But it is only relatively recently that they have become a model for current practice. Their revival is predicated on a hope, or a desire, that certain concepts and beliefs held in the seventeenth and early eighteenth centuries might be made available to the present. Wilson, creator of the Museum of Jurassic Technology, speaks of 'wonder', and Dion is enthralled by what he takes to be aesthetic properties of collections. Both artists could be criticized for projecting their desires onto past materials. Curiosity cabinets were in many ways very different from Wilson's creation. (The Museum of Jurassic Technology trades on late twentieth-century aesthetic categories such as the creepy, the cute and the campy. Dion is romantic in an Edwardian sense, and projects his desires onto material by draining its scientific content.)

The revival of the curiosity cabinet can be interpreted in a Benjaminian fashion: it is run by a *desire* for a less rigid, compartmentalized, instrumentalized, secular, technophilic sense of science and the world. When Endt-Jones conjures a time when 'the meanings of objects and the boundaries between different fields of knowledge and modes of creative production were flexible and permeable rather than set in stone', she is right as far as the genealogy of the cyborg goes. But the cultural context in which those boundaries were permeable is no longer available to us. Scholars who study curiosity cabinets, and artists who recreate them, are, I think, applying an activist historiography to an episode in history that is effectively inaccessible to us.

Maria Fabricius Hansen and Camilla Skovbjerg Paldam argue that a 'striking similarity exists between the art of the sixteenth century and the Surrealist art of the twentieth century'. This similarity has been a trope in the reception of Surrealism, which has sometimes excused Surrealism from its position as an early twentieth-century art movement, and expanded it to the status of a recurring theme in the Western visual imagination. Medieval bestiaries, the *studiolo* of Cosimo I, the paintings of Arcimboldo, the school of Fontainebleau and the court of Rudolph II, have all been proposed as Surrealist *avant la lettre*. In the history of art, the reception of Hal Foster's scholarship is an interesting example: he does not study premodernist senses of Surrealism such as Arcimboldo or the 'Surreal' qualities of curiosity cabinets, but his psychoanalytic interpretive framework is compatible with such readings, in that the psychoanalytic categories he develops can be extended beyond the temporal and cultural settings in which they were developed. 'Part object', prosthesis, the uncanny, the fetish, the phobia and other concepts seem to fit a wide range of practices that lead up to Surrealism – as Breton's range of allusions makes clear.

I am sceptical of these connections, and so I am not entirely convinced by the claim, in Hansen and Paldam's essay, that the 'imagery of metamorphosis, of flux in the passages between figurations, leading to monstrous hybridity – and of flux in the passages of meaning, resulting in ambiguities and paradoxes in the pictorial content', are found both in the sixteenth century and in the twentieth. Flux and metamorphosis, ambiguity and paradox, are there to be found in any number of traditions, but the terms themselves are very much the inheritance of Surrealism, and more specifically of late twentieth-century scholarship on Surrealism. It is difficult to know when concepts are anachronistic: metamorphosis certainly has its ancient roots, and so do the grotesque and the monstrous. But the 'hybrid', another term used by Hansen and Paldam, has a determinedly modern origin: in the humanities, it dates principally from Homi Bhabha's elaboration on Frantz Fanon's postcolonial critique. And as Barbara Stafford has shown, terms like *monstrum* had measurably different meanings in past centuries.

Hansen and Paldam also discuss the idea of generative nature, using the examples of Leonardo's instructions to consider 'stains of walls, or the ashes of a fire, or clouds, or mud or like things', and Mantegna's paintings with anthropomorphic and theriomorphic forms in clouds and landscapes. Here I think it is possible to make a fairly strong distinction between the Renaissance practices and modern interests, because the literature emphasizing and exploring Leonardo and Mantegna is a late twentieth-century literature. It's not that nature was not thought of as generative back then: it's that the cultural moment that recognized it, and began building the bridges to the past, is our own postmodern academy. (I explore this using the same examples as Hansen and Paldam in *Why Are Our Pictures Puzzles?*) An activist historiography reactivates a past for the present, and enables it to speak to present practices. Stephen Bann does the same with similar premodern examples in *Ways Around Modernism*, and the larger revival of the culture of wonder and the accompanying expansion of Surrealism are related phenomena. To the extent that they speak to the present, and work to help us understand contemporary practices, these activist historiographies are performing the crucial role of any engaged historical practice. Writers from Nietzsche to Benjamin to Hayden White, Dominick LaCapra and Georges Didi-Huberman have said similar things: history is what makes the past live for the present. But against that there should also be the balancing claim that some historical practices are actually of the past, and that they cannot be brought forward without altering, abbreviating, redefining, translating, altering or otherwise changing them. For me, the court of Rudolph II, the culture of the Wunderkammer and the idea of the *monstrum*, are sunken in history. They are not available to us, except if we are willing to change them beyond what their makers would have recognized. Breton and other theorists of Surrealism needed to believe that their project was linked in deep ways to deep pasts; but for me, Surrealism is a Western European practice of the 1920s and 1930s and is not linked in any helpful way to the deeper pasts. I offer this only as a counterbalance to Hansen and Paldam and Endt-Jones, not as a critique: each reader needs to decide for herself how to draw the line between activist and archaeological historiography.

Concluding Questions

I want to end on an abstract theme, which is agnostic about the large problems of nature, art and technology. I have been trying to get near to an especially difficult issue, which I think this book raises in an exemplary fashion. I could put it like this: under what circumstances are large-scale histories optimal? This book has an ambitious Introduction, driven by a relatively small number of concepts (art, nature, technology, *poiesis, techné, Gestell* and some others). It has an unusual analytic clarity, paired with an equally unusual historical reach. Most of the essays in this book don't try to match or answer that agenda, and some contributors use their concepts in decidedly local ways.

That sets up a contrast between the editors' Introduction and some of the essays – a contrast that is unusually productive for thinking about some of the perennial problems of historiography: how to bind conceptual analyses to the uneven ground of history, how to tie philosophic concerns to cultural contexts that do not call on philosophy, how to understand what it means to apply philosophy to history.

Work by the historian of science Peter Galison, cited in Kristine Nielsen's essay, also embodies these extremes, although in a different philosophic register. I wondered, in relation to Nielsen's essay, how it might be possible to imagine a conceptual unclarity that could complement and embody Galison's conceptual clarity. Wilson's cloud chambers, as Nielsen argues, are more complex and conflicted. I am not convinced that Nielsen's emphasis on complexity is an adequate response to what she perceives as an excessive schematization, but it throws the problem into a useful relief.

Clear dichotomies (art and technology, or image and logic) and other geometric conceptual schemata (such as the triad that organizes this book) become available when history presents itself as a subject (or an arena, discipline or domain), effectively disjunct from the historian, whose function is to find or create order and significance. From the inside, things look different. As I articulate the limbs of a Renaissance wooden doll, or work the controls of a modern cloud chamber or contemplate a curiosity cabinet, I become unsure about my leading concepts. I may mention Latour, but I'm not sure what he does for me, and perhaps he's best off in a footnote. I am uncertain, because I am inside history. It's only an illusion, but it's different from the illusion that history is something outside, before or otherwise apart from us, pictured for us by large concepts like nature, art and technology. Both illusions, together, comprise historical writing.

Bibliography

Abrams, M.H. *The Mirror and the Lamp: Romantic Theory and the Critical Tradition*. New York: Oxford University Press, 1953.

Accolti, Pietro. *Lo inganno de gl'occhi, prospettiva pratica*. Florence: P. Cecconcelli, 1625.

Achilles-Syndram, Katrin, ed. *Die Kunstsammlung des Paulus Praun. Die Inventare von 1616 und 1719*. Nuremberg: Stadtrat, 1994.

Adorno, Theodor W. *Aesthetic Theory*. London and New York: Continuum, 1997 [1970].

Adorno, Theodor W., Gretel Adorno and Rolf Tiedemann. *Ästhetische Theorie*. Frankfurt am Main: Suhrkamp, 1970.

Aitken, John. 'On the Formation of Small Clear Spaces in Dusty Air' [1883–84]. *Transactions of the Royal Society of Edinburgh* 32 (1887): pp. 239–72.

Aitken, John. 'On the Number of Dust Particles in the Atmosphere of Certain Places in Great Britain and on the Continent with Remarks on the Relation between the Amount of Dust and Meteorological Phenomena' [1892]. *Transactions of the Royal Society of Edinburgh* 37 (1892–95).

Aitken, John. 'Report on Atmospheric Dust'. Meteorology of the Ben Nevis Observatory. *Transactions of the Royal Society of Edinburgh* 42 (1902): pp. 479–89.

Alberti, Leon Battista. *De pictura*. Basel: Westheimer, 1540.

Alberti, Leon Battista. *Della pittura*. Edited by Luigi Mallé. Florence: Sansoni, 1950 [1540].

Alberti, Leon Battista. *Della pittura* [1540].Translated by Cosimo Bartoli. In Leon Battista Alberti, *Opuscoli morali*. Venice: Franceschi, 1568.

Alberti, Leon Battista. *Della pittura*. Translated by Lodovico Domenichi. Venice, 1547 [1540].

Alberti, Leon Battista. *On Painting and On Sculpture. The Latin Texts of* De Pictura *and* De Statua. Edited and translated by Cecil Grayson. London: Phaidon, 1972.

Alberti, Romano. *Origine e progreso dell'Accademia de' Pittori, Scultori ed Architetti di Roma*. Rome: Pavia, 1604.

Albert of Saxony. *Tractatus de proportionibus*. Published with Walter Burleigh, *De intentione et remissione formarum*. Venice, 1496.

Albertus Magnus. *Book of Minerals.* (*De Mineralibus*, 1491.) Translated by Dorothy Wyckoff. Oxford: Clarendon Press, 1967.

Allori, Alessandro. *Il primo libro de' ragionamenti delle regole del disegno*. Biblioteca Nazionale Centrale di Firenze BNCF, c. 1565.

Ammanati, Bartolomeo. *Lettera agli accademici del disegno*. Florence: B. Sermartelli, 1582.

Andersen, Christian Ulrik, and Søren Pold, eds. *Interface Criticism – Aesthetics Beyond Buttons*. Aarhus, Denmark: Aarhus University Press, 2011.

Andres, Mirjam. *Die Antikensammlung. Griechische, Römische, Altorientalische Puppen und Verwandtes*. Hanau-Wilhelmsbad: Hessisches Puppenmuseum, 2000.

Antennae 3: *Insect Poetics*. 2 vols (Autumn 2007).

Antennae 11: *Insecta* (Autumn 2009).

Anthenaise, Claude d', Louis-Albert de Broglie and Patrick Mauriès. *Nature fragile: le cabinet Deyrolle*. Paris: Beaux Arts Editions, 2008.

Anthes, Emily. 'The Race to Create "insect-cyborgs"'. *Observer*, February 17, 2013. www.guardian.co.uk/science/2013/feb/17/race-to-create-insect-cyborgs. Accessed February 20, 2013.

Aquinas, Thomas. *Commentary on the Metaphysics of Aristotle*. Translated by John P. Rowan. 2 vols. Chicago: Henry Regnery, 1961.

Aquinas, Thomas. *The Division and Methods of the Sciences*. 4th edition. Edited by Michel Doyne. Translated by W.H. Kane. Montréal, 1956.

Aragon, Louis. *Le paysan de Paris*. Paris: Folio, Éditions Gallimard, 1998 [1926].

Arce, Celedonio de. *Conversaciones sobre la escultura*. Pamplona: Joseph Longa, 1786.

Aristotle. *Generation of Animals*. Translated by A.L. Peck. Loeb Classical Library. Cambridge, MA: Harvard University Press; London: William Heinemann, 1953.

Aristotle. *Meteorologica*. Translated by H.D.P. Lee. Cambridge, MA: Harvard University Press, 1952.

Aristotle. *Parts of Animals*. Translated by A.L. Peck. Loeb Classical Library. Cambridge, MA: Harvard University Press; London: William Heinemann, 1961.

Aristotle. *Poetics*. Edited and translated by Stephen Halliwell. Loeb Classical Library. Cambridge, MA: Harvard University Press; London: William Heinemann, 1995.

Aristotle. *Politics*. Translated by H. Rackham. Loeb Classical Library. Cambridge, MA: Harvard University Press, London: William Heinemann, 1990 [1944].

Armstrong, Carol, and Catherine de Zegher, eds. *Ocean Flowers: Impressions from Nature*. New York: The Drawing Center; Princeton: Princeton University Press, 2004.

Arp, Jean. *On My Way: Poetry and Essays, 1912–1947*. Edited by Robert Motherwell. English translation by Ralph Manheim. New York: Wittenborn, Schultz, 1948.

Artaud, Antonin. 'L'Activité du Bureau de Recherches Surréalistes' [1925]. In *Bureau de recherches surréalistes: Cahier de la permanence octobre 1924–avril 1925*. Edited by Paule Thévenin. Paris: Gallimard, 1988.

Augustine, St. *City of God*. In *A Select Library of the Nicene and Post-Nicene Fathers of the Christian Church.* Vol. 2. Edited by Philip Schaff. New York: Christian Literature Publishing Co., 1890. http://www.ccel.org/ccel/schaff/npnf102.toc.html.

Austin, J.L., J.O. Urmson and Marina Sbisà. *How to Do Things with Words: The William James Lectures Delivered at Harvard University in 1955*. Oxford: Oxford University Press, 1962.

Bachmann, Manfred, and Claus Hansmann. *Das große Puppenbuch*. Leipzig: Edition Leipzig, 1971.

Badt, Kurt. *John Constable's Clouds*. London: Routledge and Kegan Paul, 1950.

Baker, Steve. *The Postmodern Animal*. London: Reaktion Books, 2000.

Bakkum, Douglas J. and Alexander C. Shkolnik, Guy Ben-Ary, Phil M. Gamblen, Thomas B. DeMarse and Steve M. Potter, 'Removing some "A" from AI: Embodied Cultured Networks'. In *Proceedings of the Dagstuhl Conference on Embodied Artificial Intelligence*. Edited by Luc Steels and Rolf Pfeiffer, pp. 130–45. Berlin: Springer, 2004.

Bakkum, Douglas J. and Philip M. Gamblen, Guy Ben-Ary, Zenas C. Chao and Steve Potter. 'MEART: The Semi-Living Artist'. *Frontiers in Neurorobotics* 1 (November 2007): pp. 1–10.

Bal, Mieke. 'Light Politics'. In *Take Your Time: Olafur Eliasson*. Edited by Madeleine Grynsztejn. New York and London: San Francisco Museum of Modern Art in association with Thames and Hudson, 2007.

Baldinucci, Filippo. *Notizie de' professori del disegno da Cimabue* ... Florence: S. Franchi, 1681.

Baltrušaitis, Jurgis. *Anamorphic Art*. Cambridge: Chadwyck-Healey, 1977 [1969].

Bange, Ernst Friedrich. *Die Bildwerke in Holz, Stein und Ton: Kleinplastik*. Leipzig and Berlin: De Gruyter, 1930.

Bange, Ernst Friedrich. *Die Kleinplastik der deutschen Renaissance in Holz und Stein*. Leipzig: Schmidt and Günther, 1928.

Bann, Stephen. *Ways Around Modernism*. New York and London: Routledge, 2007.

Barasch, Moshe. *Icon: Studies in the History of an Idea*. New York: New York University Press, 1995.

Baring, Anne, and Jules Cashford. *The Myth of the Goddess: Evolution of an Image*. New York: Viking Arkana, 1991.

Barocchi, Paola, ed. *Scritti d'arte del cinquecento*, I–III. Milan and Naples: Riccardo Ricciarcli, 1961–77.

Barrow, John D., and Frank J. Tipler. *The Anthropic Cosmological Principle*. Oxford and New York: Oxford University Press, 1986.

Barzman, Karen-edis. *The Florentine Academy and the Early Modern State*. Cambridge: Cambridge University Press, 2000.

Bataille, Georges. 'Œil'. *Documents* 4 (September 1929): p. 216.

Batchen, Geoffrey. *Burning with Desire: The Conception of Photography*. Cambridge, MA: MIT Press, 1997.

Baxandall, Michael. *The Limewood Sculptors of Renaissance Germany*. New Haven: Yale University Press, 1980.

Beckett, R.B., ed. *John Constable's Correspondence*. Vol. 3. Ipswich: Suffolk Records Society, 1950.

Beckmann, Poul. *Living Jewels: The Natural Design of Beetles*. Munich, Berlin, London and New York: Prestel, 2001.

Beckmann, Poul. *Living Jewels 2: The Magical Design of Beetles*. Munich, Berlin, London and New York: Prestel, 2007.

Bellori, Giovanni Pietro. 'The Idea of the Painter, Sculptor and Architect, Superior to Nature by Selection from Natural Beauties'. In Erwin Panofsky, *Idea: A Concept in Art Theory*. Translated by Joseph J.S. Peake. Columbia: University of Southern California Press, 1968.

Belozerskaya, Marina. *Luxury Arts of the Renaissance*. London: Thames and Hudson, 2005.

Bensaude-Vincent, Bernadette, and William R. Newman, eds. *The Artificial and the Natural: An Evolving Polarity*. London/Cambridge, MA: MIT Press, 2007.

Bermingham, Ann. *Landscape and Ideology: The English Rustic Tradition, 1740–1860*. Berkeley: University of California Press, 1986.

Bertalanffy, Ludwig von. *General System Theory: Foundations, Development, Applications*. New York: Braziller, 1968.

Biesenbach, Klaus, and Roxana Marcoci. 'Toward the Sun: Olafur Eliasson's Protocinematic Vision'. In *Take Your Time: Olafur Eliasson*. Edited by Madeleine Grynsztejn. New York and London: San Francisco Museum of Modern Art in association with Thames and Hudson, 2007.

Bijvoet, Marga. *Art as Inquiry: Toward New Collaborations between Art, Science and Technology*. American University Studies, Ser. 20, Fine Arts, vol. 32. New York: Peter Lang Publishing, 1997.

Black, Max. *Models and Metaphors: Studies in Language and Philosophy*. Ithaca: Cornell University Press, 1962.

Blackett, P.M.S. 'Charles Thomson Rees Wilson'. *Biographical Memoirs of Fellows of the Royal Society* 6 (November 1960).

Blunt, Wilfrid, and William T. Stearn. *The Art of Botanical Illustration*. Revised ed. London: Antique Collectors Club, 1995.

Boas Hall, Marie, ed. *The Pneumatics of Hero of Alexandria*. Facsimile of London, 1851. London, 1971. http://www.history.rochester.edu/steam/hero/. Accessed November 21, 2011.

Bois, Yve-Alain, and Rosalind E. Krauss. *Formless: A User's Guide*. New York: Zone Books; Cambridge, MA: MIT Press, 1997.

Bonacina, L.C. 'John Constable's Centenary: His Position as a Painter of Weather'. *Quarterly Journal of the Royal Meteorological Society* 63/272 (1937): pp. 483–90.

Bottari, Giovanni, and Stefano Ticozzi. *Raccolta di Lettere sulla Pittura, Scultura, e Architettura*. Vol. VI. Rome: Eredi Barbiellini, 1754.

Bourdieu, Pierre. *The Field of Cultural Production: Essays on Art and Literature*. New York: Colombia University Press, 1993.

Bredekamp, Horst. *Antikensehnsucht und Maschinenglauben. Die Geschichte der Kunstkammer und die Zukunft der Kunstgeschichte*. Berlin: Verlag Klaus Wagenbach, 1993.

Bredekamp, Horst. *The Lure of Antiquity and the Cult of the Machine: The Kunstkammer and the Evolution of Nature, Art and Technology*. Translated by Allison Brown. Princeton: Markus Wiener Publishers, 1997 [1993].

Bredekamp, Horst. *Theorie des Bildakts. Frankfurter Adorno-Vorlesungen 2007*. Berlin: Suhrkamp, 2010.

Breton, André. 'Le cadavre exquis, son exaltation' [1948]. In *Le cadavre exquis (1925–1936)*. Edited by Arturo Schwarz, pp. 5–18. Saint Etienne: Musée d'art et d'industrie, 1975.

Breton, André. *Mad Love*. Translated by Mary Ann Caws. Lincoln (NE) and London: University of Nebraska Press, 1987 [1937].

Breton, André. *Manifestoes of Surrealism* [1962]. Translated by Richard Seaver and Helen R. Lane. Ann Arbor: University of Michigan Press, 1972.

Breton, André, *Nadja*. Paris: Folio, Gallimard, 1998 [1928].

Brown, Charles. *Life of John Keats*. London: Oxford University Press, 1987.

Brown, Eric C., ed. *Insect Poetics*. Minneapolis and London: University of Minnesota Press, 2006.

Bruck, Robert. *Das Skizzenbuch von Albrecht Dürer in der königlichen öffentlichen Bibliothek zu Dresden*. Strasbourg: Heitz, 1905.

Buchan, Alexander. 'The Meteorology of Ben Nevis'. *Transactions of the Royal Society of Edinburgh* 34 (1890).

Buchloh, Benjamin H.D. 'Hans Haacke: Memory and Instrumental Reason' [1988]. In Buchloh, *Neo-Avantgarde and Culture Industry: Essays on European and American Art from 1955 to 1975*, pp. 202–41. Cambridge, MA: MIT Press, 2000.

Bucklow, Spike. *The Alchemy of Paint: Art, Science and Secrets from the Middle Ages*. London: Marion Boyars Publishers, 2009.

Bürger, Peter. *Theorie der Avantgarde*. Frankfurt am Main: Suhrkamp Verlag, 1996 [1974].

Bürger, Peter. *Theory of the Avant-Garde*. Minneapolis: University of Minnesota Press, 1984 [1980, 1974].

Burnham, Jack. 'Hans Haacke Wind and Water Sculpture'. Northwestern University. *Tri-Quarterly Supplement* 1 (Spring 1967).

Burnham, Jack. 'Introduction'. In *Great Western Salt Works: Essays on the Meaning of Post-Formalist Art*. New York: Braziller, 1974.

Burnham, Jack. 'Real-Time Systems'. In *Great Western Salt Works: Essays on the Meaning of Post-Formalist Art*. New York: Braziller, 1974.

Burnham, Jack. 'Systems Esthetics'. *Artforum* (September 1968). Reprinted in Burnham, *Great Western Salt Works*. New York: Braziller, 1974.

Butler, John M. *Fundamentals of Forensic DNA Typing*. London: Academic Press, 2010.

Cabinet 25: *Insects* (Spring 2007).

Cage, John. 'Clouds over Europe'. In *Constable's Clouds: Paintings and Cloud Studies by John Constable*. Edited by Edward Morris. Edinburgh and Liverpool: National Galleries of Scotland, 2000.

Caillois, Roger. 'La Mante religieuse'. *Minotaure* 5 (May 1934): pp. 23–26.

Caillois, Roger. 'Mimétisme et psychasthénie légendaire'. *Minotaure* 7 (June 1935): pp. 5–10.

Caillois, Roger. 'The Natural Fantastic' [1970]. In *The Edge of Surrealism: A Roger Caillois Reader*. Edited by Claudine Frank. Durham and London: Duke University Press, 2003.

Caillois, Roger. 'The Praying Mantis: From Biology to Psychoanalysis' [1934]. In *The Edge of Surrealism: A Roger Caillois Reader*. Edited by Claudine Frank. Durham and London: Duke University Press, 2003.

Cambefort, Yves. 'A Sacred Insect on the Margins: Emblematic Beetles in the Renaissance'. In *Insect Poetics*. Edited by Eric. C. Brown. Minneapolis and London: University of Minnesota Press, 2006.

Campbell, Stephen J. '"Fare una Cosa Morta Parer Viva": Michelangelo, Rosso, and the (Un)Divinity of Art'. *Art Bulletin* 84/2 (December 2002): pp. 596–620.

Camper, Fred. 'Blowin' in the Wind Films by Chris Welsby'. http://www.fredcamper.com/Film/Welsby.html. 2001.

Carducho, Vincencio. *Dialogo de la pintura. Su defensa, origin, essencia, definicion, modos y diferencias*. Madrid: Fr. Martínez, 1633.

Castroviejo, Santiago, and Pilar Coello. 'Sobre la "estampación natural" y noticias de los "libros de plants …" de Antonio Palau, Celedonio de Arce y Manuel Muñoz y Matarranz'. *Anales del Jardín Botánico de Madrid* 54/1 (July 1996): pp. 595–607.

Cataldi, Pietro Antonio. *Due lettioni di Pietr' Antonio Cattaldi fatte nell' Academia del dissegno di Perugia*. Bologna: Giouanni Rossi, 1577.

Cataldi, Pietro Antonio. *Trattato geometrico di Pietro Antonio Cataldi lettore delle scienze matematiche nello Studio di Bologna*. Bologna: Sebastiano Bonomi, 1620.

Cave, Roderick. *Impressions of Nature: A History of Nature Printing*. Revised ed. London: British Library and Mark Batty Publishers, 2010.

Cecchini, Giovanni. *L'Accademia delle belle arti di Perugia*. Florence: Le Monnier, 1954.

Cennini, Cennino. *Il libro dell'arte*. 2 vols. English translation: *The Craftsman's Handbook*. Translated by Daniel V. Thompson, Jr. New York: Dover Publications, 1960 [1933].

Chapuis, Alfred, and Edouard Gélis. *Le Monde des Automates: Etude Historique et Technique*. Geneva: Slatkine, 1984.

Chastel, André. *La Grottesque*. Paris: Le Promeneur, 1988.

Chene, Dennis Des. 'Forms of Art in Jesuit Aristotelianism (with a Coda on Descartes)'. In *The Artificial and the Natural: An Evolving Polarity*. Edited by Bensaude-Vincent and Newman. Cambridge, MA, and London: MIT Press, 2007.

Chiarini, Marco, and Cristina Acidini Luchinat. *Bizzarrie di pietre dipinte dalle collezioni dei Medici*. Florence: Silvana Editoriale, 2000.

Christian, David. *Maps of Time: An Introduction to Big History*. Berkeley, Los Angeles and London: University of California Press, 2004.

Ciamberlano, Luca. *Scuola perfetta per imparare a disegnare tutto il corpo humano*. Rome, c. 1600.

Clairvaux, Bernard de. In *Early Medieval Art 300–1150: Sources and Documents*. Edited by Caecilia Davis-Weyer. Toronto: University of Toronto Press, 1986 [1971].

Clark, J.F.M. *Bugs and the Victorians*. New Haven and London: Yale University Press, 2009.

Clottes, Jean, and David Lewis-Williams. *The Shamans of Prehistory: Trance and Magic in the Painted Caves*. Translated by Sophie Hawkes. New York: Harry N. Abrams, 1998 [1996].

Cole, Michael. *Cellini and the Principles of Sculpture*. Cambridge: Cambridge University Press, 2006.

Cole, Michael. 'The Demonic Arts and the Origin of the Medium'. *Art Bulletin* 84 (December 2002).

Conihout, Isabelle de. *De Botanica in originali: Livres de botanique réalisés en impression naturelle du XVIe au XIXe siècle*. Paris: Bibliothèque Nationale, 1993.

Crary, Jonathan. *Techniques of the Observer: On Vision and Modernity in the Nineteenth Century*. Cambridge, MA: MIT Press, 1990.

Dacos, Nicole. *La Découverte de la Domus Aurea et la formation des grotesques a la renaissance*. London: Warburg Institute; Leiden: E.J Brill, 1969.

Damisch, Hubert. *A Theory of /Cloud/: Toward a History of Painting*. Translated by Janet Lloyd. Stanford: Stanford University Press, 2002 [1972].

Danti, Egnazio. *La prospettiva di Euclide*. Florence: Stamperia de' Giunti, 1573.

Danti, Egnazio, ed. *Le due regole della prospettiva pratica di M. Iacomo Barozzi da Vignola con I comentarij del R.P.M. Egnatio Danti*. Bologna: Francesco Zannetti, 1583.

Danti, Vincenzo. *Trattato delle perfette proporzioni*. Florence, 1567.

Daston, Lorraine. 'Nature by Design'. In *Picturing Science, Producing Art*. Edited by Caroline A. Jones and Peter Galison. New York and London: Routledge, 1998.

Daston, Lorraine, and Katharine Park. *Wonders and the Order of Nature 1150–1750*. New York: Zone Books, 2001 [1998].

Daston, Lorraine, and Peter Galison. 'The Image of Objectivity'. *Representations* 40 (Autumn 1992): pp. 81–128.

Deleuze, Gilles, and Félix Guattari. *A Thousand Plateaus: Capitalism and Schizophrenia*. Translated by Brian Massumi. London: Continuum, 1987 [1980].

Descartes, René. 'Discourse on the Method of Rightly Conducting the Reason and Seeking Truth in the Sciences' [1637]. Translated by John Veitch. In *The Rationalists*. New York: Doubleday, 1960.

Dickel, Hans. *Künstlerbücher mit Photographie*. Hamburg: Maximilian-Gesellschaft, 2008.

Dissanayake, Ellen. 'What Art Is and What Art Does: An Overview of Contemporary Evolutionary Hypotheses'. In *Evolutionary and Neurocognitive Approaches to Arts, Aesthetics and Creativity*. Edited by Colin Martindale, Paul Locher and Vladimir M. Petrov. Amityville (NY): Baywood, 2007.

D'Ooge, Martin. Introduction. In Nicomachus, *Introduction to Arithmetic*. Translated by M. D'Ooge. New York: Macmillan, 1926.

Dotzler, Bernhard J., Peter Gendolla and Jörgen Schäfer. *MaschinenMenschen. Eine Bibliographie*. Frankfurt am Main and New York: P. Lang, 1992.

Drux, Rudolf. *Menschen aus Menschenhand. Zur Geschichte der Androiden. Texte von Homer bis Asimov*. Stuttgart: Metzler, 1988.

Dunkerton, Jill. 'Leonardo in Verrocchio's Workshop: Re-examining the Technical Evidence'. *National Gallery Technical Bulletin* 32 (London, 2011).

Duprat, Hubert. *Caddis, Crystal and Company*. Norwich: Norfolk Museums and Archaeology Service, 2011.

Duprat, Hubert, and Christian Besson. 'The Wonderful Caddis Worm: Sculptural Work in Collaboration with Trichoptera'. Translated by Simon Pleasance. http://www.leonardo.info/isast/articles/duprat/duprat.html. Accessed February 20, 2013.

Dusinberre, Deke. *Avant-garde British Landscape Films in the Tate Film Theatre, 3–21 March 1975*. The Hyman Kreitman Research Centre's Archive. London: Tate Britain, 1975.

Dusinberre, Deke. 'On British Avant-Garde Landscape Films'. *Undercut* 7/8 (1983): pp. 48–55.

Dusinberre, Deke. 'St George in the Forest: The English Avant-Garde'. *Afterimage* 6 (1976): pp. 4–19.

Dutta, Arindam, et al., eds. *A Second Modernism: MIT, Architecture, and the 'Techno-Social' Moment*. Cambridge, MA: MIT Press, 2013.

Eaton, Peter, and Paul West. *Atomic Force Microscopy*. Oxford: Oxford University Press, 2010.

Eco, Umberto. *Art and Beauty in the Middle Ages*. New Haven: Yale University Press, 1986.

Ede, Siân. *Art and Science*. London and New York: I.B. Tauris, 2005.

Ede, Siân. 'Fission or Fusion: Art, Science and Society'. In *Strange and Charmed: Science and the Contemporary Visual Arts*. Edited by Siân Ede, pp. 164–83. London: Calouste Gulbenkian Foundation, 2000.

Eichberger, Dagmar. *Leben mit Kunst. Wirken durch Kunst. Sammelwesen und Hofkunst unter Margarete von Österreich, Regentin der Niederlande*. Turnhout: Brepols, 2002.

Eire, Carlos M.N. *War against the Idols: The Reformation of Worship from Erasmus to Calvin*. Cambridge: Cambridge University Press, 1986.

Elderkin, Kate McK. 'Jointed Dolls in Antiquity'. *American Journal of Archaeology* 34/4 (1930): pp. 455–79.

Eliade, Mircea. *The Forge and the Crucible*. Translated by S. Corrin. Chicago and London: Chicago University Press, 1962 [1956].

Elkins, James. 'Art History and Images That Are Not Art'. *Art Bulletin* 77/4 (December 1995): pp. 553–71.

Elkins, James. 'On the Unimportance of Alchemy in Western Painting'. *Konsthistorisk Tidskrift* 1–2 (1992): pp. 21–26.

Elkins, James. *Six Stories from the End of Representation: Images in Painting, Photography, Astronomy, Microscopy, Particle Physics, and Quantum Mechanics, 1980–2000*. Stanford: Stanford University Press, 2008.

Elkins, James, ed. *Visual Practices across the University*. Munich: Wilhelm Fink Verlag, 2007.

Elkins, James. *Why Are Our Pictures Puzzles?* New York and London: Routledge, 1999.

Elliot, William H., and Daphne C. Elliot. *Biochemistry and Molecular Biology*. 4th ed. Oxford: Oxford University Press, 2009.

Endt-Jones, Marion. 'Between *Wunderkammer* and Shop Window: Surrealist *Naturalia* Cabinets'. In *Sculpture and the Vitrine*. Edited by John Welchman. Aldershot and Burlington: Ashgate, 2013.

Ernst, Max. 'Au-delà de la peinture' [1936]. In Max Ernst, *Écritures*, pp. 237–69. Paris: Gallimard, 1970.

Ernst, Max. *Beyond Painting: And Other Writings by the Artist and His Friends*. New York: Wittenborn, Schultz, 1948.

Esser, Andrea, ed. *Autonomie der Kunst? Zur Aktualität von Kants Ästhetik*. Berlin: Akademie Verlag, 1995.

Fabre, Jean-Henri. *The Glow-Worm and Other Beetles*. Miami: University Press of the Pacific, 2002.

Farago, Claire J. *Leonardo da Vinci's 'Paragone': A Critical Interpretation with a New Edition of the Text in the Codex Urbinas*. Leiden: E.J. Brill, 1992.

Farago, Claire J., ed. *Re-Reading Leonardo: The Treatise on Painting across Europe 1550 to 1900*. Farnham and Burlington: Ashgate, 2009.

Febel, Gisela. 'Neue Forschungsliteratur zum Thema "Menschenkonstruktionen"'. In *Menschenkonstruktionen. Künstliche Menschen in Literatur, Film, Theater und Kunst des 19. und 20. Jahrhunderts*. Edited by Gisela Febel and Cerstin Bauer-Funke, pp. 249–71. Göttingen: Wallstein Verlag, 2004.

Fer, Briony. 'Surrealism, Myth and Psychoanalysis'. In Briony Fer, David Batchelor and Paul Wood, *Realism, Rationalism, Surrealism: Art between the Wars*. New Haven and London: Yale University Press, 1993.

Filarete, Antonio Averlino. *Trattato di architettura* [c. 1461]. In *Antonio Averlino Filaretes Tractat über die Baukunst nebst seinen Büchern von der Zeichenkunst und den Bauten der Medici*. Edited by Wolfgang von Oettingen. Vienna: Graeser, 1890.

Findlen, Paula. 'Jokes of Nature and Jokes of Knowledge: The Playfulness of Scientific Discourse in Early Modern Europe'. *Renaissance Quarterly* 43/2 (Summer 1990): pp. 292–331.

Fiolatti, Odoardo. *Il vero modo ed ordine per dissegnar tutte le parti et membra del corpo humano*. Venice: Justus Sadeler, 1608.

Fleury, Raymond, Paul Gayot and Thieri Foulc. *Les Très Riches Heures du Collège de 'Pataphysique*. Paris: Fayard, 2000.

Flusser, Vilém. *Für eine Philosophie der Fotografie*. Göttingen: European Photography, 1983.

Folge, Douglas. 'Diana Thater: Being Inside of a Work of Art'. *Flash Art* 198 (January/ February 1998): pp. 86–89.

Foster, Hal. *Compulsive Beauty*. Cambridge, MA, and London: MIT Press, 1993.

Foster, Hal. *Design and Crime (And Other Diatribes)*. London: Verso, 2002.

Frank, Claudine, ed. *The Edge of Surrealism: A Roger Caillois Reader*. Durham and London: Duke University Press, 2003.

Frazer, James George. *The Golden Bough: A Study in Comparative Religion*. Edinburgh: Canongate, 2004 [1922].

Fredel, Jürgen. 'Ideale Maße und Proportionen. Der konstruierte Körper'. In *Die Beredsamkeit des Leibes. Zur Körpersprache in der Kunst*. Edited by Ilsebill Barta Fliedl and Christoph Geissmar, pp. 11–42. Salzburg and Vienna: Residenz, 1992.

Freedberg, David. *The Power of Images: Studies in the History and Theory of Response*. Chicago: University of Chicago Press, 1989.

Freud, Sigmund. 'The Uncanny' [1919]. In *The Uncanny*. Translated by David McLintock, with an introduction by Hugh Haughton. London: Penguin, 2003.

Fried, Michael. 'Art and Objecthood' [1967]. In *Art and Objecthood*, pp. 148–72. Chicago: University of Chicago Press, 1998.

Frisinger, Howard. *The History of Meteorology to 1800*. New York: Science History Publications, 1977.

Fusco, Laurie. 'The Use of Sculptural Models by Painters in Fifteenth-Century Italy'. *Art Bulletin* 64/2 (1982): pp. 103–20.

Gabel, Medard. 'Buckminster Fuller and the Game of the World'. In *Buckminster Fuller: Anthology for the New Millennium*. Edited by Thomas T.K. Zung. New York: St Martin's Press, 2001.

Gage, John. *Turner: Rain, Steam and Speed*. London: Allen Lane Penguin Press, 1972.

Gagné, Joan. 'Du *Quadrivium* aux *scientiae mediae*'. In *Arts Libéraux et Philosophie au Moyen Age*, pp. 975–86. Montréal: Bellarmin; Paris: Vrin, 1969.

Galison, Peter. *Image and Logic: A Material Culture of Microphysics*. Chicago: University of Chicago Press, 1997.

Garai, Luca. *Gli automata di Leonardo*. Bologna: Bononia University Press, 2007.

Gell, Alfred. *Art and Agency: An Anthropological Theory*. Oxford: Clarendon Press, 1998.

Genovese, Valeria E. *Statue vestite e snodate. Un percorso*. Pisa: Edizioni della Normale, 2010.

Godfrey, Mark. 'Roni Horn's Icelandic Encyclopedia'. *Art History* 32/5 (2009): pp. 932–53.

Ginzburg, Carlo. *The Night Battles: Witchcraft and Agrarian Cults in the Sixteenth and Seventeenth Centuries*. Translated by Anne Tedeschi and John Tedeschi. London: Routledge and Kegan Paul, 1983 [1966].

Goethe, Johann Wolfgang. 'The Pure Phenomenon'. ['Das Reine Phänomen', 1798.] In *Goethe's Color Theory*. American ed. Translated and edited by Herb Aach. New York: Van Nostrand Reinhold, 1971.

Gonzalez Bueno, Antonio, and Raúl Rodríguez Nozal. *Plantas americanas para la España ilustrada: Génesis, desarollo y ocaso del proyecto español de expediciones botánicos*. Madrid: Complutense, 2000.

Goutier, Jean-Michel, ed. *André Breton, Je vois, j'imagine: poèmes-objets*. Paris: Gallimard, 1991.

Graham, Rodney. Exh. cat. Catalogue essay by Carolyn Christov-Bakargiev. London: Whitechapel Art Gallery, 2002.

Graham, Rodney. Exh. cat. Catalogue essay by Jeff Wall. Vancouver: Vancouver Art Gallery, 1988.

Graham, Rodney. Exh. cat. Edited by Susanne Gaensheimer. Kunstverein München/ Westfälischer Kunstverein Münster. Cologne: Oktagon, 2000.

Graham, Rodney. Exh. cat. Toronto: Art Gallery of Ontario, 2004.

Graham, Rodney. *Through the Forest*. Exh. cat. Edited by Friedrich Meschede. Museu d'Art Contemporani, Barcelona. Ostfildern: Hatje Cantz, 2010.

Gray, Hanna H. 'Renaissance Humanism: The Pursuit of Eloquence'. *Journal of the History of Ideas* 24 (1967): pp. 497–514.

Greenfield, Concetta Carestia. *Humanist and Scholastic Poetics, 1250–1500*. Lewisburg: Bucknell University Press; Toronto: Associated University Presses, 1981.

Grendler, Paul F. *The Roman Inquisition and the Venetian Press, 1540–1605*. Princeton: Princeton University Press, 1977.

Gundestrup, Bente. *Det kongelige danske Kunstkammer 1737 / The Royal Danish Kunstkammer 1737*. 2 vols. Copenhagen: Nationalmuseet/Arnold Busck, 1991–95.

Haacke, Hans. 'Haacke to Exhibit Kinetic Art'. Interview by Peter Mechsler. MIT. *The Tech*, October 17, 1967.

Habermas, Jürgen. *Theorie des kommunikativen Handelns*. Frankfurt am Main: Suhrkamp, 1981.

Hackenbroch, Yvonne. *Renaissance Jewellery*. London: Sotheby Parke Bernet Publications, 1979.

Hägele, Ingrid, Gudrun Schmidt and Gudrun Schneider, eds. *Kriegsverluste der Preußischen Akademie der Künste: Kunstsammlung und Archiv*. Berlin: Akademie der Künste, 2005.

Halleux, Robert. *Les textes alchimiques*. Turnhout: Brepols, 1979.

Hansen, Maria Fabricius. 'Maniera and the Grotesque'. In *Manier und Manierismus* (Untersuchungen zur deutschen Literaturgeschichte, Band 106). Edited by Wolfgang Braungart, pp. 251–73. Tübingen: Max Niemeyer Verlag, 2000.

Haraway, Donna [J.]. 'A Cyborg Manifesto: Science, Technology and Socialist Feminism in the Late Twentieth Century'. In *Simians, Cyborgs and Women: The Reinvention of Nature*. New York: Routledge, 1991.

Haraway, Donna J. *Modest_Witness@Second_Millenium.FemaleMan©_Meet_OncoMouse™*. New York: Routledge, 1997.

Haraway, Donna J. *Simians, Cyborgs and Women: The Reinvention of Nature*. London: Free Association, 1991.

Haraway, Donna [J.]. 'Situated Knowledges: The Science Question in Feminism and the Privilege of Partial Perspective'. *Feminist Studies* 14/3 (1988): pp. 575–99.

Harris, Steven. *Surrealist Art and Thought in the 1930s: Art, Politics, and the Psyche*. Cambridge: Cambridge University Press, 2004.

Hawkins, Gary. Notes to *The 64 Sonnets* by John Keats. Edited by Edward Hirsch. Philadelphia: Paul Dry Books, 2004.

Hegel, G.W.F. Introduction to 'The Philosophy of Fine Art' [1823]. In *Philosophies of Art and Beauty: Selected Readings in Aesthetics from Plato to Heidegger*. Edited by Albert Hofstadter and Richard Kuhns. Chicago: University of Chicago Press, 1964.

Hegel, Georg Wilhelm Friedrich. *Vorlesungen Über Die Ästhetik*. 3 vols. Frankfurt am Main: Suhrkamp, 1986 [1835].

Heidegger, Martin. *Sein und Zeit*. Tübingen: Max Niemeyer, 2001 [1927].

Heidegger, Martin. *Die Technik Und Die Kehre*. Pfullingen: Neske, 1962 [1953].

Heidegger, Martin. *Der Ursprung Des Kunstwerkes*. Edited by Hans-Georg Gadamer. Stuttgart: Philipp Reclam, 1970 [1935–36].

Hein, Jørgen. *The Treasure Collection at Rosenborg Castle*. Selskabet til Udgivelse af Danske Mindesmærker. Vols 1–3. Copenhagen: Museum Tusculanum Press, 2009.

Henderson, Cyril. *Cloud and Bubble Chambers*. London: Methuen, 1970.

Henderson, Linda Dalrymple. *Duchamp in Context: Science and Technology in the Large Glass and Related Works*. Princeton: Princeton University Press, 1998.

Hesse-Honegger, Cornelia, and Peter Wallimann. 'Malformation of True Bug (Heteroptera): A Phenotype Field Study on the Possible Influence of Artificial Low-Level Radioactivity'. *Chemistry and Biodiversity* 5 (2008): pp. 499–539.

Hill, David. *Turner in the Thames*. New Haven and London: Yale University Press, 1993.

Holte, Michael Ned. 'The Administrative Sublime, or The Center for Land Use Interpretation'. *Afterall* 13 (Spring/Summer 2006). http://.afterall.org/journal/issue.13/administrative.sublime.or.center.land.use.interpre. Accessed October 2012.

Home, Stewart, ed. *What Is Situationism? A Reader*. Edinburgh: AK Press, 1996.

Hoormann, Anne. *Diana Thater: Broken Circle*. Schriften des Museums für Gegenwartskunst Siegen, vol. 2. Siegen: Museum für Gegenwartskunst, 2004.

Hope, Charles. Review of the exhibition *Leonardo da Vinci: Painter at the Court of Milan*. *New York Review of Books*, February 9, 2012.

Horace. *Ars Poetica*. In *Satires, Epistles and Ars Poetica*. Translated by H. Rushton Fairclough. Cambridge, MA: Harvard University Press, 1970.

Horkheimer, Max, and Theodor W. Adorno. *Dialektik Der Aufklärung: Philosophische Fragmente*. Frankfurt am Main: Fischer Taschenbuch Verlag, 1971 [1944].

Horn, Roni. *Another Water (The River Thames, for Example)*. Zürich: Scalo; London: Thames and Hudson, 2000.

Horn, Roni. Interview by Claudia Spinelli, 'Hiersein soll genügen. Ein Gespräch mit Roni Horn'. *Kunst-Bulletin* 10 (October 1995): pp. 10–19.

Horn, Roni. *Vatnasafn/Library of Water*. Edited and Introduction by James Lingwood. Göttingen: Steidl; London: Thames and Hudson, 2009.

Horn, Roni. *Vatnasafn/Library of Water*. London: Artangel, 2007.

Howard, Luke. *The Climate of London*. Vols 1–3. London: Harvey and Darton, 1833.

Howard, Luke. 'On the Modifications of Clouds, and on the Principles of Their Production, Suspension, and Destruction: Being the Substance of an Essay Read before the Askesian Society in the Session 1802–1803', Pts. 1 and 2. *Philosophical Magazine* 16 (1803) and 17 (1803).

Howard, Luke. *Seven Lectures on Meteorology*. London: Harvey and Darton, 1843.

Hussey, Christopher. *The Picturesque: Studies in a Point of View*. London: F. Cass, 1967 [1927].

Ihde, Don. *Bodies in Technology*. Minneapolis: University of Minnesota Press, 2002.

Ihde, Don. *Expanding Hermeneutics: Visualism in Science*. Evanston (IL): Northwestern University Press, 1998.

Ihde, Don. *Postphenomenology and Technoscience*. Albany: State University of New York Press, 2009.

Ihde, Don. *Technology and the Lifeworld: From Garden to Earth*. Bloomington: Indiana University Press, 1990.

Iles, Chrissie. *Signs of the Times: A Decade of Film, Video and Slide – Tape Installation in Britain 1980–1990*. Oxford: Museum of Modern Art, 1990.

Jacobs, Fredrika H. *The Living Image in Renaissance Art*. Cambridge: Cambridge University Press, 2005.

Janković, Vladimir. *Reading the Skies: A Cultural History of English Weather, 1650–1820*. Chicago: University of Chicago Press, 2000.

Janson, H.W. '*The Image Made by Chance* in Renaissance Thought'. In *De artibus opuscula XL: Essays in Honor of Erwin Panofsky*. Edited by M. Meiss, pp. 254–66. New York: New York University Press, 1961.

Jarzombek, Mark. 'Haacke's Condensation Cube: The Machine in the Box and the Travails of Architecture'. *Thresholds* 30 (Summer 2005): pp. 99–103.

Jay, Martin. *Downcast Eyes: The Denigration of Vision in Twentieth-Century French Thought*. Berkeley: University of California Press, 1994.

Johansson, Hanna. 'The Revival of Landscape Art'. In *Landscape Theory*. Edited by Rachel deLue and James Elkins. London and New York: Routledge, 2007.

Jonas, Hans. *The Phenomenon of Life: Toward a Philosophical Biology*. Chicago: University of Chicago Press, 1982.

Jones, Amelia. *Irrational Modernism: A Neurasthenic History of New York Dada*. Cambridge, MA, and London: MIT Press, 2004.

Jones, Caroline A. *Hans Haacke 1967*. Cambridge, MA: List Visual Art Center, 2011.

Jones, Caroline A. 'Rendering Time'. In *Einstein for the 21st Century*. Edited by Peter Galison et al., pp. 130–49. Princeton: Princeton University Press, 2008.

Kac, Eduardo. *Telepresence and Bio Art: Networking Humans, Rabbits, and Robots*. Ann Arbor: University of Michigan Press, 2005.

Kant, Immanuel. *Critique of Judgment*, Part 1, § 45. New York: Hafner, 1951 [1790].

Kant, Immanuel. *Gesammelte Schriften*. Preußische Akademie der Wissenschaften. Berlin: Walter de Gruyter, 1902.

Kant, Immanuel. *Kritik Der Reinen Vernunft*. Frankfurt am Main: Suhrkamp, 1974 [1788].

Kant, Immanuel. *Kritik Der Urteilskraft*. Frankfurt am Main: Suhrkamp, 1992 [1790].

Kastner, Jeffrey, ed. *Nature*. Documents of Contemporary Art. London: Whitechapel Gallery and MIT Press, 2012.

Kastner, Jeffrey. 'Trichopterae'. *Cabinet* 25: *Insects* (Spring 2007): pp. 102–3.

Keith, Larry. 'Leonardo's Painting Technique'. In Luke Syson with Larry Keith, *Leonardo da Vinci: Painter at the Court of Milan*. Exh. cat. London: National Gallery, 2011.

Kelly, Kevin. *What Technology Wants*. New York: Viking, 2010.

Kemp, Martin. '"Il Concetto dell'anima" in Leonardo's Early Skull Studies'. *Journal of the Warburg and Courtauld Institutes* 34 (1971): pp. 115–34.

Kemp, Martin. 'Leonardo and the Visual Pyramid'. *Journal of the Warburg and Courtauld Institutes* 40 (1977): pp. 129–49.

Kemp, Martin. *The Science of Art*. Yale: Yale University Press, 1990.

Kemp, Wolfgang. 'Disegno. Beiträge zur Geschichte des Begriffs zwischen 1547 und 1607'. *Marburger Jahrbuch für Kunstwissenschaft* 19 (1974): pp. 219–40.

Kenny, Neil. *The Uses of Curiosity in Early Modern France and Germany*. Oxford: Oxford University Press, 2004.

Kenseth, Joy, ed. *The Age of the Marvelous*. Hanover (NH): Hood Museum of Art/ Dartmouth College, 1991.

Kepes, Gyorgy. *The New Landscape in Art and Science*. Chicago: P. Theobald, 1956.

Kieckhefer, Richard. *Magic in the Middle Ages*. Cambridge: Cambridge University Press, 2010 [1989].

Kilgour, William Thomas. *Twenty Years on Ben Nevis*. London: Alexander Gardner, 1906.

Kitchener, Gary. 'Pentagon Plans Cyber-Insect Army'. *BBC News*, March 16, 2006. news.bbc.co.uk/1/hi/world/americas/4808342.stm. Accessed February 20, 2013.

Kittler, Friedrich, ed. *Austreibung des Geistes aus den Geisteswissenschaften: Programme des Poststrukturalismus*. Paderborn, Munich, Vienna, Zürich: Schöningh, 1980.

Kittler, Friedrich. *Gramophone, Film, Typewriter*. Translated by Geoffrey Winthrop-Young and Michael Wutz. Stanford: Stanford University Press, 1999 [1986].

Klein, Robert, and Henri Zerner. *Italian Art 1500–1600: Sources and Documents*. Englewood Cliffs, N.J.: Prentice-Hall, 1966.

Kleist, Heinrich von. *Über das Marionettentheater*. In Heinrich von Kleist, *Briefe und Werke*. Edited by Siegfried Streller and Peter Goldammer. 3 vols, vol. 3, pp. 473–80. Berlin and Weimar: Aufbau Verlag, 1978.

Klubertanz, George P. *The Discursive Power: Sources and Doctrine of the Vis Cogitativa according to St Thomas Aquinas*. St Louis: Modern Schoolman, 1952.

Klüver, Billy, Julie Martin and Barbara Rose, eds. *Pavilion: Experiments in Art and Technology*. New York: E.P. Dutton, 1972.

Kniphof, Johann Hieronymus. *Botanica in originali seu Herbarium vivum …* 3 vols. Halle an der Saale: J.G. Trampe, 1757–64.

Knoppers, Laura Lunger, and Joan B. Landes, eds. *Monstrous Bodies/Political Monstrosities in Early Modern Europe*. Ithaca and London: Cornell University Press, 2004.

Koerner, Joseph Leo. *The Moment of Self-Portraiture in German Renaissance Art*. Chicago and London: University of Chicago Press, 1993.

Kovatz, Stephen, and Thomas Munz, eds. *Deep North*. Berlin: Revolver Publishing, 2009.

Krauss, Rosalind E. 'The Photographic Conditions of Surrealism'. *October* 19 (Winter 1981): pp. 3–34.

Kris, Ernst. 'Der Stil "Rustique". Die Verwendung des Naturabgusses bei Wenzel Jamnitzer und Bernard Palissy'. *Jahrbuch der Kunsthistorischen Sammlungen* 1 (1926): pp. 137–208.

Kruse, Christiane. 'Fleisch werden: Fleisch malen: Malerei als "incarnazione". Mediale Verfahren des Bildwerdens im Libro dell'Arte von Cennino Cennini'. *Zeitschrift für Kunstgeschichte* 63/3 (2000): pp. 305–25.

Kuhn, Thomas S. *The Structure of Scientific Revolutions*. 2nd ed. Chicago: University of Chicago Press, 1970 [1962].

La Beaumelle, Agnès Angliviel de, and Isabelle Monod-Fontaine, eds. *André Breton: la beauté convulsive*. Paris: Editions du Centre Pompidou, 1991.

Lacan, Jacques. 'The Mirror Stage as Formative of the I as Revealed in Psychoanalytic Experience' [1949]. In *Écrits: A Selection*. Translated by Alan Sheridan. New York and London: W.W. Norton, 1977.

Latour, Bruno. 'Atmosphère, Atmosphère'. In *Olafur Eliasson: The Weather Project*. Edited by Susan May. London: Tate Modern, 2003.

Latour, Bruno. 'On Technical Mediation – Philosophy, Sociology, Genealogy'. *Common Knowledge* 3/2 (Fall 1994): pp. 29–64.

Latour, Bruno. *The Politics of Nature: How to Bring the Sciences into Democrary*. Translated by Catherine Porter. Cambridge, MA: Harvard University Press, 2004 [1999].

Latour, Bruno. *Reassembling the Social: An Introduction to Actor-Network-Theory*. Oxford: Oxford University Press, 2007.

Latour, Bruno. *We Have Never Been Modern*. New York and London: Harvester Wheatsheaf, 1993.

Latour, Bruno. 'What Is the Style of Matters of Concern'. Spinoza Lectures at the University of Amsterdam, April and May 2005, published as an independent pamphlet. Amsterdam: Van Gorcum, 2008.

Latour, Bruno. 'Why Has Critique Run Out of Steam? From Matters of Fact to Matters of Concern'. *Critical Inquiry* 30/2 (Winter 2004): pp. 225–48.

Latour, Bruno, and Peter Weibel. *Making Things Public*. Cambridge, MA: MIT Press, 2005.

Latour, Bruno, and Steve Woolgar. *Laboratory Life: The Construction of Scientific Facts*. Princeton: Princeton University Press, 1986 [1979].

Lautréamont, Comte de. *Maldoror and the Complete Works of the Comte de Lautréamont*. Translated by Alexis Lykiard. Cambridge, MA: Exact Change, 1998.

Lautréamont, Comte de. *Œuvres complètes, Les Chants de Maldoror. Lettres. Poésies I et II*. Paris: Poésie/Gallimard, 2001.

Law, John. 'Actor Network Theory and Material Semiotics'. In *The New Blackwell Companion to Social Theory*. Edited by Bryan S. Turner, pp. 141–58. Malden, MA: Wiley Blackwell, 2009.

Le Corbusier. *Towards a New Architecture*. London: Architectural Press, 1960 [1927, 1923].

Leslie, Charles Robert. *Memoirs of the Life of John Constable*. London: Phaidon, 1995.

Lewis-Williams, David. *The Mind in the Cave: Consciousness and the Origins of Art*. London: Thames and Hudson, 2004.

Lomas, David. *The Haunted Self: Surrealism, Psychoanalysis, Subjectivity*. New Haven and London: Yale University Press, 2000.

Lorenzoni, A. *Il carteggio artistico inedito di Don Vincenzo Borghini*. Florence: Succ. B. Seeber, 1912.

Lovejoy, Arthur O., and George Boas. *Primitivism and Related Ideas in Antiquity*. Baltimore: Johns Hopkins University Press, 1935.

Ludovico, Alessandro. 'Hehe Interview'. *Neural* (2009): pp. 39–40.

Lyles, Anne. '"The immense canopy": Studies of Sky and Cloud by British Artists c. 1770–1860'. In *Constable's Clouds: Paintings and Cloud Studies by John Constable*. Edited by Edward Morris. Edinburgh and Liverpool: National Galleries of Scotland, 2000.

Malina, Roger. 'What Is a Climate Artist?'. In *Deep North*. Edited by Stephen Kovatz and Thomas Munz, pp. 98–101. Transmediale Parcours. Berlin: Revolver Publishing, 2009.

Malt, Johanna. *Obscure Objects of Desire: Surrealism, Fetishism and Politics*. Oxford: Oxford University Press, 2004.

Malvasia, C.C. *Felsina pittrice, vite de' pittori Bolognesi*. Bologna, 1678.

Marchese, Vincenzo. *Memorie dei piú insigni pittori, scultori e architetti*. Florence: A. Parenti, 1846.

Marsh, David. *Quattrocento Dialogue: Classical Tradition and Humanist Innovation*. Cambridge, MA: Harvard University Press, 1980.

Maurice, Klaus. 'The Princely Art of Turning: On the Aesthetic Significance of the Natural Sciences and Technology'. In *Gedrehte Kostbarkeiten/Turned Treasuries*. Edited by Georg Laue, pp. 17–27. Munich: Kunstkammer Georg Laue, 2004.

Mauriès, Patrick. *Cabinets of Curiosities*. London: Thames and Hudson, 2002.

Mauss, Marcel. *A General Theory of Magic*. Translated by Robert Brain. London and Boston: Routledge and Kegan Paul, 1972 [1950].

Mayr, Ernst. 'The Autonomy of Biology: The Position of Biology among the Sciences'. *Quarterly Review of Biology* 71/1 (March 1996): pp. 97–106.

McAdie, Alexander. *A Cloud Atlas*. Chicago: Rand McNally, 1923.

McLuhan, Marshall. *Understanding Media: The Extensions of Man*. New York, Toronto and London: McGraw-Hill, 1964.

Meadow, Mark A. 'Quiccheberg and the Copious Object: Wenzel Jamnitzer's Silver Writing Box'. In *The Lure of the Object*. Edited by Stephen Melville. New Haven and London: Yale University Press, 2005.

Merchant, Carolyn. *The Death of Nature: Women, Ecology, and the Scientific Revolution*. New York: Harper and Row, 1980.

Meuter, Norbert. 'Ästhetische Autonomie. Einige Gedanken über Kunst und Moral'. In *Bodies in Action and Symbolic Forms. Zwei Seiten der Verkörperungstheorie*. Edited by Horst Bredekamp, Marion Lauschke and Alex Arteaga, pp. 219–36. Berlin: Akademie Verlag, 2012.

Mitchell, W.J.T. *Picture Theory: Essays on Verbal and Visual Representation*. Chicago: University of Chicago Press, 1995.

Mitchell, W.J.T. 'There Are No Visual Media'. *Journal of Visual Culture* 4/3 (August 2005): pp. 257–66.

Möbius, Hanno. *Montage und Collage: Litteratur, bildende Künste, Film, Fotografie, Musik, Theater bis 1933*. Munich: Wilhelm Fink Verlag, 2000.

Morel, Philippe. *Les Grotesques. Les figures de l'imaginaire dans la peinture italienne de la fin de la Renaissance*. Paris: Flammarion, 1997.

Morris, Edward, ed. *Constable's Clouds: Paintings and Cloud Studies by John Constable*. Edinburgh and Liverpool: National Galleries of Scotland, 2000.

Müller-Tamm, Pia, and Katharina Sykora, eds. *Puppen Körper Automaten – Phantasmen der Moderne*. Cologne: Oktagon, 1999.

Mulvey, Laura. 'Introductory essay'. In *Chris Welsby British Artists Films*. DVD Jacket. London: British Film Institute, 2005.

Myssok, Johannes. *Bildhauerische Konzeption und plastisches Modell in der Renaissance* Münster: Rhema, 1999.

National Gallery Technical Bulletin: Leonardo da Vinci: Pupil, Painter, and Master 32 (London, 2011).

Nelkin, Dorothy, and M. Susan Lindee. *The DNA Mystique: The Gene as a Cultural Icon*. New York: W.H. Freeman and Company, 1995.

Neri, Janice. *The Insect and the Image: Visualizing Nature in Early Modern Europe, 1500–1700*. Minneapolis and London: University of Minnesota Press, 2011.

Neri, Louise, Lynne Cooke and Thierry de Duve. *Roni Horn*. London: Phaidon 2000.

Neumann, Erich. *The Great Mother: An Analysis of the Archetype*. Translated by R. Manheim. Princeton: Princeton University Press, 1963 [1955].

Neumann, Erich. *Ursprungsgeschichte des Bewusstseins*. Zurich: Rascher, 1949.

Newman, William R. 'Art, Nature, Alchemy, and Demons'. In *The Artificial and the Natural: An Evolving Polarity*. Edited by Bensaude-Vincent and Newman. London and Cambridge, MA: MIT Press, 2007.

Newman, William R. *Promethean Ambitions: Alchemy and the Quest to Perfect Nature*. Chicago and London: University of Chicago Press, 2005 [2004].

Newman, William R., and Anthony Grafton, eds. *Secrets of Nature: Astrology and Alchemy in Early Modern Europe*. Cambridge, MA, and London: MIT Press, 2006 [2001].

Niemetz, Anne. 'Singing Cells, Art, Science, and the Noise in Between', MFA Thesis, UCLA Department of Design Media Arts, 2004.

Nolte, John. *The Human Brain: An Introduction to Its Functional Anatomy*. 6th ed. Philadelphia: Mosby Elsevier, 2009.

Obergruber-Boerner, Carlos. 'Menschenpaare'. In *Nackt. Die Ästhetik der Blöße*. Edited by Wilhelm Hornbostel and Nils Jockel, pp. 35–44. Munich: Prestel, 2002.

Oubré, Alondra Yvette. *Instinct and Revelation: Reflections on the Origins of Numinous Perception*. London and New York: Taylor and Francis, 1997.

Paldam, Camilla Skovbjerg. 'Woman as Cohesive Agent: Surrealist Perception and Montage as Reconstructive Strategy in Louis Aragon's *Le paysan de Paris*'. In *From Homer to Hypertext: Studies in Narrative, Literature and Media*. Edited by Hans Balling and Anders Klinkby Madsen, pp. 95–110. Odense, Denmark: Syddansk Universitetsforlag, 2002.

Palissy, Bernard. *The Admirable Discourses*. Translated by Aurèle la Rocque. Urbana: University of Illinois Press, 1957 [1580].

Panofsky, Erwin. *Idea: A Concept in Art Theory*. Translated by Joseph J.S. Speake. Columbia: University of Southern California Press, 1968 [1924].

Panofsky, Erwin. *Perspective as Symbolic Form*. New York: Zone Books, 1991 [1927].

Parkinson, Gavin. *Surrealism, Art and Modern Science*. New Haven and London: Yale University Press, 2008.

Parris, Leslie, and Ian Fleming-Williams. *Constable*. London: Tate Gallery, 1991.

Payo Hernanz, René Jesús. *El artista burgalés en la época ilustrada*. Burgos: Institución Fernán González, 2005.

Pederson, Jill. 'Henrico Boscano's *Isola beata*: New Evidence for Academia Leonardi Vinci in Renaissance Milan'. *Renaissance Studies* 22/4 (2008): pp. 450–75.

Peirce, Charles Sanders. 'The Icon, Index, and Symbol'. In *Collected Papers II: Elements of Logic*. Edited by C. Hartshorne and P. Weiss. Cambridge, MA: Harvard University Press, 1965.

Phillips, L.R. *D.W. Griffith: Titan of the Film Art*. New York: Gordon Press, 1976.

Pietz, William. 'Fetish'. In Richard Shiff and Robert S. Nelson, *Critical Terms for Art History*. Chicago (IL): University of Chicago Press, 1996.

Plato. *Republic*. Translated by Paul Shorey. In *The Collected Dialogues of Plato including the Letters*. Edited by Edith Hamilton and Huntington Cairns. Bollingen series LXXI. Princeton: Princeton University Press, 1989 [1938].

Pliny the Elder. *Natural History*. Translated by H. Rackham (vols 1–5 and 9), W.H.S. Jones (vols 6–8) and D.E. Eichholz (vol. 10). Loeb Classical Library (10 vols). Cambridge, MA: Harvard University Press; London: William Heinemann 1947–80.

Pochat, Götz. 'Imitatio und Superatio in der bildenden Kunst'. In *Imitatio. Von der Produktivität künstlerischer Anspielungen und Mißverständnisse*. Edited by Paul Naredi-Rainer. Berlin: Reimer, 2001.

Poliziano, Angelo. *Panepistemon*. In *Omnia Opera* [Venice, 1498]. Edited by I. Maier. 2 vols. Turin: Bottega d'Érasmo, 1971.

Pomian, Krzysztof. *Collectors and Curiosities: Paris and Venice, 1500–1800*. Translated by Elizabeth Wiles-Portier. Cambridge: Polity, 1990.

Pressly, William L. 'The Praying Mantis in Surrealist Art'. *Art Bulletin* 55/4 (December 1973): pp. 600–615.

Preziosi, Donald, and Claire Farago. *Art Is Not What You Think It Is*. Oxford: Wiley-Blackwell, 2012.

Principe, Lawrence M., and William R. Newman. 'Some Problems with the Historiography of Alchemy'. In *Secrets of Nature*. Edited by William R. Newman and Anthony Grafton. Cambridge, MA, and London: MIT Press, 2006 [2001].

Puerto Sarmiento, Francisco Javier. *Ciencia de Cámara: Casimiro Gomez Ortega (1741–1818), el científico cortesano*. Madrid: CSIC, 1992.

Raffles, Hugh. *Insectopedia*. New York: Vintage, 2010.

Randall, John Hermann. 'The Development of Scientific Method in the School of Padua' [1940]. In *Renaissance Essays from the Journal of the History of Ideas*. Edited by P.O. Kristeller and P.D. Weiner, pp. 217–51. New York: Harper and Row, 1968.

Rasmussen, Jörg. *Deutsche Kleinplastik der Renaissance und des Barock*. Hamburg: Museum für Kunst und Gewerbe Hamburg, 1975.

Rath, Markus. 'Die Haptik der Bilder. *Rilievo* als Verkörperungsstrategie der Malerei'. In *Das haptische Bild. Körperhafte Bilderfahrung in der Neuzeit*. Edited by Markus Rath, Jörg Trempler and Iris Wenderholm, pp. 3–29. Berlin: Akademie Verlag, 2013.

'Real Academia de las tres nobles artes'. *Memorial literaria* (May 1784): p. 19.

Reisinger-Weber, Jutta. *Der Monogrammist IP und sein Umkreis*. Passau: Dietmar Klinger Verlag, 2007.

Reti, Ladislao. 'The Library of Leonardo'. In Leonardo da Vinci, *The Madrid Codices* [1491–1505], 5 vols., vol. III: pp. 91–108. New York: McGraw-Hill, 1974.

Reynolds, Graham. *Constable: The Natural Painter*. London: Cory, Adams and Mackay, 1965.

Ritter, Joachim. 'Landschaft. Zur Funktion des Ästhetischen in der modernen Gesellschaft' [1962]. In Joachim Ritter, *Subjektivität: Sechs Aufsätze*. Frankfurt am Main: Suhrkamp, 1974.

Roberts, Donna. 'Surrealism and Hybrid (Psycho-) Morphologies'. *Henry Moore Institute Online Papers and Proceedings*. March 19, 2008. http://www.henry-moore.org/docs/donna_roberts_surrealist_hybrids_0.pdf.

Rodner, William S. *J.M.W. Turner: Romantic Painter of Industrial Revolution*. Berkeley, Los Angeles and London: University of California Press, 1997.

Rosen, Charles, and Henri Zerner. *Romanticm and Realism: The Mythology of Nineteenth Century Art*. London and Boston: Faber and Faber, 1984.

Rosenblum, Robert. *Modern Painting and the Northern Romantic Tradition: Friedrich to Rothko*. London: Thames and Hudson, 1975.

Rosenthal, Michael. *Constable: The Painter and His Landscape*. New Haven: Yale University Press, 1983.

Ross, Michael H., Gordon I. Kaye and Wojciech Pawlina. *Histology: A Text and Atlas*. 4th ed. Baltimore: Lippincott Williams and Wilkins, 2003.

Ruby, Ilka, and Andreas Ruby: 'Take-Away Landscape'. In *Your Black Horizon Art Pavilion: Olafur Eliasson and David Adjaye*. Cologne: Verlag der Buchhandlung Walther König, 2007.

Rupprich, Hans. *Albrecht Dürer. Schriftlicher Nachlass*. 3 vols. Berlin: Deutscher Verein für Kunstwissenschaft, 1956–69.

Ruskin, John. *Modern Painters: Part Three.* Whitefish: Kessinger Publishing, 2005 [1856].

Ruskin, John. *Modern Painters*. Vol. 4, *Of Mountain Beauty*. Orpington, Kent: George Allen, 1888.

Ruskin, John. 'The Storm-Cloud of the Nineteenth Century' [1884]. In *The Works of John Ruskin*. Edited by E.T. Cook and Alexander Wedderburn. London: George Allen, 1908.

Rutherford, Lord. 'C.T.R. Wilson – Bibliography'. In *Nobel Lectures, Physics 1922–1941*. Amsterdam: Elsevier Publishing, 1965. http://nobelprize.org/nobel/nobel-foundation/publications/lectures/index.htmlhttp://nobelprize.org/nobel_prizes/physics/laureates/1927/wilson-bio.html.

Sánchez Cantón, Francisco Javier. 'Un maniquí del siglo XVI'. *Archivo Español de Arte* XXV (1952): pp. 101–9.

Saunders, J.B. de C.M., and Charles D. O'Malley. *The Illustrations from the Works of Andreas Vesalius of Brussels: with annotations and translations, a discussion of the plates and their background, authorship and influence, and a biographical sketch of Vesalius*. Cleveland and New York: Dover, 1950.

Schaaf, Larry J. *Out of the Shadows: Herschel, Talbot, and the Invention of Photography*. New Haven and London: Yale University Press, 1992.

Schiebinger, Londa, 'Naming and Knowing: The Global Politics of Eighteenth-Century Botanical Nomenclatures'. In *Making Knowledge in Early Modern Europe: Practices, Objects, and Texts, 1400–1800*. Edited by Pamela H. Smith and Benjamin Schmidt, pp. 90–105. Chicago and London: University of Chicago Press, 2007.

Schiller, Friedrich. 'On Naive and Sentimental Poetry' [1795]. Translated by Julius A. Elias. In *German Aesthetic and Literary Criticism: Winckelmann, Lessing, Hamann, Herder, Schiller, Goethe*. Edited by H.B. Nisbet. Cambridge: Cambridge University Press, 1985.

Schinzel, Britta. 'Recognisability and Visual Evidence in Medical Imaging versus Scientific Objectivity'. In *Dynamics and Performativity of Imagination: The Image between the Visible and the Invisible*. Edited by Bernd Huppauf and Christoph Wulf, pp. 339–56. New York: Routledge, 2009.

Schlosser, Julius von. 'Aus der Bildnerwerkstatt der Renaissance'. *Jahrbuch der kunsthistorischen Sammlungen des Allerhöchsten Kaiserhauses* XXXI (1913/14): pp. 67–135.

Schoch, Rainer, Matthias Mende and Anna Scher, eds. *Albrecht Dürer. Das druckgraphische Werk*. Munich: Prestel, 2001–2004.

Schulze Altcappenberg, Heinrich Th., and Michael Thimann, eds. *Disegno. Der Zeichner im Bild der Frühen Neuzeit*. Berlin and Munich: Deutscher Kunstverlag, 2007.

Schwarz, Arturo, ed. *Le Cadavre exquis (1925–1936)*. Saint Etienne: Musée d'art et d'industrie, 1975.

Sciolla, Gianni Carlo. 'Il disegno dal manichino in legno e dal modello in terra e cera nella tradizione dal Quattrocento al Seicento. Appunti per una ricerca'. In *Conservazione dei materiali librari archivistici e grafici*. Edited by Marina Regni and Piera Giovanna Tordella, pp. 209–26. Turin: U. Allemandi, 1996.

Scribner, Robert W. 'Vom Sakralbild zur sinnlichen Schau. Sinnliche Wahrnehmung und das Visuelle bei der Objektivierung des Frauenkörpers in Deutschland im 16. Jahrhundert'. In *Gepeinigt, begehrt, vergessen: Symbolik und Sozialbezug des Körpers im späten Mittelalter und in der frühen Neuzeit*. Edited by Klaus Schreiner and Norbert Schnitzler, pp. 309–30. Munich: Wilhelm Fink Verlag, 1992.

Seel, Martin. 'Das Unsichtbare sichtbar machen. Zur aktuellen Lage des Austauschs von Kunst und Natur'. In *Ferne Nähe. 'Natur' in der Kunst der Gegenwart*. Exh. cat. Edited by Volker Adolphs. Kunstmuseum Bonn. Cologne: Wienand, 2009.

Seigel, Jerrold E. *Rhetoric and Philosophy in Renaissance Humanism: The Union of Eloquence and Wisdom from Petrarch to Valla*. Princeton: Princeton University Press, 1968.

Seipel, Wilfried. *Alle Wunder dieser Welt: Die kostbarsten Kunstwerke aus der Sammlung Erzherzog Ferdinands II (1529–1595)*. Vienna: Kunsthistorischer Museum Wien, 2001.

Sframeli, Maria, and Yvonne Hackenbroch. *I gioielli dell'Elettrice Palatina al Museo degli Argenti*. Florence: Centro Di, 1988.

Shanken, Edward. 'The House That Jack Built: Jack Burnham's Concept of "Software" as a Metaphor for Art'. *Leonardo Electronic Almanac* 6/10 (November 1998). http://www.artexetra.com/House.html. Accessed October 2012.

Shiner, Larry. *The Invention of Art: A Cultural History*. Chicago: University of Chicago Press, 2001.

Sidler, Michelle. 'The Rhetoric of Cells: Understanding Molecular Biology in the Twenty-First Century'. *Rhetoric Review* 25/1 (2006): pp. 58–75.

Skulptur Projekte in Münster 1997. Exh. cat. Edited by Klaus Bussmann. Westfälisches Landesmuseum Münster. Ostfildern-Ruit: Hatje, 1997.

Smail, Daniel Lord. *On Deep History and the Brain*. Berkeley, Los Angeles and London: University of California Press, 2008.

Smith, Pamela H. *The Body of the Artisan: Art and Experience in the Scientific Revolution*. Chicago: University of Chicago Press, 2004.

Smith, Pamela [H.]. 'Science on the Move: Recent Trends in the History of Early Modern Science'. *Renaissance Quarterly* 62 (2009): pp. 345–75.

Snyder, Joel. 'Visualization and Visibility'. In *Picturing Science, Producing Art*. Edited by Caroline A. Jones and Peter Galison. New York and London: Routledge, 1998.

Spiteri, Raymond, and Donald LaCoss, eds. *Surrealism, Politics and Culture*. Aldershot and Burlington: Ashgate, 2003.

Suchin, Peter. 'Forces of the Small: Painting as Sensuous Critique'. In *Cornelia Hesse-Honegger: The Future's Mirror*. Newcastle-upon-Tyne: Locust, 1997.

Summers, David. 'Contrapposto: Style and Meaning in Renaissance Art'. *Art Bulletin* 59 (1977): pp. 336–61.

Summers, David. *The Judgment of Sense: Renaissance Naturalism and the Rise of Aesthetics*. Cambridge: Cambridge University Press, 1987.

Summers, David. 'Michelangelo on Architecture'. *Art Bulletin* 54 (1972): pp. 146–57.

Summers, David. *Michelangelo and the Language of Art*. Princeton: Princeton University Press, 1981.

Syson, Luke, and Dora Thornton. *Objects of Virtue: Art in Renaissance Italy*. London: British Museum Press, 2001.

Syson, Luke, with Larry Keith. *Leonardo da Vinci: Painter at the Court of Milan*. Exh. cat. London: National Gallery, 2011.

Tauber, Gesine, and Johannes Tauber. 'Mittelalterliche Kruzifixe mit schwenkbaren Armen. Ein Beitrag zur Verwendung von Bildwerken in der Liturgie'. *Zeitschrift des deutschen Vereins für Kunstwissenschaft* 23 (1969): pp. 79–121.

Taylor, Janelle S. 'Image of Contradiction: Obstetrical Ultrasound in American Culture'. In *Reproducing Reproduction: Kinship, Power and Technological Innovation*. Edited by Sarah Franklin and Helena Ragoné. Philadelphia: University of Pennsylvania Press, 1998.

Teale, Edwin Way, ed. *The Insect World of J. Henri Fabre*. Translated by Alexander Teixeira de Mattos. Boston: Beacon Press, 1991.

Thater, Diana. *Delphine*. Exh. cat. Vienna: Wiener Secession, 2000.

Thater, Diana. *Keep the Faith: A Survey Exhibition*. Exh. cat. Edited by Barbara Engelbach and Wulf Herzogenrath. Kunsthalle Bremen, Museum für Gegenwartskunst Siegen. Cologne: Salon, 2004.

Thater, Diana. *Parkett* 60 (2000): pp. 76–116.

Theuerkauff, Christian. 'Gliederpuppe'. In *Der Mensch um 1500. Werke aus Kirchen und Kunstkammern*. Berlin, 1977.

Thévenin, Paule, ed. *Bureau de recherches surréalistes: Cahier de la permanence octobre 1924–avril 1925*. Paris: Gallimard, 1988.

Thompson, Clive. 'Sonocytology'. *Collision Detection – Science, Technology, Culture*, May 12, 2004.

Thornes, John. *The Accurate Dating of Certain of John Constable's Cloud Studies 1821/22 Using Historical Weather Records*, Occasional Papers, 34. London: Department of Geography, University College London, 1978.

Toulmin, Stephen. *The Return to Cosmology: Postmodern Science and the Theology of Nature*. Berkeley and Los Angeles: University of California Press, 1982.

Toulmin, Stephen, and June Goodfield. *The Fabric of the Heavens*. London: Hutchinson, 1960.

Trinkhaus, Charles. *In Our Image and Likeness: Humanity and Divinity in Italian Humanist Thought*. 2 vols. Chicago: University of Chicago Press, 1970.

Tripps, Johannes. *Das handelnde Bildwerk in der Gotik. Forschung zu den Bedeutungsschichten und der Funktion des Kirchengebäudes und seiner Ausstattung in der Hoch- und Spätgotik*. Berlin: Mann Verlag, 1998.

Tzara, Tristan. 'Recette du "cadavre exquis" écrit' and 'Recette du "cadavre exquis" dessiné'. In *Le Cadavre exquis (1925–1936)*. Edited by Arturo Schwarz. Saint Etienne: Musée d'art et d'industrie, 1975.

Ubl, Ralph. 'Autonomie'. In *Metzler Lexikon Kunstwissenschaft*. Edited by Ulrich Pfisterer, pp. 45–49. Stuttgart and Weimar: Metzler, 2011.

Valla, Lorenzo. *Dialectica libri tres*. In *Dialectical Disputations*. Edited and translated by Brian P. Copenhaver and Lodi Nauta. Cambridge, MA: Harvard University Press, 2012.

van den Berg, Hubert. 'Poetry as a Reification of Nature: On the Role of Nature in the Poetry of Hans Arp'. In *Nature: Literature and Its Otherness/La littérature et son autre*. Edited by Svend Erik Larsen, pp. 124–41. Odense, Denmark: Odense University Press, 1997.

Vanouse, Paul. 'Discovering Nature, Apparently: Analogy, DNA Imaging, and the Latent Figure Protocol'. In *Tactical Biopolitics: Art, Activism, and Technoscience*. Edited by Beatriz da Costa and Kavita Philip. Cambridge, MA: MIT Press, 2008.

Vanouse, Paul. 'The Relative Velocity Inscription Device'. In *Signs of Life: Bio Art and Beyond*. Edited by Eduardo Kac. Cambridge, MA: MIT Press, 2007.

Vasari, Giorgio. *Lives of the Painters, Sculptors and Architects*. Translated by Gaston du C. de Vere, introduction and notes by David Ekserdjian. New York: Everyman's Library, 1996.

Vasari, Giorgio. *Le vite de' più eccellenti pittori, scultori e architettori nelle redazioni del 1550 e 1568*. Edited by Paola Barocchi and Rosanna Bettarini. 6 vols. Florence: Sansoni, 1966–87.

Vasoli, Cesare. *La dialettica e la retorica dell'Umanesimo: 'Invenzione' e 'Metodo' nella cultura del XV e XVI secolo.* Milan: Feltrinelli, 1968.

Verbeek, Peter-Paul. 'Moralizing Technology: On the Morality of Technological Artifacts and Their Design'. Paper for workshop, pp. 1–25. Ethics Institute, Utrecht University, March 24–25, 2006.

Verbeek, Peter-Paul. *What Things Do: Philosophical Reflections on Technology, Agency, and Design*. University Park: Pennsylvania State University Press, 2005.

Vernant, Jean-Pierre. 'Arbeit und Natur in der griechischen Antike'. In *Seminar: Die Entstehung von Klassengesellschaften*. Frankfurt am Main: Suhrkamp, 1973.

Vitruvius. *Ten Books on Architecture*. Translation by Ingrid D. Rowland. Commentary and illustrations by Thomas Noble Howe. Cambridge and New York: Cambridge University Press, 1999.

Völker, Klaus, ed. *Künstliche Menschen. Dichtungen und Dokumente über Golems, Homunculi, lebende Statuen und Androiden*. Franfurt am Main: Suhrkamp, 1994.

Wamberg, Jacob, ed. *Art and Alchemy*. Copenhagen: Museum Tusculanum, 2006.

Wamberg, Jacob. 'Dehumanizing Danto and Fukuyama: Toward a Post-Hegelian Role for Art in Evolution'. In *The Posthuman Condition: Ethics, Aesthetics and Politics of Biotechnological Challenges*. Edited by Kasper Lippert-Rasmussen, Mads Rosendahl Thomsen and Jacob Wamberg, pp. 141–54. Aarhus, Denmark: Aarhus University Press, 2012.

Wamberg, Jacob. *Landscape as World Picture: Tracing Cultural Evolution in Images*. 2 vols. Aarhus, Denmark: Aarhus University Press, 2009 [2005].

Wamberg, Jacob. 'Det monstrøse'. In *Monster*. Exh. cat., pp. 7–43. Albertslund Rådhus, 2002.

Wamberg, Jacob. 'Toward a (Re)constructed Endosemiotics? Art, Magic, and Augmented Reality'. In *Throughout: Art and Culture Emerging with Ubiquitous Computing*. Edited by Ulrik Ekman, pp. 976–1002. Cambridge, MA: MIT Press, 2013.

Wamberg, Jacob. 'Wounded Working Heroes: Seeing Millet and Van Gogh through the Cleft Lens of Totalitarianism (Adding Reflections from Kiefer and Baseliz)'. In *Totalitarian Art and Modernity*. Edited by Mikkel Bolt Rasmussen and Jacob Wamberg, pp. 26–104. Aarhus, Denmark: Aarhus University Press, 2010.

Wazbinsky, Zygmunt. *L'Accademia medicea del disegno a Firenze nel Cinquecento.* 2 vols. Florence: L.S. Olschki, 1987.

Weeks, Andrew, ed. and trans. *Paracelsus (Theophrastus Bombastus von Hohenheim, 1493–1541): Essential Theoretical Writings*. Leiden and Boston: Brill, 2008.

Weisheipl, John. 'Classification of the Sciences in Medieval Thought'. *Medieval Studies* 27 (1965): pp. 54–90.

Weixlgärtner, Arpad. 'Dürer und die Gliederpuppe'. In *Beiträge zur Kunstgeschichte. Franz Wickhoff gewidmet*, pp. 80–90. Vienna, 1903.

Weixlgärtner, Arpad. *Dürer und Grünewald. Ein Versuch, die beiden Künstler zusammen – in ihren Besonderheiten, ihrem Gegenspiel, ihrer Zeitgebundenheit – zu verstehen*. Gothenburg: Elanders Boktryckeri, 1949.

Weixlgärtner, Arpad. 'Von der Gliederpuppe'. *Göteborgs Konstmuseum Åstryck* (1954): pp. 37–71.

Wiener, Norbert. *Cybernetics: Or, Control and Communication in the Animal and the Machine*. 2nd ed. New York: MIT Press, 1961.

Wilson, C.T.R. 'Banquet Speech'. Nobel Banquet in Stockholm (December 10, 1927). In *Les Prix Nobel en 1927*. Edited by Carl Gustaf Santesson. Stockholm: Nobel Foundation, 1928. nobelprize.org/ nobel_prizes/physics/laureates/1927/wilson-speech.html.

Wilson, C.T.R. 'Condensation of Water Vapour in the Presence of Dust-Free Air and Other Gases'. *Philosophical Transactions of the Royal Society of London, A* 189, 1897.

Wilson, C.T.R. 'On the Cloud Method of Making Visible Ions and the Tracks of Ionizing Particles'. Nobel Lecture (December 12, 1927), 194. In *Nobel Lectures, Physics 1922–1941*. http://www.nobel.se /physics/laureates/1927/wilson-lecture.pdf.

Wilson, C.T.R. 'Reminiscences of My Early Years'. *Notes and Records of the Royal Society of London* 14/2 (1960): pp. 163–73.

Wilson, Stephen. *Information Arts: Intersections of Art, Science, and Technology*. Cambridge, MA: MIT Press, 2002.

Witzke, Anne Sophie, and Sune Hede. 'Rethink – Contemporary Art and Climate Change'. Edited by the Alexandra Institute. Aarhus, Denmark: Alexandra Institute, 2009.

Wollen, Peter. 'Chris Welsby – Films/Photograps/Writings'. Originally published by the Arts Council of Great Britain, 1980. http://www.sfu.ca/~welsby/Wollen.htm.

Wood, Gillen D'Arcy. 'Constable, Clouds, Climate Change'. *Wordsworth Circle* 38/1–2 (2007): pp. 25–34.

Worms, Jeannine. *Entretiens avec Roger Caillois*. Paris: La Différence, 1991.

Zamperini, Alessandra. *Le Grottesche. Il sogno della pitura nella decorazione parietale*. San Giovanni Lupatoto (Verona): Arsenale Editrice, 2007.

Zhou, Weilie, Robert P. Apkarian, Zhong Lin Wang and David Joy. 'Fundamentals of Scanning Electron Microscopy'. In *Scanning Microscopy for Nanotechnology: Techniques and Applications*. Edited by Weilie Zhou and Zhong Lin Wang, pp. 1–40. New York: Springer, 2007.

Zingerle, Ignaz. *Das deutsche Kinderspiel im Mittelalter*. Innsbruck: Wagner, 1873.

Zuccaro, Federico. *L'idea dei pittori*. Turin: A. Disserolio, 1607.

Zwijnenberg, Robert. *The Writings and Drawings of Leonardo da Vinci: Order and Chaos in Early Modern Thought*. Translated (from the Dutch original) by Caroline A. van Eck. Cambridge and New York: Cambridge University Press, 1999.

Zwirner, Dorothea. *Rodney Graham*. Friedrich Christian Flick Collection, 1. Cologne: DuMont 2004.

Index

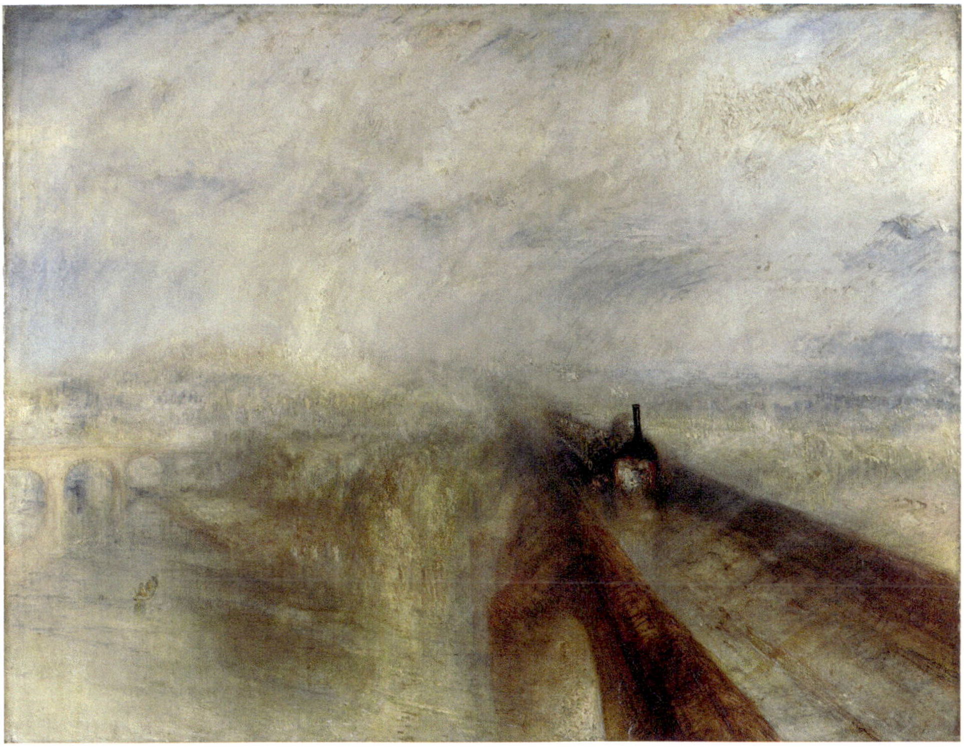

1 J.M.W. Turner, *Rain, Steam and Speed: The Great Western Railway* (1844), oil on canvas. National Gallery, London. Photo credit: National Gallery, London

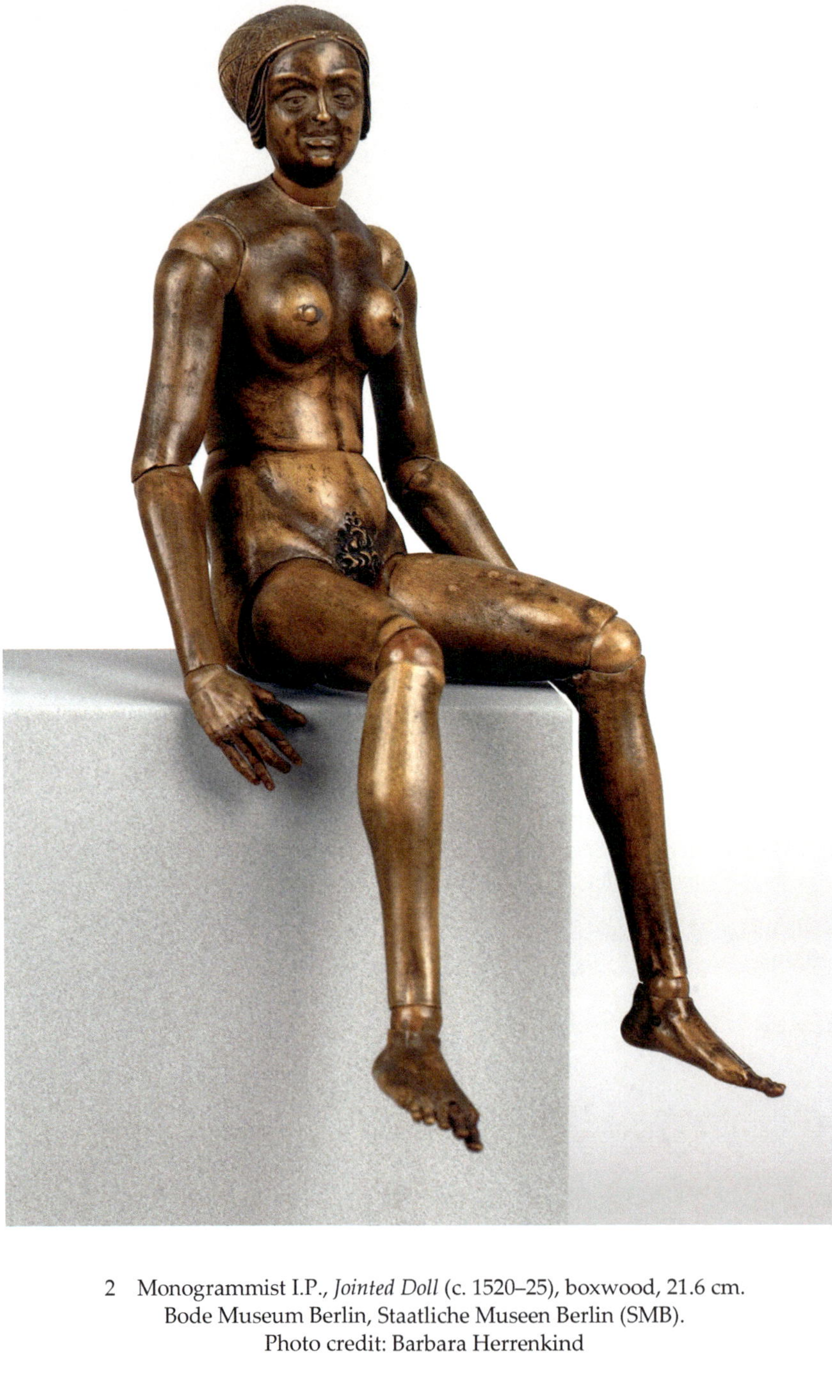

2 Monogrammist I.P., *Jointed Doll* (c. 1520–25), boxwood, 21.6 cm. Bode Museum Berlin, Staatliche Museen Berlin (SMB). Photo credit: Barbara Herrenkind

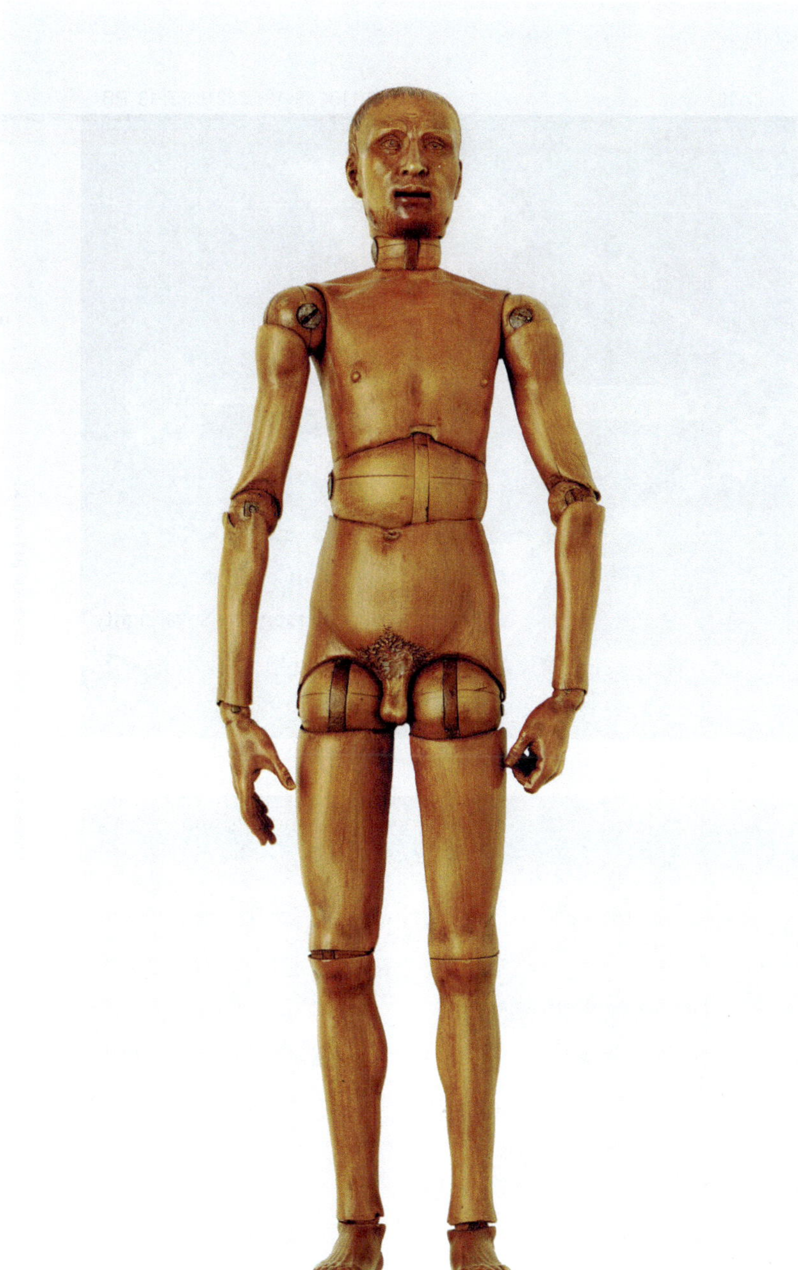

3 *Jointed Doll* (sixteenth century), boxwood, 21.6 cm.
National Museum of Denmark, Copenhagen, Denmark.
Photo credit: Nationalmuseet

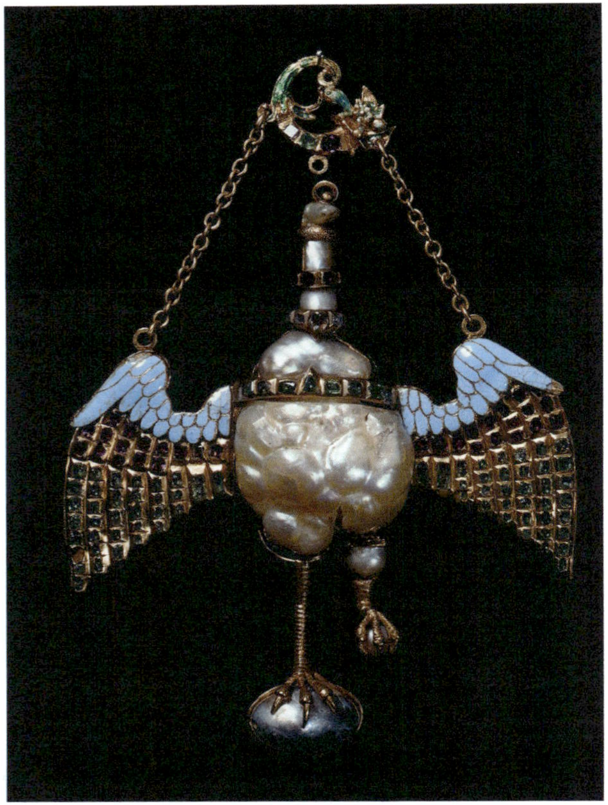

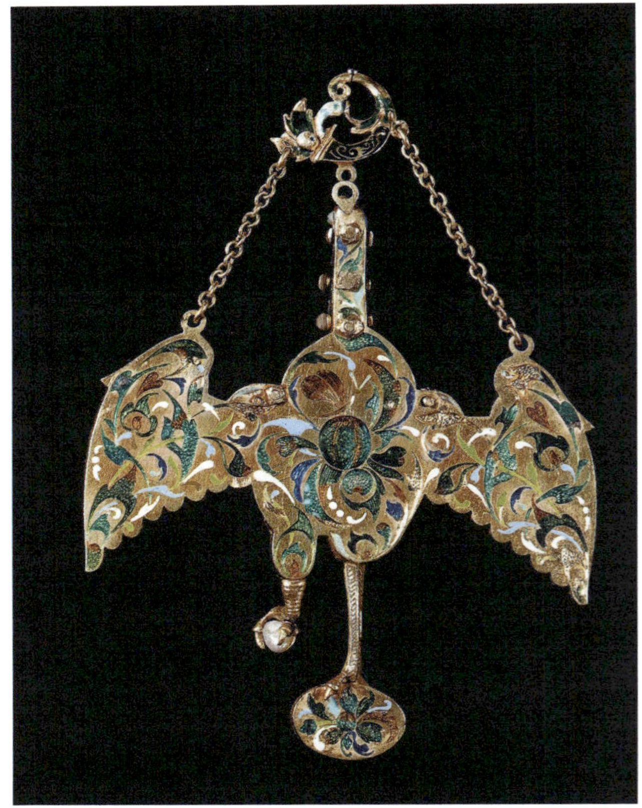

4a (*left*) front of artwork; 4b (*right*) back of artwork.
Crane (1610), 8 pearls, gold, enamel, 76 emeralds (1 missing), 36 rubies, 5 diamonds, 7.8 x 6.5 cm.
Northern Europe. Rosenborg Castle, Copenhagen.
Photo credit: Danish Royal Collections

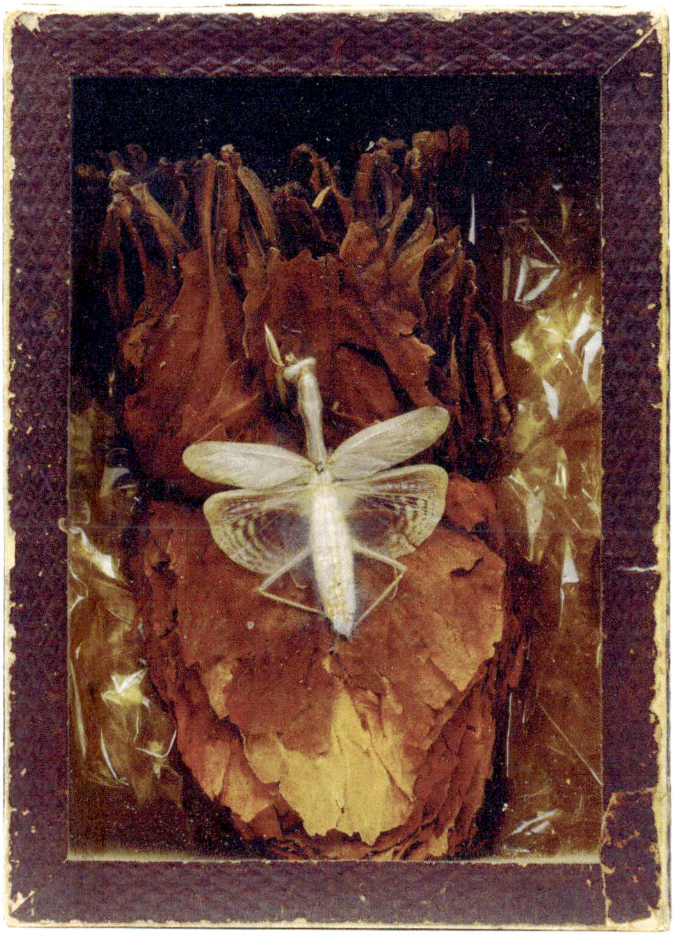

5 André Breton and Jacqueline Lamba, *Le Petit Mimétique* (1936), insect, dried leaves, cellophane. Musée national d'art moderne, Paris. © André Breton/billedkunst.dk. © Jacqueline Lamba/billedkunst.dk

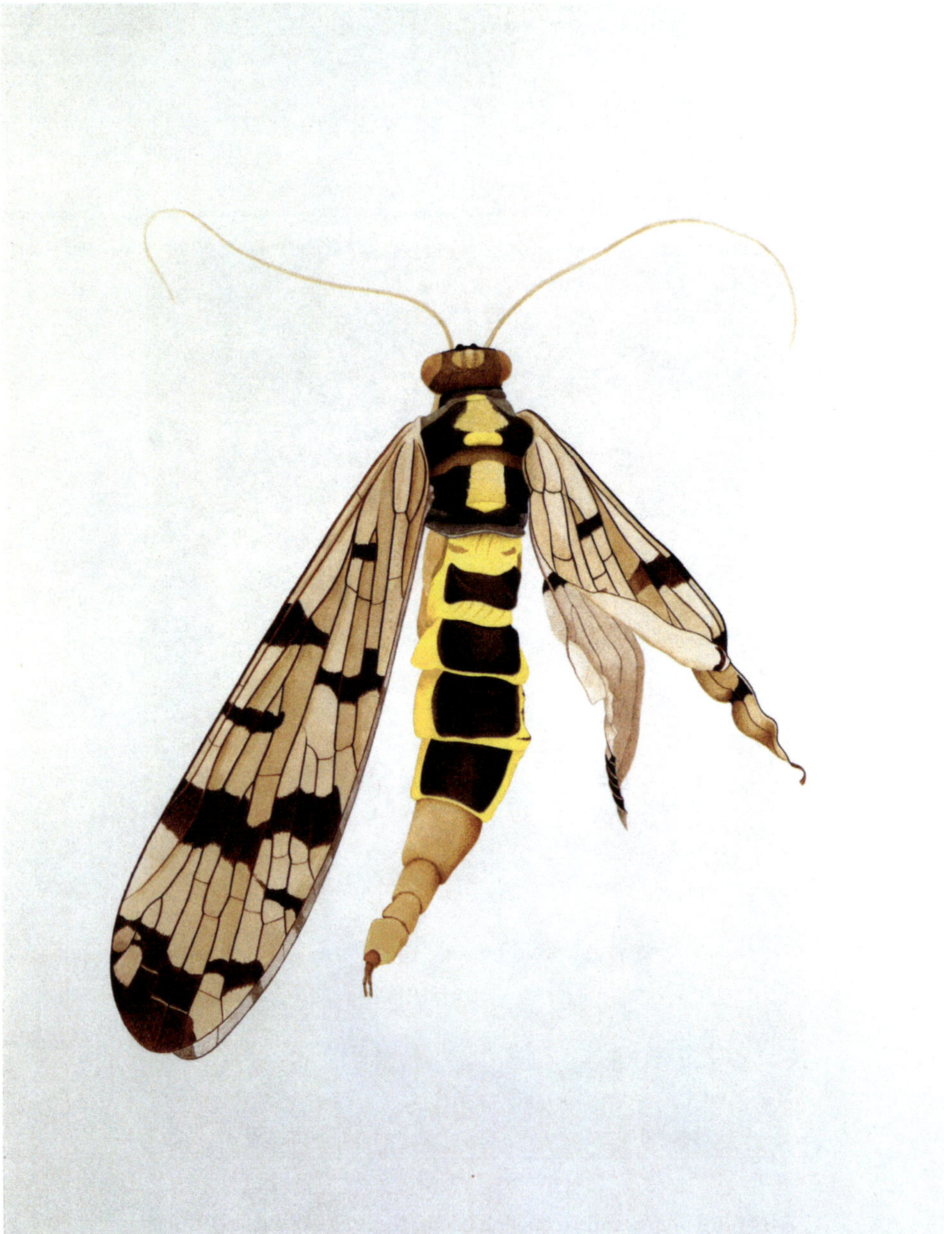

6 Cornelia Hesse-Honegger, *Scorpion Fly* (1988), watercolour, *Parnorpa communis* from Reuenthal, found near the nuclear power plant Leibstadt, Zürich, Switzerland. © Cornelia Hesse-Honegger/billedkunst.dk/Photo credit: Cornelia Hesse-Honegger

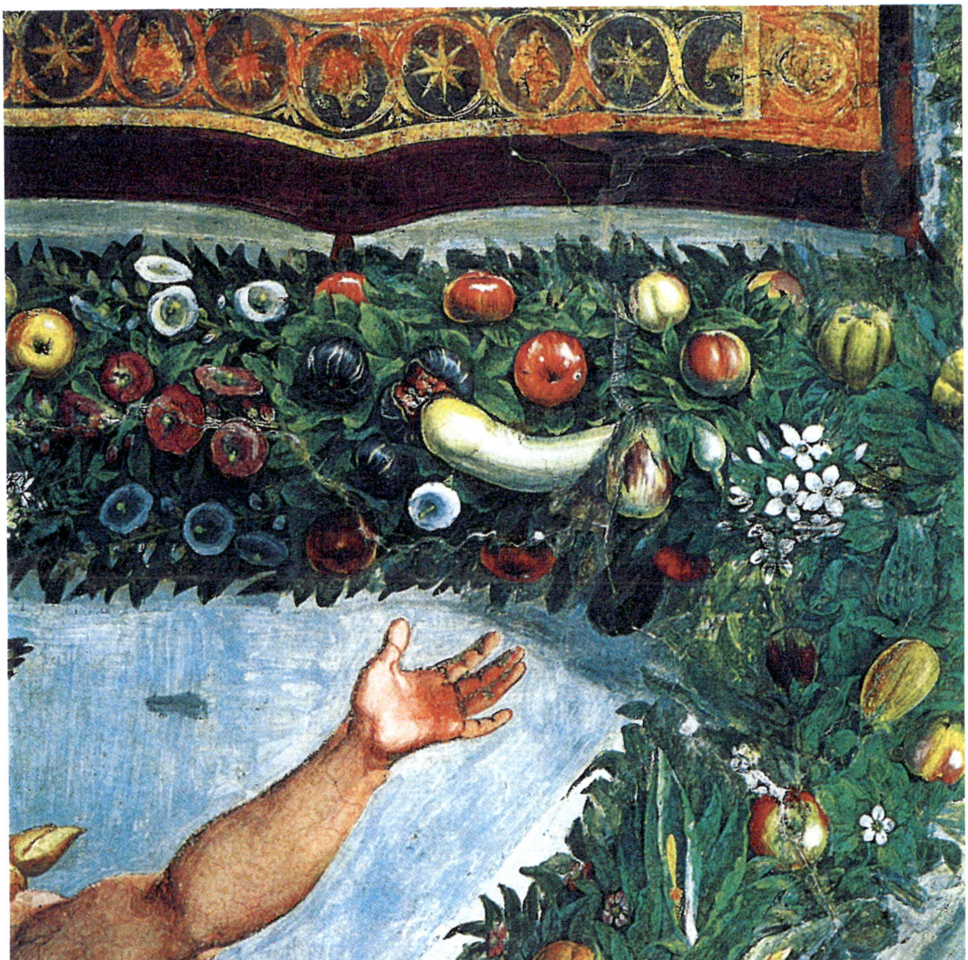

7 Giovanni da Udine, detail with fig, squash and eggplant as female and male genitals (1518–19), frescoes in the ceiling of the Loggia di Psiche, Villa Farnesina, Rome. Nicole Dacos and Caterina Furlan, *Giovanni da Udine, 1487–1561*, Udine [Italy]: Casamassima, 1987, p. 26

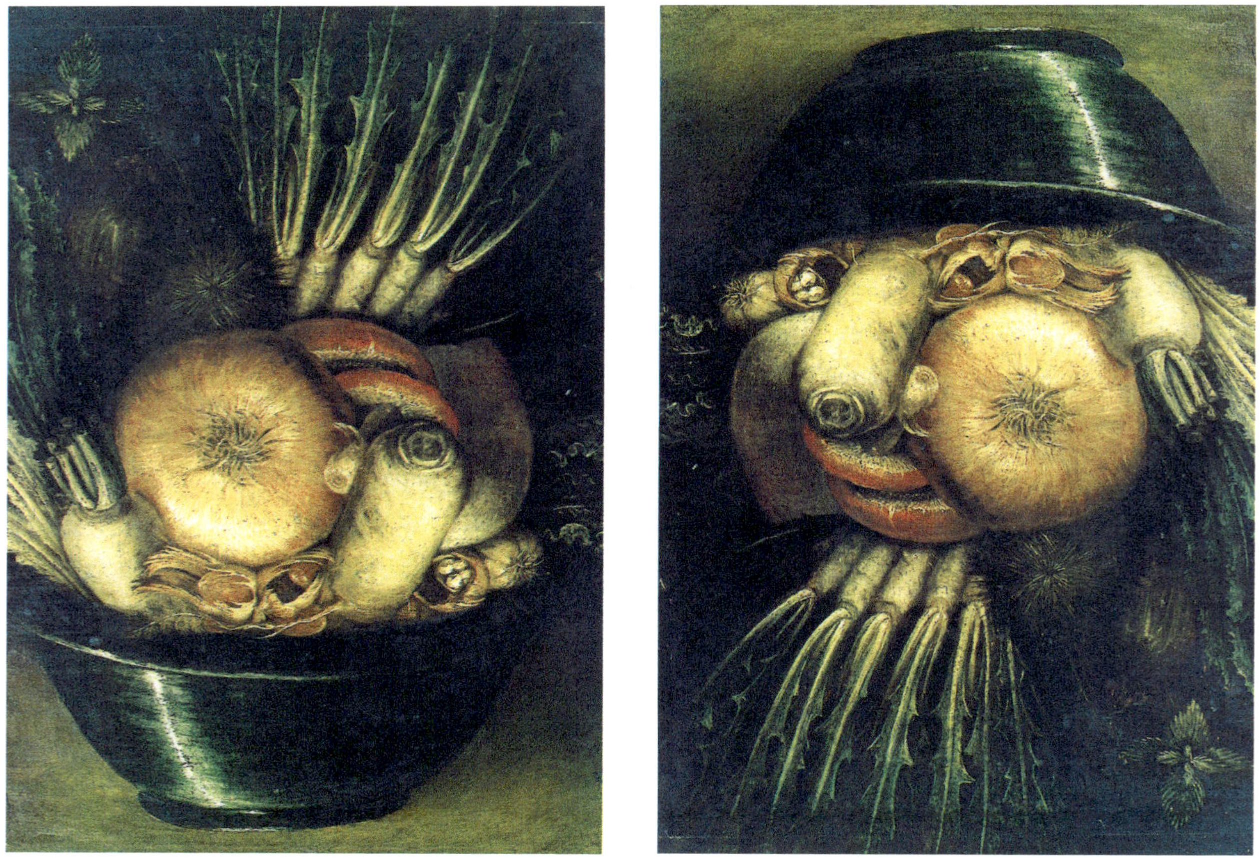

8 Arcimboldo, *Still Life/The Gardener* (c. 1590), in normal view and upside down to show the face, oil on wood. Museo Civico ala Ponzone, Cremona

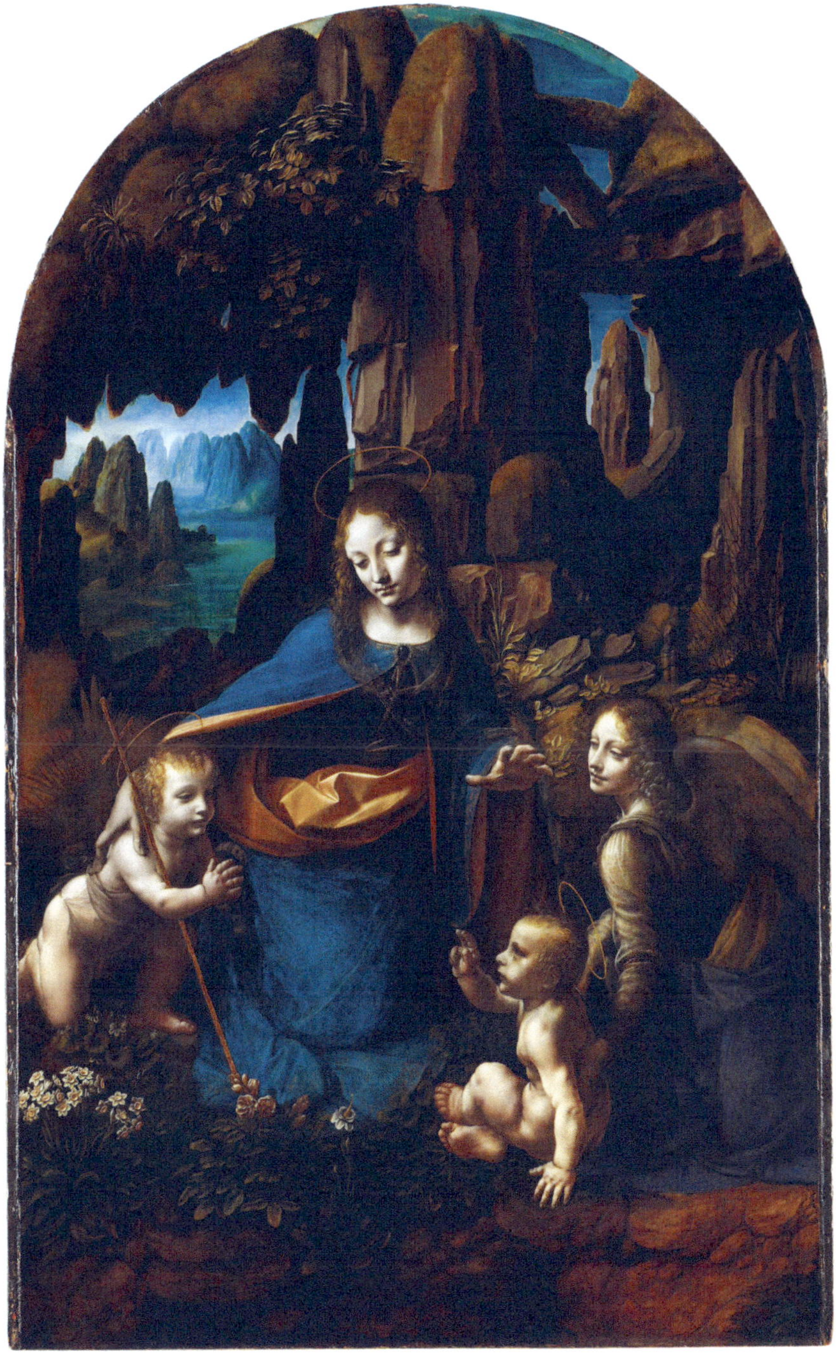

9 Leonardo da Vinci, *Virgin of the Rocks* (1490–1508), oil on wood. National Gallery, London

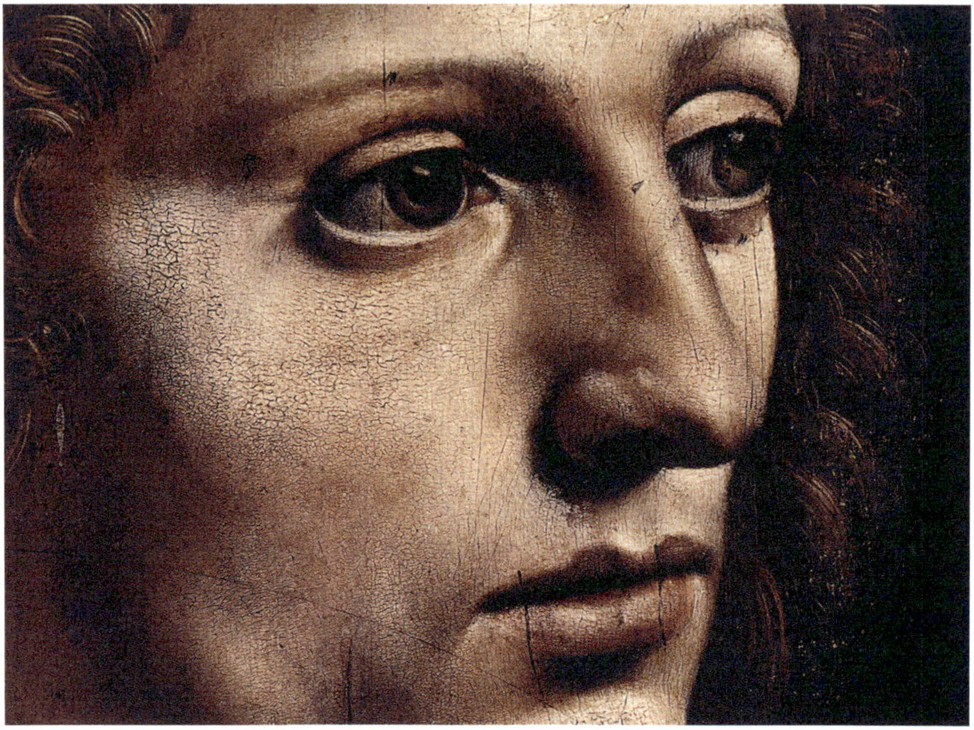

10 Leonardo da Vinci, *Portrait of a Musician* (c. 1489–90), unfinished, tempera and oil on panel (inv. 99). © Veneranda Biblioteca Ambrosiana, Milan. Photo credit: De Agostini/ Veneranda Biblioteca Ambrosiana

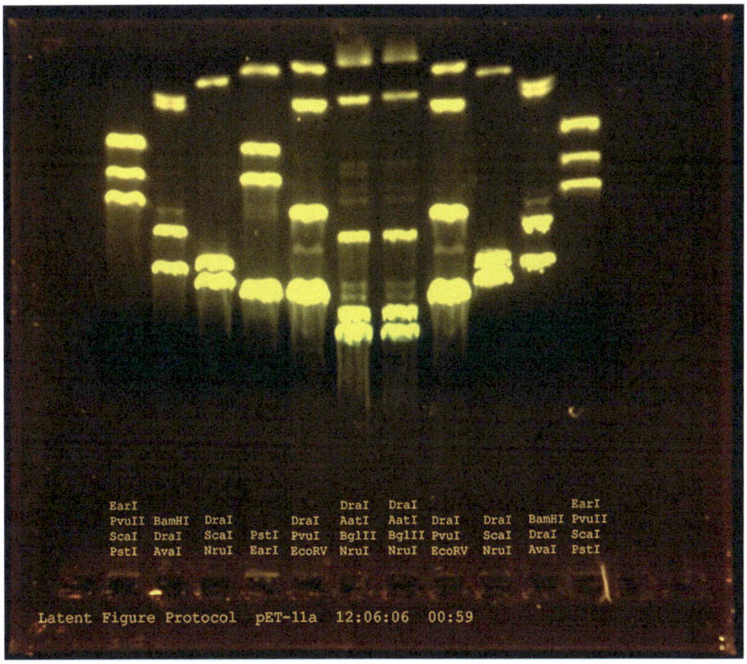

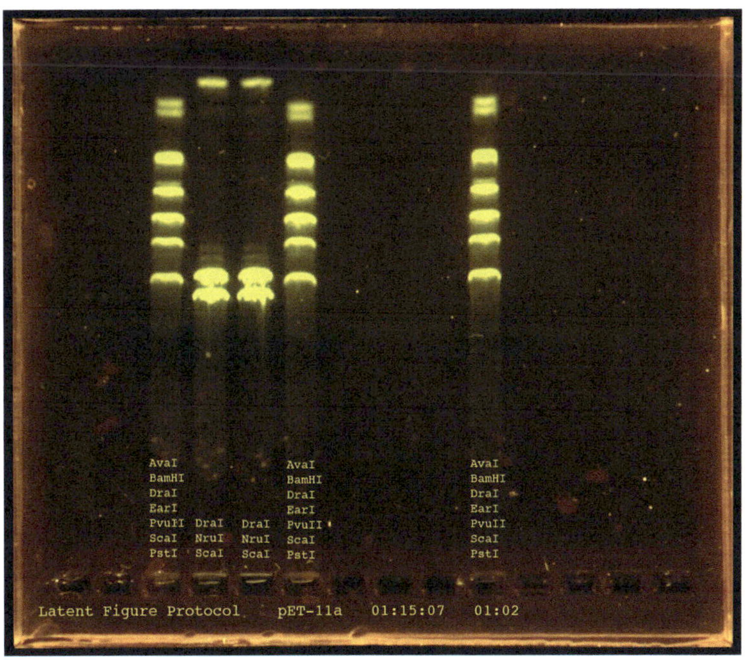

11a (*top*) Produced December 6, 2006; 11b (*bottom*) Produced January 11, 2007. Paul Vanouse, *Latent Figure Protocol*. Figures were produced with the DNA of bacterial plasmid pET-11a. Enzymes used to process the DNA are listed in each column. © Paul Vanouse/Photo credit: Paul Vanouse

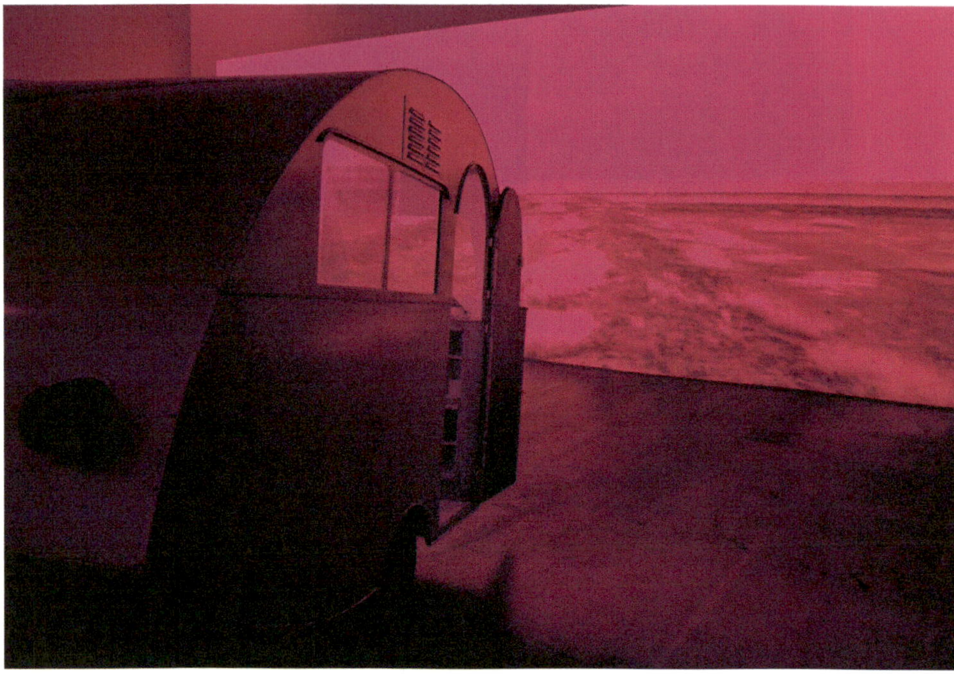

12 Petko Dourmana, *Post Global Warming Survival Kit* (2008), infrared photography.
The project was produced by Edith Russ Site for Media Art.
© Petko Dourmana/Photo credit: Petko Dourmana

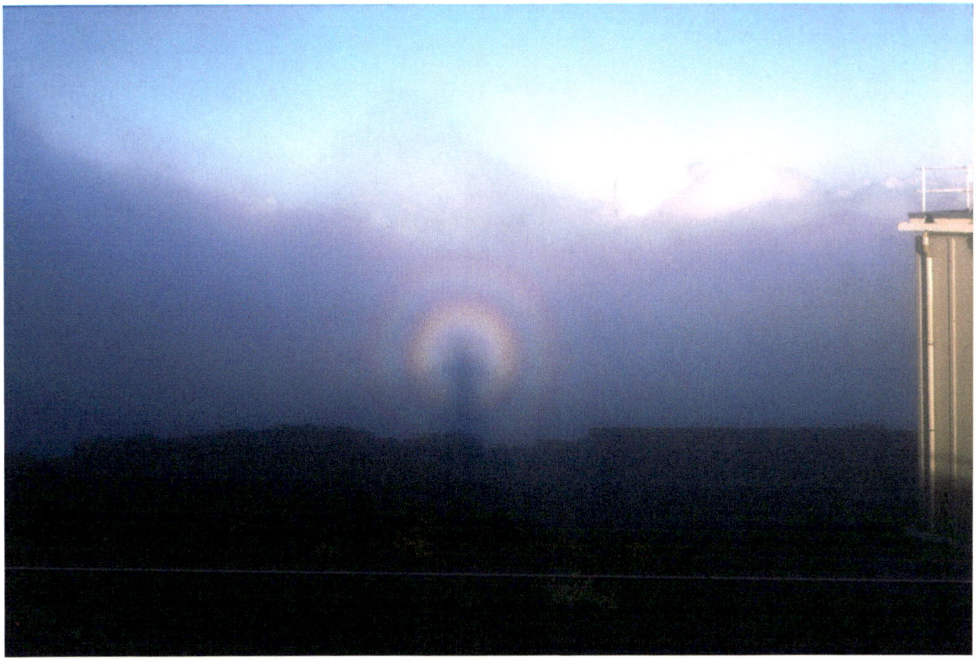

13 Nik Szymanek, *The Glory*, photograph.

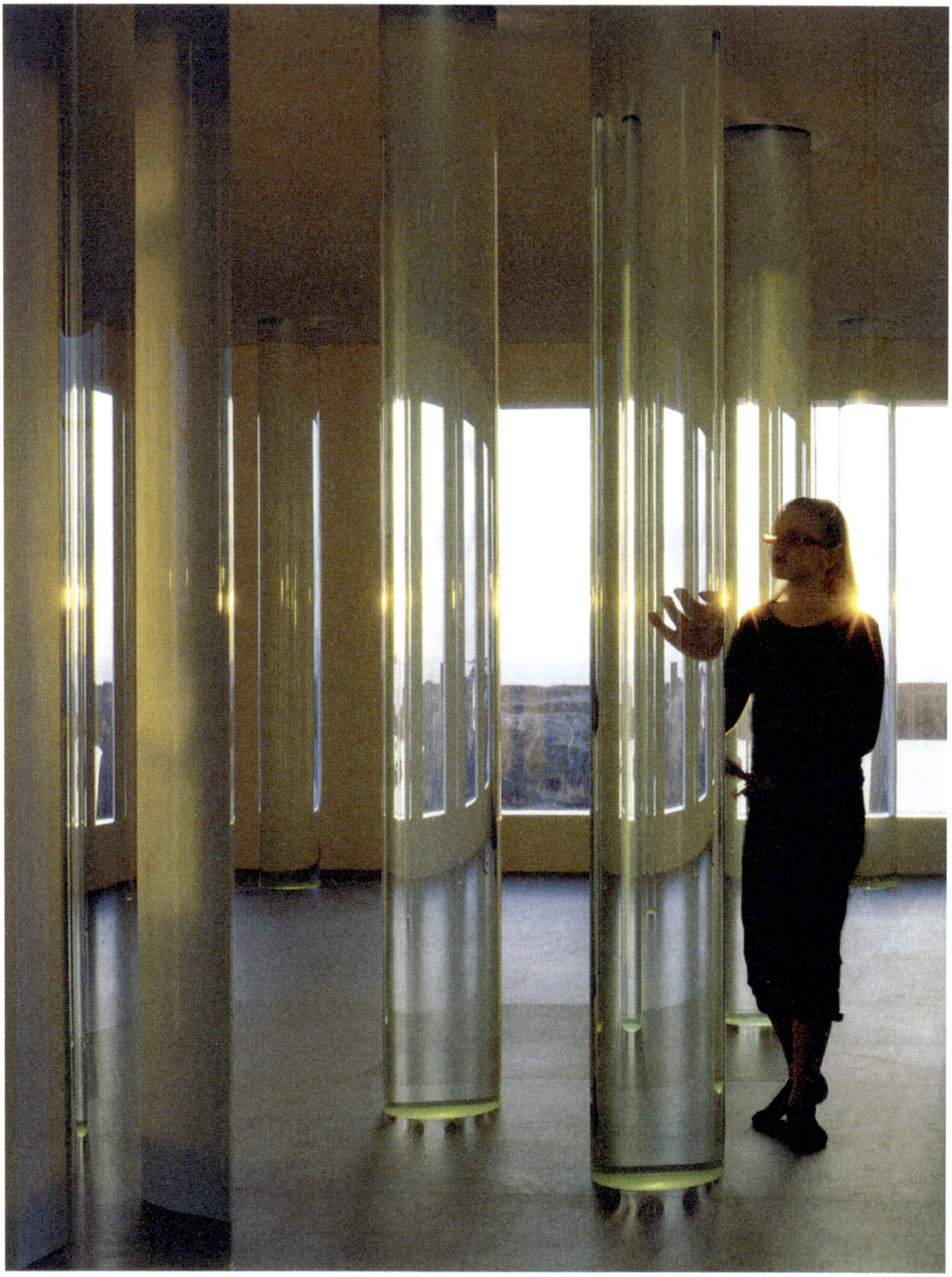

14 Roni Horn, *Vatnasafn/Library of Water* (2007), 24 floor-to-ceiling columns filled with water from unique glacial sources. Source: Roni Horn, *Vatnasafn/Library of Water*, ed. James Lingwood, Göttingen: Steidl, 2009. © Roni Horn/Photo credit: Roni Horn

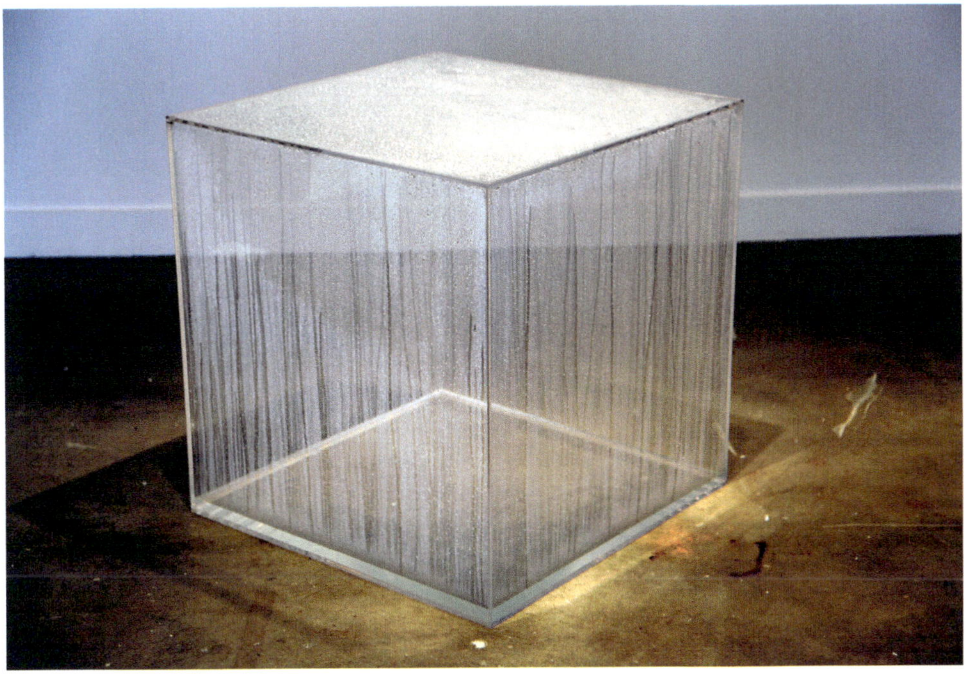

15 Hans Haacke, *Condensation Cube* (originally exhibited as *Weather Cube*) (1963–65), Plexiglas and water.
© Hans Haacke/billedkunst.dk

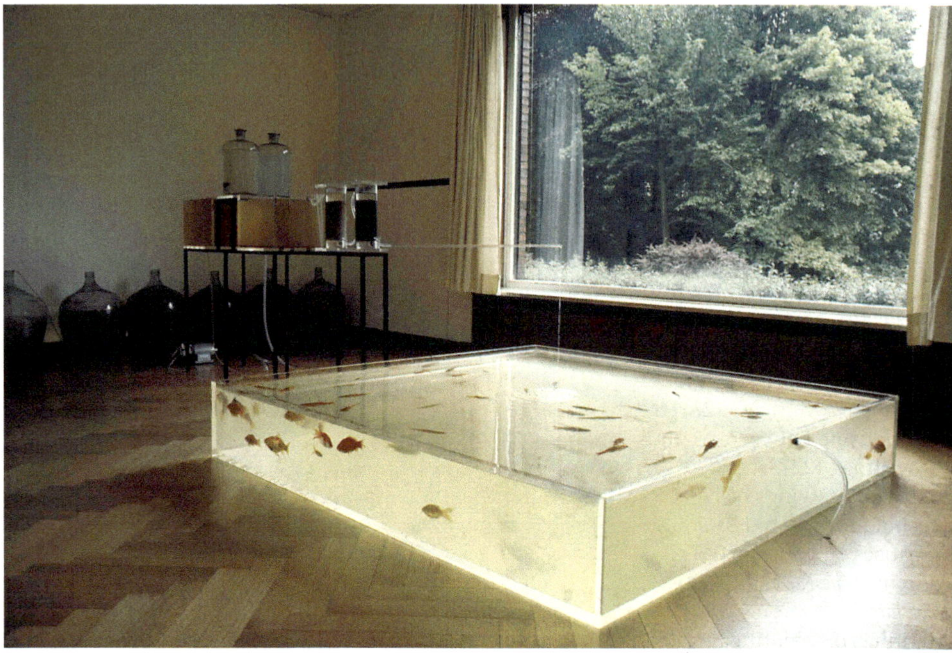

16 Hans Haacke, *Rhine Water Purification Plant* (1972), contaminated water, purification system, goldfish. Museum Haus Lange, Krefeld, Germany. © Hans Haacke/billedkunst.dk

Batch number: 10726526

Printed by Printforce, the Netherlands